T0391192

Trees in a Changing Environment

Plant Ecophysiology

Volume 9

Series Editors:

Luit J. De Kok

University of Groningen, The Netherlands

Malcolm J. Hawkesford

Rothamsted Research, United Kingdom

Aims & Scope:

The Springer Series in Plant Ecophysiology comprises a series of volumes that deals with the impact of biotic and abiotic factors on plant functioning and physiological adaptation to the environment. The aim of the Plant Ecophysiology series is to review and integrate the present knowledge on the impact of the environment on plant functioning and adaptation at various levels: from the molecular, biochemical and physiological to a whole plant level. This series is of interest to scientists who like to be informed of new developments and insights in plant ecophysiology, and can be used as advanced textbooks for biology students.

The titles published in this series are listed at the end of this volume.

Trees in a Changing Environment

Ecophysiology, Adaptation, and Future Survival

Edited by

Michael Tausz
Forest and Ecosystem Science
University of Melbourne
Victoria
Australia

and

Nancy Grulke
Pacific NW Research Station WWETAC
USDA Forest Service
Prineville, Oregon, USA

 Springer

Editors
Michael Tausz
Forest and Ecosystem Science
University of Melbourne
Victoria
Australia

Nancy Grulke
Pacific NW Research Station WWETAC
USDA Forest Service
Prineville, Oregon, USA

ISSN 1572-5561   ISSN 1572-5561 (electronic)
ISBN 978-94-017-9099-4   ISBN 978-94-017-9100-7 (eBook)
DOI 10.1007/978-94-017-9100-7
Springer Dordrecht Heidelberg New York London

Library of Congress Control Number: 2014946587

Printed on acid-free paper

Springer is part of Springer Science+Business Media (www.springer.com)

Editor's Note

For a variety of reasons, this book project took much longer to completion than planned. It was commenced 4 years ago, and some contributions were received as long as 3 years ago. Fortunately, this volume compiles excellent overviews of current topics in tree environmental physiology and ecosystem process aspects, and the delay does not diminish their value in any way. I am convinced that the contributions in this book will prove of long lasting value in an otherwise fast moving field of research. I would like to take the opportunity to thank all authors, editors, and publishers involved for their excellent contributions and, particularly, for their patience.

Creswick, Australia
March 2014

Michael Tausz

Preface

The Earth's climate is continuously changing and has always changed through time. These changes are based on complex, oscillating cycles that occur on decadal, century, and millennial time scales. Climate shifts are common, marked by ice ages as well as long, warm periods.

There is by now overwhelming evidence that human activities have altered natural climatic cycles (Stocker et al. 2013). Although atmospheric chemistry changes (in CO_2, CO, O_3, CH_4) have occurred in the past due to natural causes, the current and expected future atmospheric composition is unlike any in the past due to anthropogenically generated air pollution (in addition to the above: NO_x and tropospheric O_3, double the concentration of the pre-industrial era).

In computer-based models (general circulation models, GCMs), rising concentrations of greenhouse gases have resulted in an increase in air temperature and instabilities in weather. Warmer air holds more water, and it evaporates from all surfaces: soil, vegetation, and open water. In other areas, there will be flooding, just as deleterious as drought to maladapted species. Because every component of ecosystems responds to temperature and water, current ecosystems are and will continue to change in response to increases in temperature, increases in evaporation, and weather instabilities (extremes in temperature and precipitation, its form, and when the extremes occur). Evidence for climate change has already been reported in thousands of publications, in locations distributed throughout the globe. These changes, as well as predicted future changes, are predicted with high confidence on a global scale, yet may differ considerably from place to place (Stocker et al. 2013).

An increase in air temperature of 1–1.5 °C above the mean for 1850–1900 is highly likely by mid-century. In addition to the direct effect of increasing air temperature on water balance, global circulation models predict different amounts of precipitation (Stocker et al. 2013). The greatest threat to ecosystems is increased frequency, duration, and extremity of water availability (from drought to flooding) and temperature (unusual timing and duration of cold snaps, prolonged heat spells) that will disrupt function, survival, and distribution of plants, animals, insects, and pathogens adapted to a past, or at best the current environment. In addition, air pollution effects on ecosystems need to be considered over the long term, especially with regard to the fertilizing effects of CO_2 and nitrogen deposition, and the deleterious and CO_2-negating effects of tropospheric O_3 on carbon uptake and its allocation. Although generalized approaches to managing ecosystems for climate change may be developed, the novel combinations of atmospheric chemistry, temperature, water availability, and the instabilities and extremities in weather will require novel, place-based land management approaches for ecosystems.

At a time when much of the world seems to be discussing climate change, one might ask, 'why another book anticipating effects of climate change?' Firstly, because trees are such long-lived organisms they depend on the acclimation potential of the individuals throughout their lifetime for their survival. Adaptive evolutionary change is slow in species with long generation cycles; hence trees are particularly vulnerable to rapid environmental changes. It is therefore even more important to understand the life functions of trees and the function of forests to underpin possible adaptive management strategies, and these will most likely be different from strategies under consideration for annual or short cycle natural or cropping systems. Secondly, most treatises on climate change effects on biological systems are CO_2-centric: they emphasize CO_2 fertilization and CO_2-induced increased temperatures, accompanying decreases in water availability (in general), but increased plant water use efficiency. We have included the interactive effects of elevated CO_2, the physical environmental effects of greenhouse gas accumulation, and the *source* of the CO_2: atmospheric chemical changes of air pollution (CO_2, O_3, NO_x, and nitrogen deposition). This is a fundamental consideration that many of the discussions on climate change have ignored, or considered only in isolation (with some noted exceptions, see Emberson et al. 2000, who advocated integration of the effects of these components in a process-based model). Due to the difficulty and the magnitude of experimental studies with multiple factors, there are few field studies that have accomplished two abiotic interactive factors (such as CO_2 x temperature, Kellomäki et al. 2000, CO_2 x N amendment, Pääkkönen and Holopainen 1995, O_3 x CO_2 Karnosky et al. 2003, or O_3 x N amendment, Watanabe et al. 2006), let alone many environmental and biological factors over the lifetime of trees and within the complexity of forest ecosystems. Some studies along environmental gradients with carefully matched sites (e.g., high N deposition, drought stress, and moderate O_3 exposure vs. high N deposition and moderate O_3 exposure alone, Miller and McBride 1999) can provide an insight into multiplicative effects. However, we are still restricted to the *current* range in conditions and responses of extant trees that established in a past climate: 80–250+ years ago. The future holds an unprecedented combination and quantity of atmospheric chemicals, and it is as yet unclear whether and which current species or populations of trees are sufficiently equipped to cope with such conditions.

Our ecosystems are already and unequivocally (Stocker et al. 2013) experiencing environmental and climate change, and forests and other tree dominated ecosystems are likely to be severely affected. In this book, the authors have thoughtfully reviewed and described constituent functions and processes that will help us understand tree responses to the complex, concurrent effects of environmental stresses imposed by climate change, and its ultimate source, air pollution. In many cases they have challenged current theory on expected responses, and in all cases they have contributed their expert knowledge on tree and forest ecosystem response to environmental change: an integrated, qualitative assessment. We offer this comprehensive analysis of tree responses and their capacity to respond to environmental changes to give us better insight as to how to plan for the future.

Creswick, Australia   Michael Tausz
Prineville, Oregon   Nancy Grulke

References

Emberson LD, Simpson D, Tuovinen J-P, Ashmore MR, Cambridge HM (2000) Modeling and mapping ozone deposition in Europe. Water Air Soil Pollut 130: 577–582

Karnosky DF, Zak DR, Pregitzer KS, Awmack CS, Bockheim JG et al (2003) Tropospheric O_3 moderates responses of temperate hardwood forests to elevated CO_2: a synthesis of molecular to ecosystem results from the Aspen FACE project. Funct Ecol 17: 289–394

Kellomäki S, Wang K-Y, Lemettinen M (2000) Controlled environment chambers for investigating tree response to elevated CO_2 and temperature under boreal conditions. Photosynthetica 38: 69–81

Miller PR, McBride JR (1999) Oxidant air pollution impacts in the Montane forests of Southern California: a case study of the San Bernardino Mountains. Ecological Studies 134, Springer, New York

Pääkkönen E, Holopainen T (1995) Influence of nitrogen supply on the response of clones of birch (*Betula pendula* Roth) to ozone. New Phytol 129: 595–603

Stocker TF, Qin D, Plattner G-K, Tignor M, Allen SK, Boschung J, Nauels A, Xia Y, Bex V, Midgley PM (eds) (2013) Climate Change 2013: the physical science basis. Contribution of Working Group I to the Fifth Assessment Report of the Intergovernmental Panel on Climate Change. Cambridge University Press, Cambridge, New York

Watanabe M, Yamaguchi M, Iwasaki M, Matsuo N, Naba J, Tabe C, Matsumura H, Kohno Y, Izuta T (2006) Effects of ozone and/or nitrogen load on the growth of *Larix kaempferi*, *Pinus densiflora* and *Crytomeria japonica* seedlings. J Jpn Soc Atmos Environ 41: 320–334

Contents

Chapter 1
Resource Allocation and Trade-Offs in Carbon Gain of Leaves Under Changing Environment

Kouki Hikosaka, Yuko Yasumura, Onno Muller, and Riichi Oguchi

Abstract In leaf canopies, environmental conditions such as light availability and temperature vary spatially and temporally. Plants change leaf traits such as leaf nitrogen content, leaf mass per area, leaf anatomy, photosynthetic capacity, and organization of the photosynthetic apparatus in response to the change in conditions. These changes occur because a trait that is optimal under a certain condition is not advantageous under others. When growth irradiance is high or air temperature is low, plants invest more nitrogen into ribulose-1,5-bisphosphate carboxylase (Rubisco) rather than photosystems. Leaf nitrogen content is high under such conditions because nitrogen content that maximizes nitrogen use efficiency of daily carbon gain is higher under higher irradiance or lower temperature conditions. Leaf anatomy constrains the maximal rate of photosynthesis: leaves with higher photosynthetic rate should be thicker to allot more chloroplasts on mesophyll surface. To increase maximal photosynthetic rate after gap formation, shade leaves of some species are thicker than the minimum required for the photosynthetic rate, allowing further increase in chloroplast volume.

1.1 Introduction

In leaf canopies, there are spatial and temporal variations in photosynthetically active photon flux density (PFD) and temperature. Air temperature and PFD change seasonally especially in higher latitudes. PFD decreases with depth within canopies

K. Hikosaka (✉) • Y. Yasumura • R. Oguchi
Graduate School of Life Sciences, Tohoku University, Aoba, Sendai 980-8578, Japan
e-mail: hikosaka@m.tohoku.ac.jp

O. Muller
Graduate School of Life Sciences, Tohoku University, Aoba, Sendai 980-8578, Japan

Institute for Bio- and Geosciences IBG-2: Plant Sciences, Forschungszentrum Jülich GmbH, 52425 Jülich, Germany

M. Tausz and N. Grulke (eds.), *Trees in a Changing Environment*,
Plant Ecophysiology 9, DOI 10.1007/978-94-017-9100-7_1,
© Springer Science+Business Media Dordrecht 2014

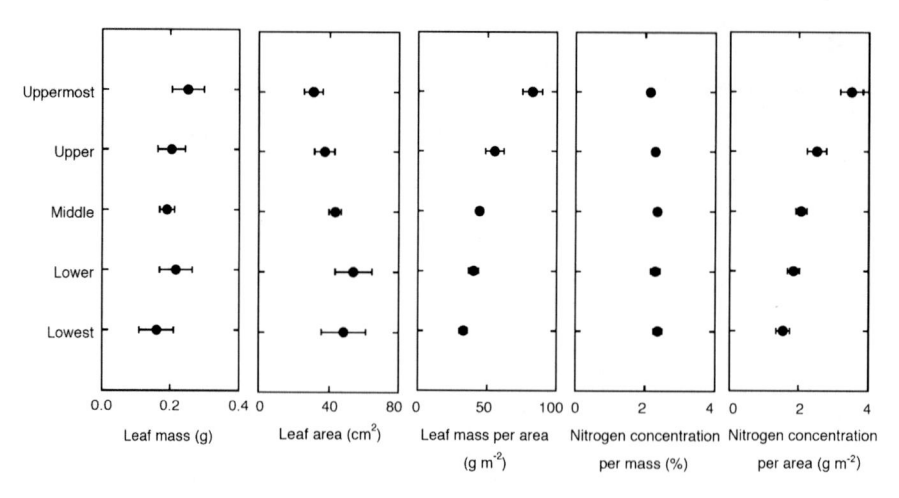

Fig. 1.1 Gradient of leaf traits in a canopy of *Fagus crenata* (Drawn from data shown by Yasumura et al. (2005) and unpublished data (Y. Yasumura))

and often varies by two orders between the top and bottom of dense canopies (Monsi and Saeki 1953). Gap formation, which is an important event in forest ecosystems, greatly increases PFD in understorey. Various leaf traits exhibit significant changes in response to such environmental changes. For example, leaf nitrogen content is highest in the leaves at the top of canopy (Fig. 1.1). As a result of these variations, photosynthetic activity of leaves varies greatly across canopy layers and seasonal environment.

Responses in leaf traits to environmental change is an important information for correct prediction of carbon flow in forest ecosystems (Baldocchi and Harley 1995; Wilson et al. 2001; Ito 2010). Why do leaf traits change in response to environmental changes? This may be because a leaf that is adapted to a certain environment is not necessarily adapted to other environments. If a resource is allocated to improve one function, it inevitably causes a reduction in other functions, i.e., there are trade-offs in resource allocation.

Here is a review of photosynthetic acclimation to spatial and temporal heterogeneity in environment. We particularly focus on light and temperature as important environmental factors. We discuss trade-offs in resource allocation and its relation to optimization of photosynthetic performance.

1.2 Trade-Off in Nitrogen Allocation Among Photosynthetic Components

Nitrogen is one of the most important factors that limit plant growth in many ecosystems (Aerts and Chapin 2000). Even under non-limiting conditions, nitrogen acquisition requires carbon costs, which are utilized for development and maintenance of root systems and uptake, assimilation, and translocation of nitrogen. Therefore efficient use of nitrogen is an important strategy to survive, grow and reproduce under natural environments (Aerts and Chapin 2000).

The photosynthetic apparatus is the largest sink of nitrogen in plants; approximately half of leaf nitrogen is invested in photosynthetic proteins (Evans and Seemann 1989; Hikosaka 2010). The photosynthetic apparatus consists of various proteins. Photons are absorbed by chlorophylls (chl) associated with photosystems I and II (PSI and PSII) and the excitation energy is utilized for electron transport from water to NADPH and for proton transport across thylakoid membranes to produce ATP. NADPH and ATP are utilized in Calvin cycle to produce sugars. The first step of CO_2 fixation is catalyzed by ribulose-1,5-bisphosphate carboxylase oxygenase (Rubisco), where CO_2 is bound to ribulose-1,5-bisphosphate (RuBP) forming 3-phosphoglyceric acid (PGA). Triose phosphate (TP) is then produced by using ATP and NADPH. Some of TP is transported to the cytosol and used for sucrose synthesis, the remainder is used for the regeneration of RuBP. From the viewpoint of energy utilization, the photosynthetic apparatus can be divided into light harvesting (photosystems) and light use (other parts). Under low light, light harvesting limits photosynthesis, while light use is the limiting process under high light.

The organization of the photosynthetic apparatus changes depending on growth photon flux density (PFD). For example, the ratios of Rubisco to chl and of chl a to chl b increase with increasing growth PFD. Such acclimation has been reported not only for herbaceous species (Boardman 1977; Anderson 1986; Terashima and Evans 1988) but also for woody species (Hikosaka et al. 1998; Fig. 1.2). Within-canopy variation in the photosynthetic apparatus has also been shown along light gradients (Niinemets 1997; Niinemets et al. 1998; Warren and Adams 2001; Laisk et al. 2005; Turnbull et al. 2007; Fig. 1.3).

Changes in the organization of the photosynthetic apparatus are related to the role of each component in photosynthesis. Photosynthetic rates exhibit a saturating curve against PFD. When PFD is low, photosynthetic rate linearly increases with increasing light, whereas it saturates at high PFD. The initial slope of the curve is the product of quantum yield and light absorption, the latter of which increases with increasing chl content of the leaf (Gabrielsen 1948). The light-saturated rate of photosynthesis (P_{max}) is, on the other hand, related to the content of other photosynthetic proteins (von Cammerer and Farquhar 1981; Evans 1983; Makino et al. 1983). Particularly, Rubisco content is important because it catalyzes the limiting step of photosynthesis when CO_2 concentration is low under saturating light. Thus nitrogen should be invested more into the light harvesting part under low PFD and to the light use part such as Rubisco under high PFD (Evans 1989).

Hikosaka and Terashima (1995) developed this idea further and constructed a comprehensive model to predict the optimal nitrogen partitioning among photosynthetic components. In this model photosynthetic components were categorized into five groups: Group I, Rubisco; Group II, electron carriers, ATP synthetase, and Calvin cycle enzymes other than Rubisco; Group III, core complex of PSII (PSII core); Group IV, core complex and light harvesting chl-protein complex of PSI, and

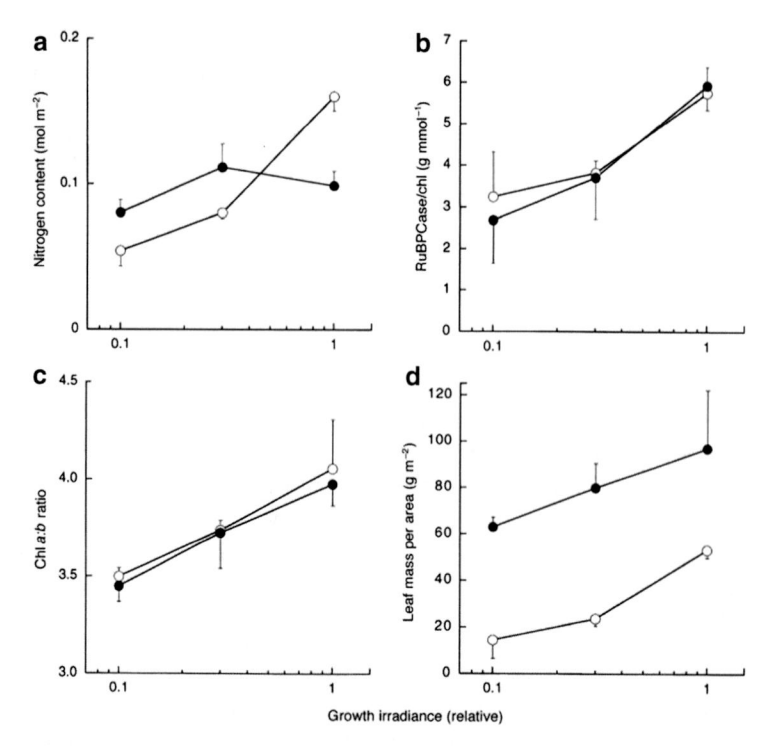

Fig. 1.2 Photosynthetic acclimation in *Chenopodium album* (annual herb; *open symbols*) and *Quercus myrsinaefolia* (evergreen tree; *closed symbols*) grown at different light regimes. RuBCase = Rubisco (Redrawn from Hikosaka et al. 1998)

Group V, light harvesting chl-protein complex of PSII (LHCII). The nitrogen cost for each group was calculated from published data. P_{max} was assumed to be a function of amount of proteins belonging to Group I, II and III. The initial slope was assumed to be a function of chl content. Optimal nitrogen partitioning that maximizes daily carbon gain was calculated. It considerably changed with PFD conditions. Under high PFD, daily carbon gain increases with increasing nitrogen allocation to proteins related to P_{max}, while under low PFD carbon fixation is high when nitrogen is allocated more to photosystems (Fig. 1.4). These results are consistent with the observations that leaves allocate more nitrogen to Rubisco than to chl at higher PFD (Fig. 1.2). Optimal nitrogen investment is higher in PSII core than in LHCII at higher PFD. This is because more PSII core is necessary for higher P_{max}. On the other hand, greater amount of LHCII is only advantageous under low PFD because nitrogen cost of chl (the ratio of chl to N in each group) is higher in LHCII than in PSII. Because most of chl *b* is associated with LHCII, this result explains why the chl *a/b* ratio increases with increasing growth PFD (Fig. 1.2).

Hikosaka and Terashima (1996) applied this model to plants of a sun (*Chenopodium album*) and a shade (*Alocasia macrorrhiza*) species grown under

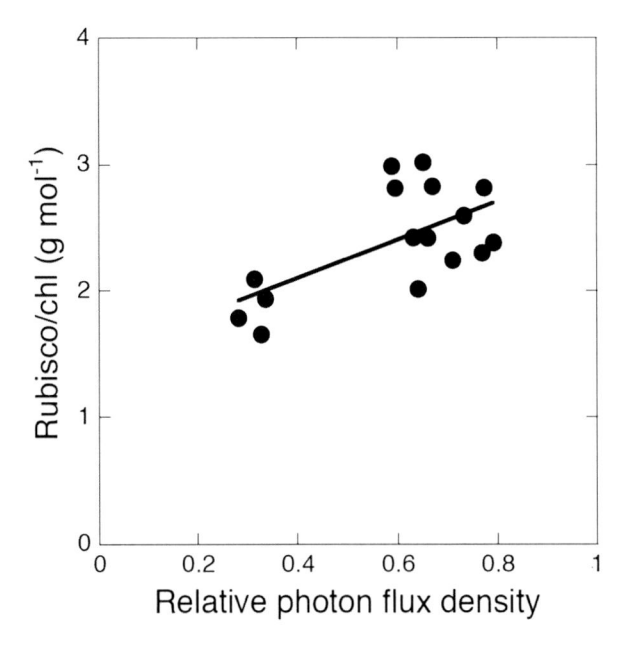

Fig. 1.3 The ratio of Rubisco to chl as a function of the intercepted irradiance in canopy leaves of *Quercus crispula* (Unpublished data (O. Muller))

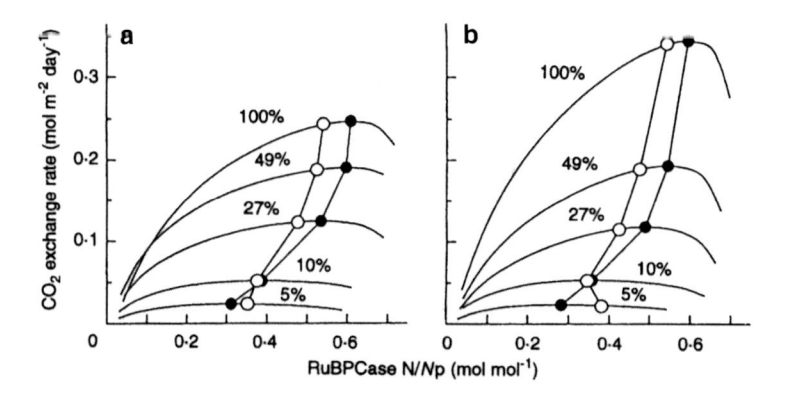

Fig. 1.4 Daily carbon gain as a function of N partitioning in *Alocasia macrorrhiza* (perennial herb; (**a**)) and *Chenopodium album* (annual herb; (**b**)) grown under various PFD (5–100 % of full sunlight). *Open* and *closed symbols* denote actual and optimal nitrogen partitioning (Redrawn from Hikosaka and Terashima 1996)

various PFD conditions. Figure 1.4 shows the effect of nitrogen partitioning on daily carbon gain. There was an optimal nitrogen allocation to Rubisco and it increased with increasing growth PFD (closed circles). Similar to the optimum, actual nitrogen partitioning to Rubisco also increased with increasing growth PFD (open circles). Difference between optimal and actual nitrogen partitioning was small except for *C. album* plants grown at the lowest PFD (5 % of full sunlight). These results suggest that plants can adjust nitrogen partitioning among photosynthetic components nearly optimally to different light regimes, though sun species might be unable to do so in very low PFD.

Similar changes in nitrogen partitioning occur when growth temperature changes. With decreasing growth temperature, for example, the Rubisco/chl and chl *a/b* ratios increase (Hikosaka 2005; Yamori et al. 2009). This may be because of the difference in temperature dependence between the light harvesting and light use parts. As with other enzyme activities, activity of Calvin cycle enzymes is sensitive to temperature and is generally lower at low temperatures. In contrast, photochemical reactions are insensitive to temperature, and consequently the initial slope of light response curve is less affected by temperature. At low temperatures, therefore, proteins related to the light use part should be enhanced to keep the balance between the light harvesting and use.

In temperate climates at mid-latitudes, temperature and light climate vary strongly during the year (Fig. 1.5). In winter, the air temperature is around freezing point and the PFD is lower with shorter day lengths than in summer when air temperature is around 25 °C (Röhrig 1991). In the understorey of deciduous forests, large changes in PFD occur due to sprouting and fall of canopy leaves in spring and autumn, respectively. Leaves of evergreen species in the understorey of such forests are exposed to large changes in light and temperature conditions over the year, which may affect leaf functioning.

Muller et al. (2005) investigated seasonal change in the photosynthetic traits of leaves of an evergreen understory shrub *Aucuba japonica* grown at three different light regimes: gap, understory of deciduous forest, and understory of evergreen forest. They applied multiple regression to evaluate quantitative contribution of temperature and PFD to the photosynthetic acclimation (Fig. 1.5). The Rubisco/chl ratio was significantly correlated both with air temperature and PFD as well as the chl *a/b* ratio. Across sites PFD had stronger effects than air temperature, while within a site temperature had stronger effects on photosynthetic acclimation. It was concluded that the photosynthetic apparatus is strongly affected by the prevailing PFD at the time of leaf development. Within a given light regime, however, the balance between Rubisco and chl responds mainly to temperature and to a lesser extent to PFD.

Apart from the trade-off mentioned above, there is another trade-off between nitrogen allocation between two processes, carboxylation and regeneration of RuBP. At low CO_2 concentrations under saturated light, RuBP carboxylation is the limiting step of photosynthesis, while RuBP regeneration limits photosynthesis at high CO_2 concentrations. Thus to increase photosynthetic rates at low CO_2 concentrations nitrogen should be more allocated to Rubisco, whereas it should be more to RuBP regeneration processes (Group II and III) at high CO_2 concentrations (Hogan et al. 1991; Sage 1994; Webber et al. 1994; Medlyn 1996; Hikosaka and Hirose 1998; Fig. 1.6). Hikosaka and Hirose (1998) theoretically showed that elevated CO_2 (from 350 to 700 μmol mol^{-1}) increased daily carbon gain by 40 %

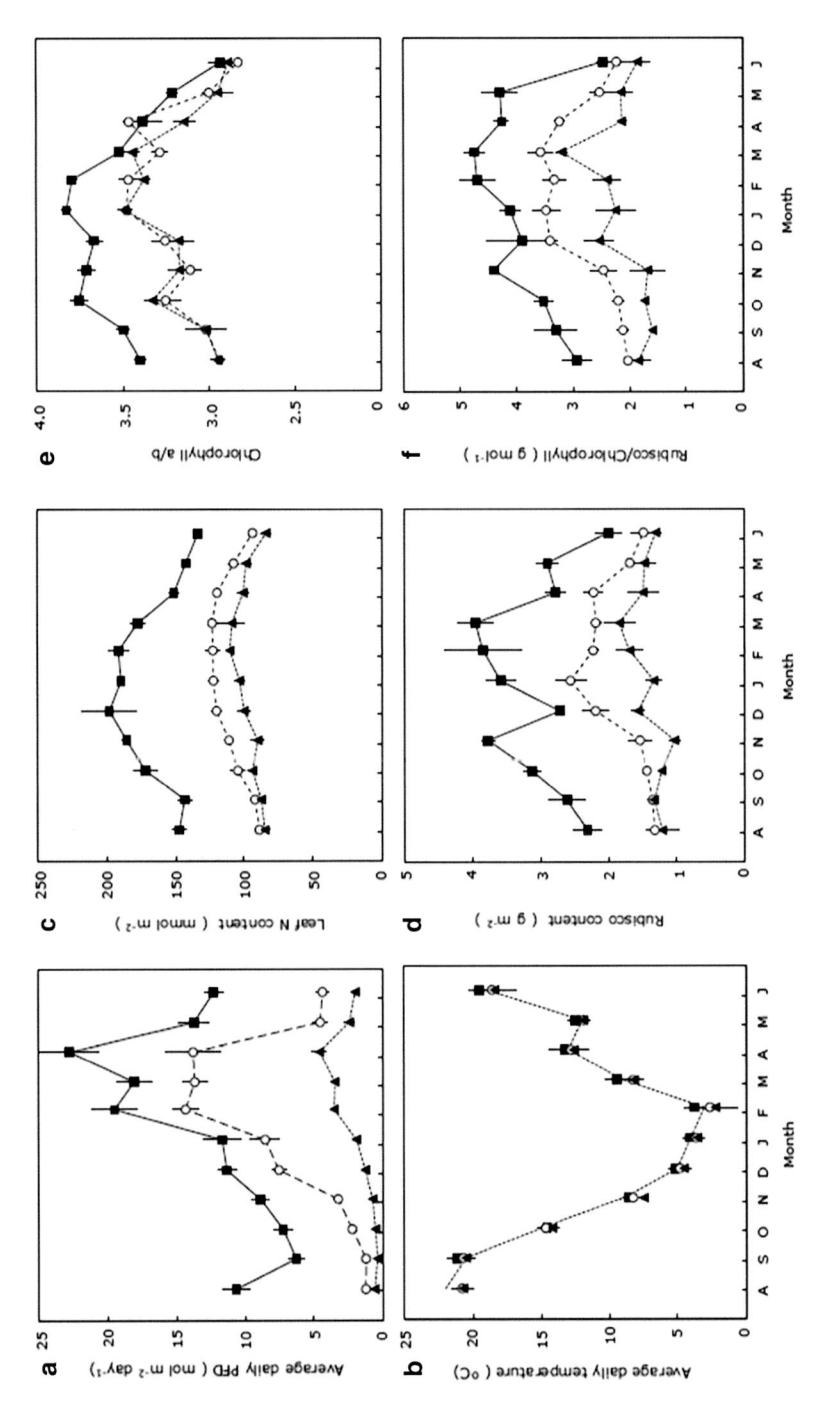

Fig. 1.5 Seasonal changes in PFD (**a**), temperature (**b**), leaf nitrogen content per area (**c**), Rubisco content (**d**), chl a/b ratio, and Rubisco/chl ratio in *Aucuba japonica*, an evergreen shrub, growing three light regimes (*squares*, gap; *circles*, deciduous understory; *triangle*, evergreen understory) (Redrawn from Muller et al. 2005)

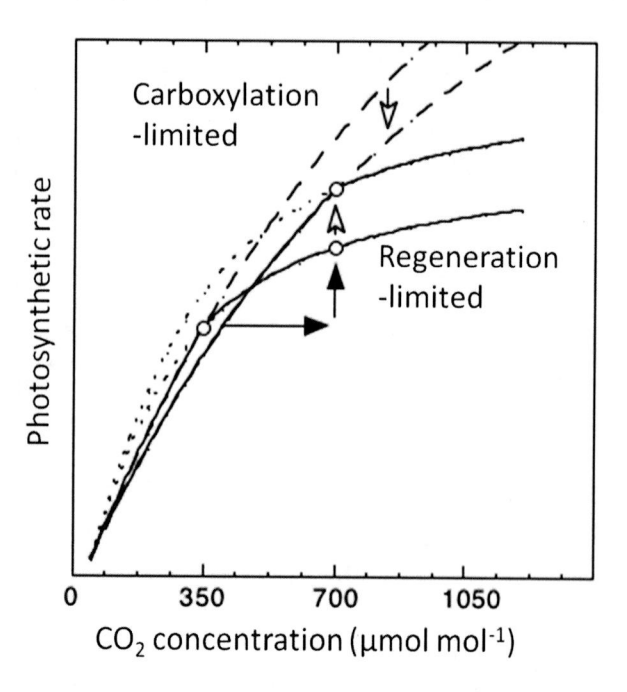

Fig. 1.6 Effects of nitrogen allocation in the photosynthetic apparatus on CO_2 dependence of photosynthesis. See text for detail

when nitrogen partitioning is optimal for 350 μmol mol^{-1} CO_2, while it increased by 60 % when nitrogen is reallocated to maximize photosynthesis at 700 μmol mol^{-1}. This prediction is consistent with the result on transgenic rice with decreased Rubisco content (Makino et al. 1997). When leaves with similar nitrogen content were compared, the transgenic leaves had lower photosynthetic rates than those of wild type at low CO_2 concentrations, but the opposite was the case at high CO_2 concentrations.

In experimental studies, nitrogen allocation to Rubisco and RuBP regeneration processes has been evaluated as V_{cmax} (maximum carboxylation rate) and J_{max} (maximum electron transport rate, Farquhar et al. 1980), respectively. In the 1990s, it was believed that the balance between J_{max} and V_{cmax} was not significantly affected by growth CO_2 concentrations (e.g., Sage 1994; Medlyn et al. 1999). However, recent meta-analyses of FACE (free air CO_2 enrichment) studies have indicated that the J_{max}/V_{cmax} ratio changed significantly with growth CO_2 concentration (Long et al. 2004). Osada et al. (2010) studied photosynthetic traits of *Polygonum sachalinense* plants (a perennial herb) growing around natural CO_2 springs where plants had been exposed to high CO_2 concentrations for the long term and found a significant effect of CO_2 concentration on the J_{max}/V_{cmax} ratio. These results suggest that these plants can alter the balance between carboxylation and regeneration of RuBP depending on growth CO_2 concentration. However, the alteration in actual plants does not seem to be optimal in a quantitative sense. For example, in the study of Osada et al. (2010), the J_{max}/V_{cmax} ratio increased by only 5–6 % when the current CO_2 concentration doubled, and increased by 5 % in FACE experiments at ambient CO_2 + 200 μmol mol^{-1} CO_2 (Ainsworth and Long 2005).

These values are much smaller than the theoretical prediction of a 40 % increase in the J_{max}/V_{cmax} ratio with a doubling of the current CO_2 concentration (Medlyn 1996).

The balance between carboxylation and regeneration of RuBP changes also with temperature. Potential rate of RuBP regeneration exponentially increases with increasing temperature, whereas that of RuBP carboxylation depends less on temperature because of Rubisco kinetics (Fig. 1.7). When temperature is low, therefore, increased nitrogen allocation to RuBP regeneration processes is beneficial (Hikosaka 1997; Hikosaka et al. 2006; Fig. 1.7). Hikosaka et al. (1999a) found that *Quercus myrsinaefolia*, an evergreen tree, realized such changes in the balance between carboxylation and regeneration of RuBP. Hikosaka (2005) found that *Plantago asiatica*, a perennial herb, invested more nitrogen in RuBP regeneration at low growth temperature (Fig. 1.8). However, it has been indicated that some species alter the balance but the others do not (Hikosaka et al. 2006). For example, Hikosaka et al. (2007) studied temperature dependence of photosynthesis in canopy leaves of *Quercus crispula*, a deciduous tree, which did not show seasonal change in the J_{max}/V_{cmax} ratio. Onoda et al. (2005) showed that the J_{max}/V_{cmax} ratio exhibited a seasonal change in seedlings of *Polygonum cuspidatum*, a perennial herb, but not in those of *Fagus crenata*, a deciduous tree. Recently, Yamori et al. (2010) compared temperature acclimation in cold-sensitive and tolerant crop species, the latter of which tended to show greater changes in the J_{max}/V_{cmax} ratio depending on growth temperatures.

1.3 Nitrogen Use Efficiency of Daily Carbon Gain at Leaf and Canopy Levels

In many canopies, there is a vertical gradient of leaf nitrogen content per unit area (N_{area}) (De Jong and Doyle 1985; Hirose and Werger 1987b; Hollinger 1989; Evans 1993; Ellsworth and Reich 1993; Anten et al. 1998; Niinemets 1997; Niinemets et al. 2001; Kikuzawa 2003; Wright et al. 2006; Migita et al. 2007; Yasumura et al. 2005; Fig. 1.1). This gradient is formed mainly in response to the gradient of light availability. This has been proved mainly using herbaceous canopies. For example, the gradient of N_{area} is steeper in a denser than in a scarce canopy (Hirose et al. 1988). N_{area} in vine species where PFD was manipulated, changes depended on PFD (Hikosaka et al. 1994). The gradient of N_{area} is steeper in canopies that have steeper light gradients (Anten et al. 1995, 2000; Ackerly and Bazzaz 1995).

Because almost half of leaf nitrogen is invested in the photosynthetic apparatus, photosynthetic rate is related to N_{area} (Evans 1989; Evans and Seemann 1989; Hikosaka 2010). In particular, there is a strong correlation between P_{max} and N_{area} (Hirose and Werger 1987a; Evans 1989; Hikosaka et al. 1998; Hikosaka 2004; Niinemets et al. 2001; Warren and Adams 2001; Fig. 1.9a). Dark respiration rate is also positively correlated with N_{area} (Hirose and Werger 1987a; Anten

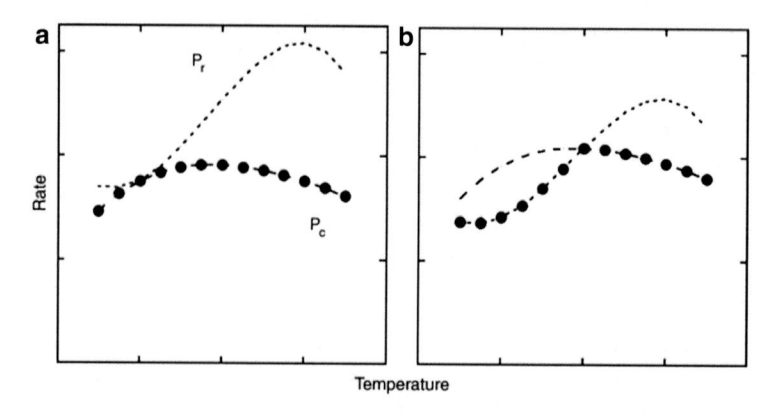

Fig. 1.7 Effect of nitrogen allocation on temperature dependence of photosynthesis. *Broken* and *dotted lines* represent potential rate of Rubisco- (P_c) and RuBP regeneration-limited (P_r) photosynthesis, respectively. *Closed circles* are realized photosynthetic rate, defined as the lower rate of Rubisco- and RuBP regeneration-limited photosynthesis (Redrawn from Hikosaka et al. 2006)

et al. 1995; Hikosaka et al. 1999b). On the other hand, correlation between photosynthetic rate at low light and N_{area} is generally weak (Hirose and Werger 1987a). Weak but significant correlation has been observed between the initial slope and N_{area} in some studies (Hirose and Werger 1987a; Hikosaka et al. 1999b), but not in others (Anten et al. 1995).

As a result of these correlations, daily carbon gain depends on N_{area}; daily carbon gain increases with increasing N_{area} until its optimum and gradually declines due to increasing respiration rate (Hirose and Werger 1987a; Fig. 1.9b). Each curve has two optima. One is the N_{area} that maximizes daily carbon gain (N_{optCER}; A in Fig. 1.9b) and the other is the N_{area} that maximizes nitrogen use efficiency of carbon gain (daily carbon gain per unit leaf nitrogen, daily NUE) (N_{optNUE}; B in Fig. 1.9b, c) (Hirose 1984; Hirose and Werger 1987a; Hikosaka and Terashima 1995). N_{area} values at both optima increase with increasing PFD (Fig. 1.9b, c), which explains why N_{area} is higher in upper leaves.

The optimal N_{area} that maximizes daily NUE (N_{optNUE}) implicitly assumes a trade-off between leaf area and N_{area}. When the amount of nitrogen for a leaf is limited, plants have two choices: one is increasing leaf area, which inevitably reduces N_{area}, and the other is increasing N_{area} at the expense of leaf area. N_{optNUE} is truly optimal when photosynthesis is limited only by nitrogen. However, it is not the case if there are other limitations such as carbon supply. Hikosaka and Terashima (1995) discussed that N_{area} will be closer to the N_{optCER} when nitrogen is more available, while it will be closer to N_{optNUE} when nitrogen is more limited. This is consistent with the experimental results of leaves of spinach (*Spinacia oleracea*) (Hikosaka and Terashima 1995; Terashima and Hikosaka 1995).

N_{area} exhibits seasonal change. In deciduous trees, N_{area} increases after unfolding and reaches maximum in mid summer (Wilson et al. 2000, 2001; Hikosaka et al. 2007; Fig. 1.10). Some of leaf nitrogen is resorbed and others

Fig. 1.8 Nitrogen partitioning in *Plantago asiatica* (perennial herb) leaves grown at high-light with low-temperature (*closed circle*), high-light with high-temperature (*open circle*), and low-light with low-temperature (*closed square*). FBPase (stroma fructose-1,6-bisphosphatase) activity represents nitrogen investment in the RuBP regeneration process. RuBPCase = Rubisco

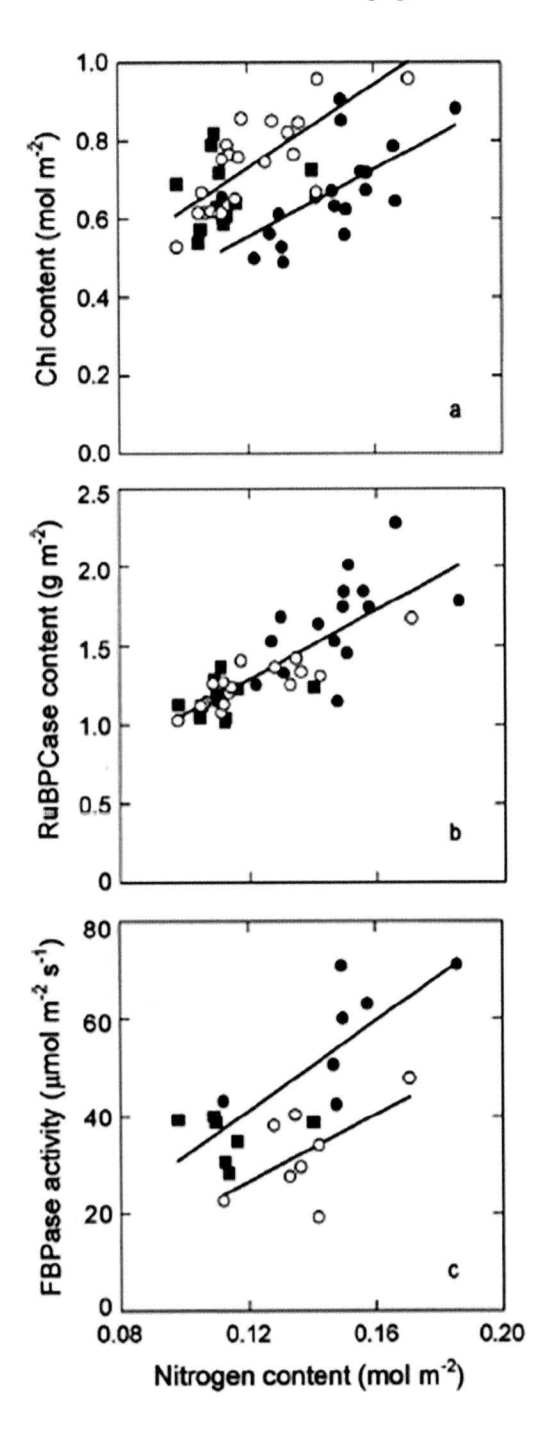

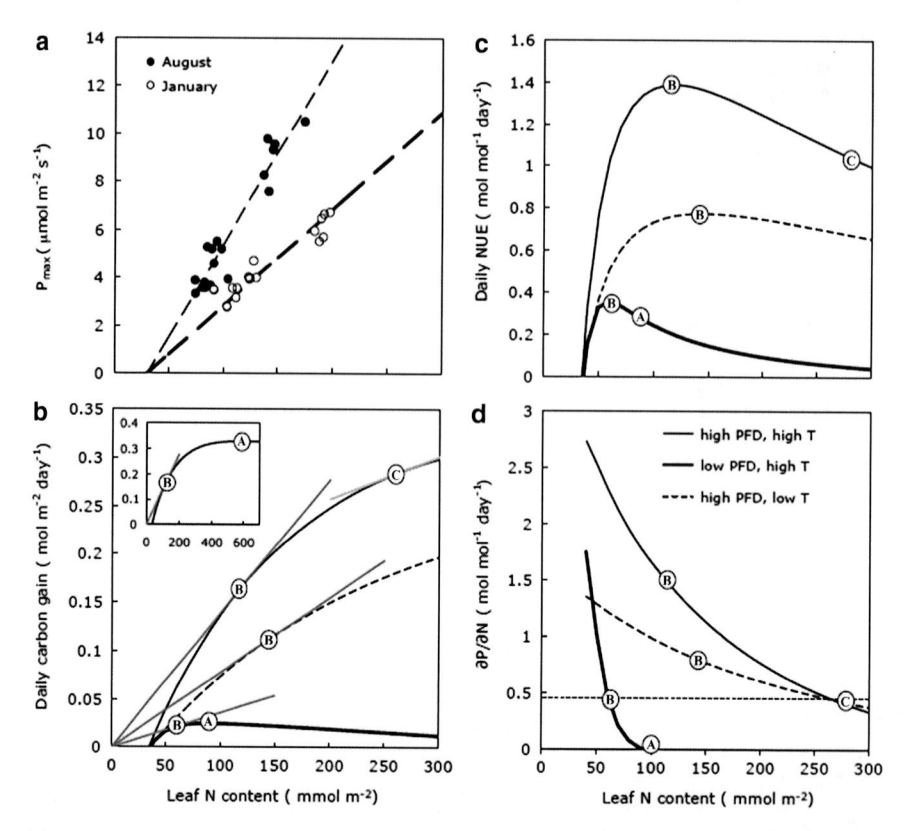

Fig. 1.9 Dependence on leaf nitrogen content (N_{area}) of the light saturated rate of photosynthesis (P_{max}; **a**), daily carbon exchange rate (CER; **b**), daily nitrogen use efficiency (NUE, CER per leaf nitrogen; and slope of the tangent of daily CER ($\partial P/\partial N$; **d**) in *Aucuba japonica*. *Closed* and *open symbols* in a are data obtained in summer (August) and winter (January) at the growth temperature. Daily CER was calculated based on the data shown in (**a**). *Continuous, dotted*, and *thick lines* denote values at summer under high light conditions, those at winter under high light conditions, and those at summer under low light conditions, respectively. The *circle A* and *B* denote the N_{area} that maximizes daily CER and daily NUE, respectively. In the *circle "C"*, $\partial P/\partial N$ of high light leaves is identical to that of low light leaves in *"B"*, indicating optimal allocation of nitrogen between these two leaves. Calculated with data in Muller et al. (2011)

drop with dead leaves (Yasumura et al. 2005). Yasumura et al. (2005) showed that nitrogen resorption efficiency in leaves was not different among layers, though N_{area} was very different.

In evergreen trees, N_{area} is generally highest in winter (Fig. 1.5). Muller et al. (2005) applied multiple regression analysis to analyze effect of PFD and temperature on N_{area} and showed that both PFD and temperature significantly affected N_{area}; leaf N_{area} was high when PFD was high and temperature was low. Experimental studies have also shown that N_{area} is higher at lower temperature regimes (Hikosaka 2005; Yamori et al. 2009).

Fig. 1.10 Seasonal changes in (**a**) mean leaf mass per unit area (LMA), (**b**) leaf nitrogen concentration per unit mass (N_{mass}) and (**c**) leaf nitrogen concentration per unit area (N_{area}) in canopy leaves of *Quercus crispula* in 2001 (*closed circle*) and 2002 (*open circle*). Bars are standard deviations. Polynomial curves are fitted for (**a**) ($r^2 = 0.95$, $P < 0.05$) and (**c**) ($r^2 = 0.99$, $P < 0.05$) (Redrawn from Hikosaka et al. 2007)

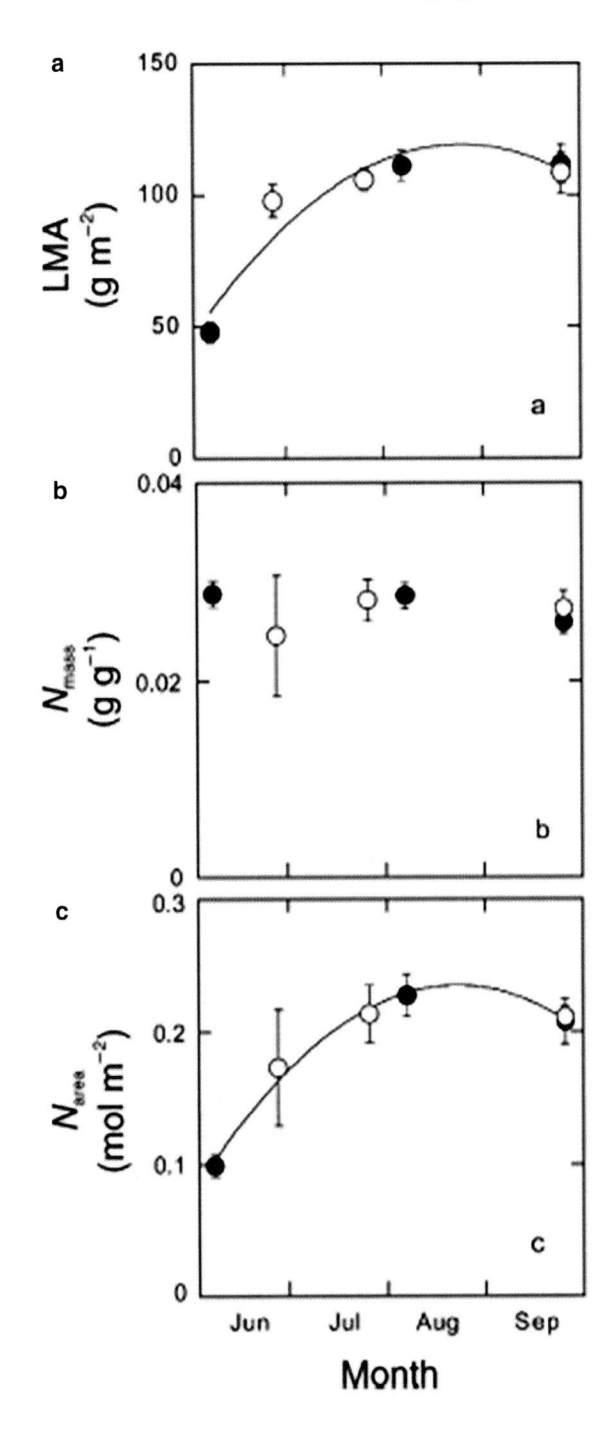

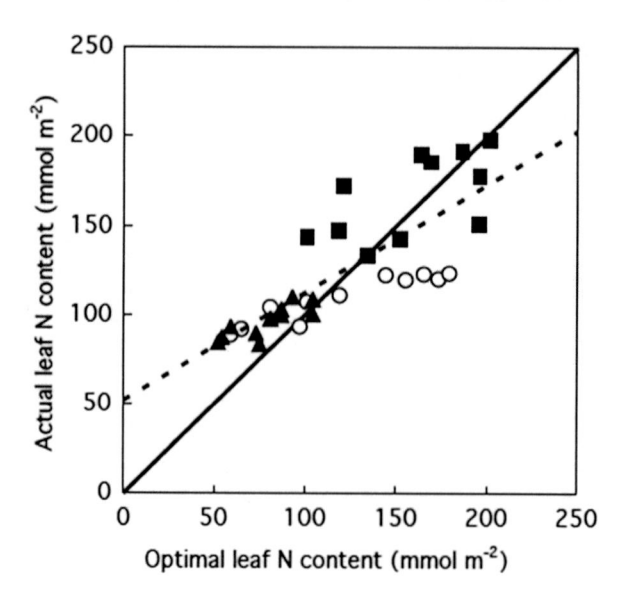

Fig. 1.11 Optimal leaf nitrogen content in relation to the actual leaf nitrogen content in *Aucuba japonica* growing in a gap (*closed squares*), under a deciduous canopy (*open circles*) and evergreen canopy (*closed triangles*) with linear regression line (*broken line*; $r^2 = 0.65$) and 1:1 line (*solid line*) (Redrawn from Muller et al. 2011)

Muller et al. (2011) addressed a question why N_{area} increases in winter. They determined photosynthetic rate and N_{area} in leaves of *Aucuba japonica* plants growing under three light regimes. From nitrogen dependence of daily carbon gain, they calculated optimal N_{area} that maximizes nitrogen use efficiency of daily carbon gain (N_{optNUE}; B in Fig. 1.9). Both increasing PFD and decreasing temperature increases the N_{optNUE}. There was a strong correlation between the N_{optNUE} and actual N_{area}, which was close to the 1:1 relationship (Fig. 1.11). Sensitivity analyses showed that both temperature and PFD had comparable contribution to the change in the variation in the N_{optNUE}.

Vertical gradient of N_{area} has been discussed with respective to maximizing canopy photosynthesis. As mentioned above, photosynthetic rate is less sensitive to N_{area} at low PFD but increases with increasing N_{area} at high PFD. Therefore canopy photosynthesis is improved when nitrogen is allocated more to leaves that receive higher PFD (Field 1983; Hirose and Werger 1987b). Field (1983) showed that canopy photosynthesis is maximized if nitrogen is allocated such that every leaf satisfies following equation:

$$\lambda = \partial P / \partial N \qquad (1.1)$$

where λ is the Lagrange multiplier, P is daily carbon gain and N is N_{area}. When compared at the same N_{area}, $\partial P / \partial N$ is higher in leaves that receive high light (Fig. 1.9d). Therefore N_{area} should be higher in upper leaves. Farquhar (1989) suggested that canopy photosynthesis is maximized if P_{max} of each leaf is proportional to light availability of each leaf. Anten et al. (1995) proved that this relationship is maintained when the initial slope and convexity of the light-response curve is constant across leaves. Hirose and Werger (1987b) calculated optimal nitrogen distribution among leaves in a canopy of *Solidago altissima*, a perennial herb. The actual nitrogen distribution was significantly different from the inferred optimal distribution, as it was less steep. Similar differences between actual and optimal nitrogen distribution were reported by other researchers (Anten et al. 1995). Anten et al. (2000) compiled data obtained from herbaceous canopies and showed that the slope of the actual nitrogen distribution was almost half of that of the optimal distribution.

It should be noted that optimal nitrogen content to maximize daily NUE (N_{optNUE}) is not necessarily consistent with the optimal nitrogen allocation among leaves to maximize canopy photosynthesis. In N_{optNUE}, the curve of daily CER-N has a tangent from the origin (Fig. 1.9b). Slope of the tangent is different depending on PFD (B in Fig. 1.9b). On the other hand, slope of a tangent of the curve, i.e. $\partial P/\partial N$, is required to be identical among leaves in a canopy that maximizes canopy photosynthesis (C for high light leaves in Fig. 1.9b, d). Nitrogen distribution may be less steep in a canopy in which every leaf has N_{optNUE} than in a canopy that maximizing canopy photosynthesis. Therefore, less steep nitrogen distribution found in actual canopies may be caused by a result of optimal regulation at a leaf level rather than that at a canopy level.

1.4 Trade-Offs in Leaf Morphology

Sun and shade leaves differ from each other in morphological traits as well as in physiological traits. In general, sun leaves are thicker and have higher leaf mass per area than shade leaves. Also in tree canopies, there is a vertical gradient in morphological traits (Ellsworth and Reich 1993; Niinemets 1997; Wright et al. 2006). Figure 1.1 shows gradients of leaf traits in a *Fagus crenata* canopy. Leaf mass per area (LMA) exhibited a large decrease from the top to the bottom. N_{area} can be expressed as a product of LMA and nitrogen concentration per mass (N_{mass}). In tree canopies, gradient of N_{area} is mainly attributed to LMA because N_{mass} is relatively constant or even higher in lower canopies (Fig. 1.1). In herbaceous canopies, in contrast, gradient of N_{area} is mainly ascribed to N_{mass} (Hirose et al. 1988). This difference reflects differences in canopy development. In herbaceous canopies, new leaves are mainly formed at the top of the canopy and light availability for each leaf declines with development of new leaves. New leaves developed as a sun leaf and N_{area} and N_{mass} gradually decrease mainly due to resorption while morphological traits are relatively constant (but LMA generally exhibits small reduction through leaf senescence). In tree canopies, on the other hand, new leaves are produced in each layer and light availability does not change greatly. Leaf thickness and LMA were altered according to the environment where the leaves developed, while N_{mass} is relatively constant (Ellsworth and Reich 1993).

Leaf morphology is an important constraint for P_{max} (Terashima et al. 2001). Large investment of photosynthetic proteins is necessary to achieve high P_{max}. Since all photosynthetic enzymes are involved in chloroplasts, sun leaves need to have a large number of chloroplasts in the mesophyll cells. CO_2 diffusion in the liquid phase is very slow and chloroplasts distribute near the cell surface. If a leaf increased the number of chloroplasts without thickening the mesophyll layer, some chloroplasts would become separated from the cell surface and any increase in the number of such chloroplasts contributes little to increasing photosynthetic capacity because they do not receive sufficient CO_2 to fix. Therefore, sun leaves are thick in order to arrange all chloroplasts along the mesophyll cell surface. Hence there are strong correlations between photosynthetic capacity and leaf thickness (McClendon 1962; Jurik 1986), between photosynthetic capacity and mesophyll cell surface area (Nobel et al. 1975), and between the internal conductance of CO_2 and chloroplast surface area facing the intercellular space (von Caemmerer and Evans 1991; Evans et al. 1994).

This constraint of P_{max} brings about a trade-off between leaf thickness and leaf area. If biomass is limited for production of a leaf, large leaf area is advantageous for light capture but it inevitably forces small leaf thickness and thus suppresses P_{max}. In fact, leaf area in *Fagus crenata* canopy was greater at lower layers (Fig. 1.1).

Gap formation abruptly increases light availability for understorey plants in a forest. This event is considered indispensable for further growth of tree seedlings and thus for regeneration of forests (Denslow 1987; Naidu and DeLucia 1997; Ryel and Beyschlag 2000). In a mixed temperate forest, gaps are formed throughout a year (Romme and Martin 1982). When irradiance increased in the growing season, plants often showed light acclimation where P_{max} increased even in already expanded leaves (Turnbull et al. 1993; Naidu and DeLucia 1998; Yamashita et al. 2000). Nevertheless, it has been shown that leaf thickness is determined by the irradiance at leaf development, and changes little after leaves have matured (Milthorpe and Newton 1963; Verbelen and De Greef 1979; Sims and Pearcy 1992). Does this imply that leaves do not have to become thick to increase their P_{max}?

Oguchi et al. (2003) found that mature shade leaves of *Chenopodium album*, an annual herb, have vacant space along the mesophyll surface which is not occupied by chloroplasts (Fig. 1.12). When the shade leaves were exposed to high irradiance, chloroplast volume increased to fill the space and P_{max} increased without an increase in leaf thickness. However, these leaves had vacant space and consequently were thicker than the minimum required to arrange all chloroplasts to fill the mesophyll cell surface.

Oguchi et al. (2005, 2006) investigated leaf anatomy of various deciduous tree species in a growth cabinet (Oguchi et al. 2005) and in the field where an artificial gap was formed (Oguchi et al. 2006). They found that the response of existing leaves to increasing PFD was different among species. Shade leaves of *Betula ermanii*, *Kalopanax pictus*, *Magnolia obovata*, and *Quercus crispula* had the vacant space in mesophyll cells and increased chloroplast volume after exposure to high light, similar to the results on *C. album* (Fig. 1.13). Three *Acer* species, *A. rufinerve*, *A. mono*, and *A. japonicum* extended not only chloroplast volume but also mesophyll cell surface after exposure to high light, suggesting that *Acer* species have

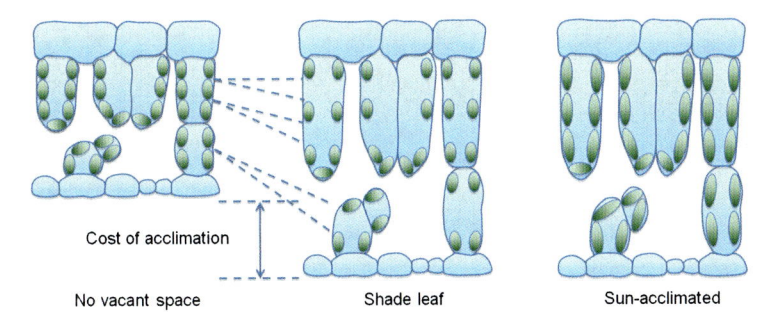

Fig. 1.12 Diagram of anatomical acclimation in shade leaves that are exposed to a sunny condition

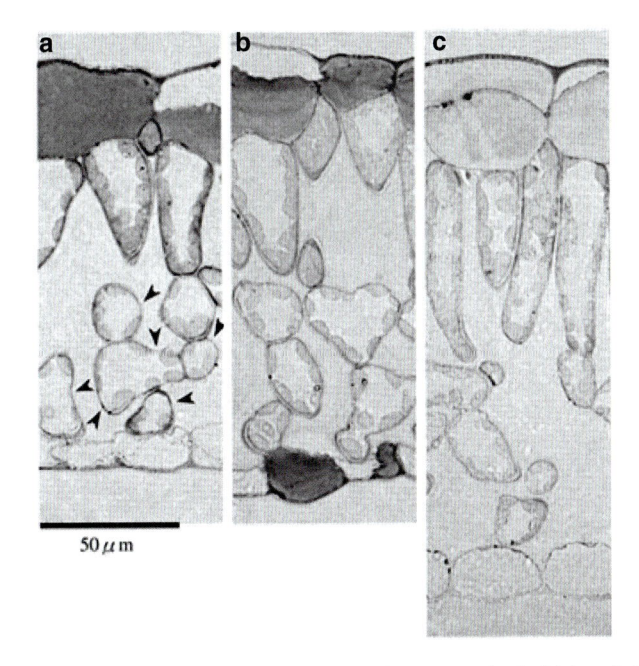

Fig. 1.13 Leaf anatomy of *Betula ermanii*. (**a**) Low-light grown leaf, (**b**) low-light grown leaf after transfer to high light, and (**c**) high-light grown leaf. *Arrows*: vacant space (Redrawn from Oguchi et al. 2005)

plasticity in leaf anatomy even after full expansion (Fig. 1.14). On the other hand, *F. crenata* had little mesophyll cell surface unoccupied by chloroplasts and leaf anatomy was not changed after exposure to high light (Fig. 1.15). Consequently, it did not increase P_{max}. These results suggest that light acclimation potential is primarily determined by the availability of unoccupied cell surface into which chloroplasts expand, as well as by the plasticity of the mesophyll that allows an increase in its surface area.

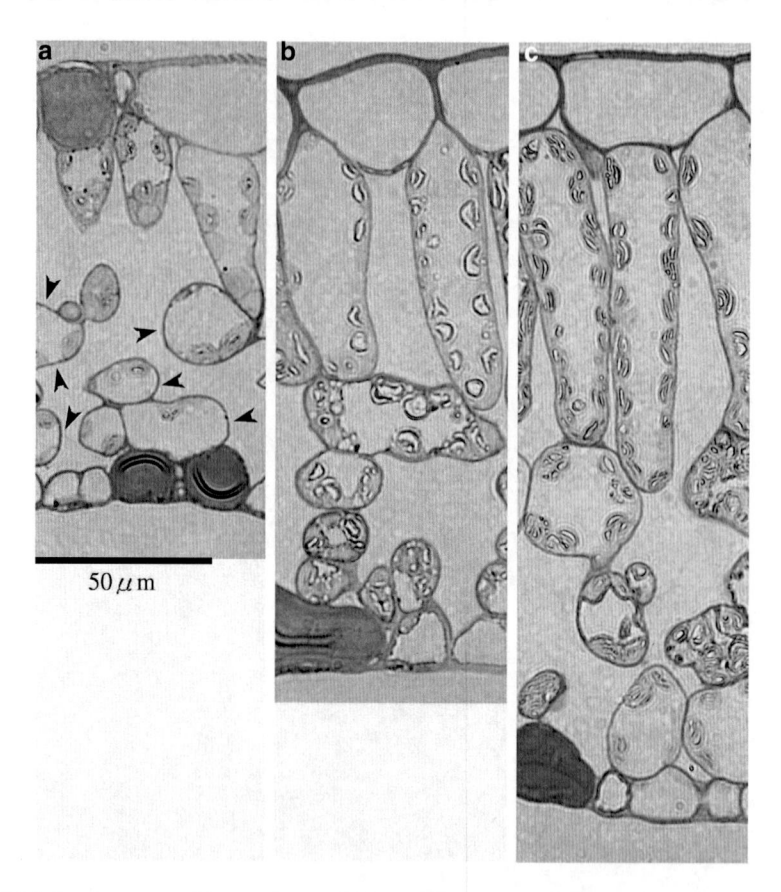

Fig. 1.14 Leaf anatomy of *Acer rufinerve*. (**a**) Low-light grown leaf, (**b**) low-light grown leaf after transfer to high light, and (**c**) high-light grown leaf. *Arrows*: vacant space (Redrawn from Oguchi et al. 2005)

Then the question arises, why some species have vacant space while others do not? Plants need to invest more biomass into thicker leaves (Fig. 1.12). Plants could intercept more light if this biomass were used to enlarge leaf area. Oguchi et al. (2008) evaluated the cost and benefit of photosynthetic light acclimation in a natural environment. The researchers created gaps by felling canopy trees in a cool-temperate forest and evaluated the cost and benefit of light acclimation in *K. pictus*, a species that had vacant space in shade leaves. These leaves increased P_{max} by enlarging chloroplasts into this space after gap formation. An increase in carbon gain of light-acclimated leaves over non-acclimated leaves is a benefit of acclimation. The authors used a biochemical model based on Rubisco kinetics and combined it with an empirical model for stomatal conductance described as a function of environmental factors (Harley and Tenhunen 1991). The costs are the additional investment in biomass needed to construct the vacant space which would

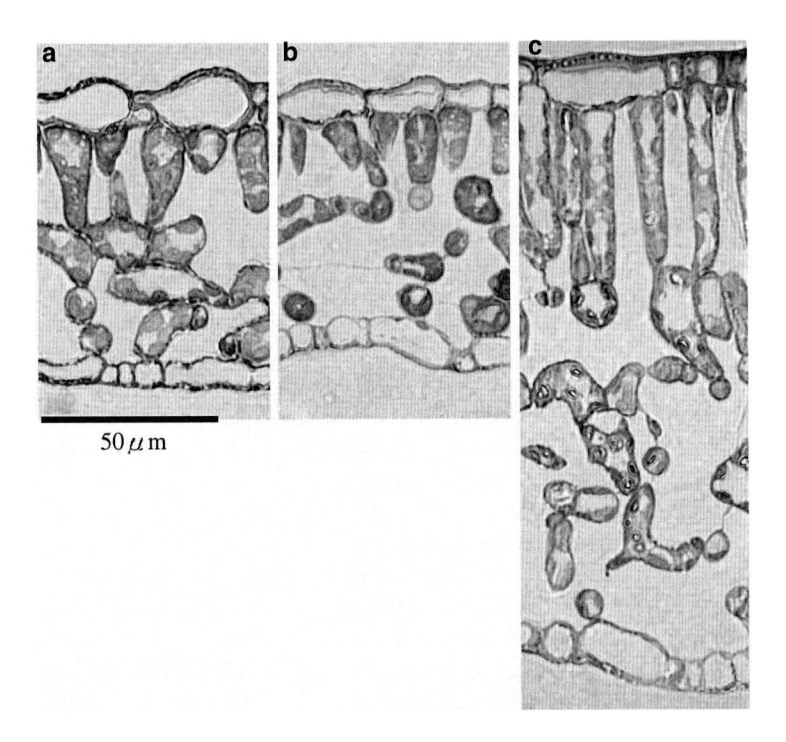

Fig. 1.15 Leaf anatomy of *Fagus crenata*. (**a**) Low-light grown leaf, (**b**) low-light grown leaf after transfer to high light, and (**c**) high-light grown leaf (Redrawn from Oguchi et al. 2005)

enable the chloroplast volume to increase for the future gap formation. When those leaves were exposed to higher irradiance after gap formation, the area of the mesophyll surface covered by chloroplasts increased by 17 % and P_{max} by 27 %. This increase in P_{max} led to an 11 % increase in daily carbon gain, which was greater than the amount of biomass additionally invested to construct thicker leaves. Thus the capacity of a plant to acclimate to light (photosynthetic acclimation) would contribute to rapid growth in response to gap formation. On the other hand, if gaps are not formed, the cost to construct thicker leaf is wasteful. This suggests that optimistic species may produce thicker leaves to allow an increase in P_{max} after gap formation, while pessimistic species produce leaves with a minimum thickness to reduce construction cost.

As discussed above, evergreen leaves in temperate forests exhibit seasonal changes in leaf nitrogen content. Muller et al. (2009) studied leaf anatomy of *Aucuba japonica* and showed that a vacant space on mesophyll surfaces in summer enabled the chloroplast volume to increase in winter. Thus, summer leaves were thicker than needed to accommodate mesophyll surface chloroplasts at this time of the year, but this allowed for increases in mesophyll surface chloroplasts in winter. The authors also performed a transfer experiment in which irradiance regimes were changed at the beginning of autumn and of spring to evaluate differential effects of winter and summer irradiance on leaf anatomy and photosynthesis. Leaf anatomical characteristics such as mesophyll surface area were significantly dependent on growth light in winter, suggesting that summer leaf anatomical characteristics help facilitate photosynthetic acclimation to winter conditions.

1.5 Conclusion

Increasing photosynthetic capacity may benefit carbon gain but entails many kinds of costs: investment of nitrogen, increased respiration, and thickening the leaf. Shade leaves may be rather advantageous if they have low nitrogen content, low respiration, and smaller thickness, which enables greater leaf area. Optimal traits are different depending on growth irradiance and temperatures. Furthermore, the strategy may be different depending on whether the species is "optimistic" or "pessimistic" for the future. If a shaded plant "expects" improvement in light environment in the future, it produces leaves that are thicker than the minimum to allow increase in chloroplast volume after the light improvement. If it is "pessimistic", it may produce leaves with minimum thickness. Plants alter their leaf traits not only in response to environmental change but also with their strategy, which may partly contribute to coexistence of various species on the earth.

Acknowledgments The study was supported by KAKENHI, the Global Environment Research Fund (F-092) by the Ministry of the Environment, Japan, and the Global COE Program j03 (Ecosystem management adapting to global change) by the MEXT.

References

Ackerly DD, Bazzaz FA (1995) Leaf dynamics, self-shading and carbon gain in seedlings of a tropical pioneer tree. Oecologia 101:289–298

Aerts R, Chapin FS III (2000) The mineral nutrition of wild plants revisited: a re-evaluation of processes and patterns. Adv Ecol Res 30:1–67

Ainsworth EA, Long SP (2005) What have we learned from 15 years of free-air CO_2 enrichment (FACE)? A meta-analytic review of the responses of photosynthesis, canopy properties and plant production to rising CO_2. New Phytol 165:351–372

Anderson JM (1986) Photoregulation of the composition, function, and structure of thylakoid membranes. Annu Rev Plant Physiol 46:161–172

Anten NPR, Schieving F, Werger MJA (1995) Patterns of light and nitrogen distribution in relation to whole canopy carbon gain in C3 and C4 mono and dicotyledonous species. Oecologia 101:504–513

Anten NPR, Miyazawa K, Hikosaka K, Nagashima H, Hirose T (1998) Leaf nitrogen distribution in relation to leaf age and photon flux density in dominant and subordinate plants in dense stands of a dicotyledonous herb. Oecologia 113:314–324

Anten NPR, Hikosaka K, Hirose T (2000) Nitrogen utilisation and the photosynthetic system. In: Marshall B, Roberts J (eds) Leaf development and canopy growth. Sheffield Academic Press, Sheffield

Baldocchi DD, Harley PC (1995) Scaling carbon dioxide and water vapour exchange from leaf to canopy in a deciduous forest II. Model testing and application. Plant Cell Environ 18:1157–1173

Boardman NK (1977) Comparative photosynthesis of sun and shade plants. Annu Rev Plant Physiol 28:355–377

De Jong TM, Doyle JF (1985) Seasonal relationships between leaf nitrogen content (photosynthetic capacity) and leaf canopy light exposure in peach (*Prunus persica*). Plant Cell Environ 8:701–706

Denslow JS (1987) Tropical rainforest gaps and tree species diversity. Annu Rev Ecol Syst 18:431–451

Ellsworth DS, Reich PB (1993) Canopy structure and vertical patterns of photosynthesis and related leaf traits in a deciduous forest. Oecologia 96:169–178

Evans JR (1983) Nitrogen and photosynthesis in the flag leaf of wheat (*Triticum aestivum* L.). Plant Physiol 72:297–302

Evans JR (1989) Photosynthesis and nitrogen relationships in leaves of C3 plants. Oecologia 78:9–19

Evans JR (1993) Photosynthetic acclimation and nitrogen partitioning within a lucerne canopy. I. Canopy characteristics. Aust J Plant Physiol 20:55–67

Evans JR, Seemann JR (1989) The allocation of protein nitrogen in the photosynthetic apparatus: costs, consequences, and control. In: Briggs WR (ed) Photosynthesis. Liss, New York

Evans JR, von Caemmerer S, Setchell BA, Hudson GS (1994) The relationship between CO_2 transfer conductance and leaf anatomy in transgenic tobacco with a reduced content of rubisco. Aust J Plant Physiol 21:475–495

Farquhar GD (1989) Models of integrated photosynthesis of cells and leaves. Philos Trans R Soc Lond B 323:357–367

Farquhar GD, von Caemmerer S, Berry JA (1980) A biochemical model of photosynthetic CO_2 assimilation in leaves of C3 species. Planta 149:78–90

Field C (1983) Allocating leaf nitrogen for the maximization of carbon gain: leaf age as a control on the allocation program. Oecologia 56:341–347

Gabrielsen EK (1948) Effects of different chlorophyll concentrations on photosynthesis in foliage leaves. Physiol Plant 1:5–37

Harley PC, Tenhunen JD (1991) Modeling the photosynthetic response of C3 leaves to environmental factors. In: American Society of Agronomy, Crop Science Society of America (eds) Modeling crop photosynthesis – from biochemistry to canopy. CSSA Special Publication, Madison

Hikosaka K (1997) Modelling optimal temperature acclimation of the photosynthetic apparatus in C3 plants with respect to nitrogen use. Ann Bot 80:721–730

Hikosaka K (2004) Interspecific difference in the photosynthesis-nitrogen relationship: patterns, physiological causes, and ecological importance. J Plant Res 117:481–494

Hikosaka K (2005) Nitrogen partitioning in the photosynthetic apparatus of Plantago asiatica leaves grown under different temperature and light conditions: similarities and differences between temperature and light acclimation. Plant Cell Physiol 46:1283–1290

Hikosaka K (2010) Mechanisms underlying interspecific variation in photosynthetic capacity across wild plant species. Plant Biotech 27:223–229

Hikosaka K, Hirose T (1998) Leaf and canopy photosynthesis of C_3 plants at elevated CO_2 in relation to optimal partitioning of nitrogen among photosynthetic components: theoretical prediction. Ecol Model 106:247–259

Hikosaka K, Terashima I (1995) A model of the acclimation of photosynthesis in the leaves of C3 plants to sun and shade with respect to nitrogen use. Plant Cell Environ 18:605–618

Hikosaka K, Terashima I (1996) Nitrogen partitioning among photosynthetic components and its consequence in sun and shade plants. Funct Ecol 10:335–343

Hikosaka K, Terashima I, Katoh S (1994) Effects of leaf age, nitrogen nutrition and photon flux density on the distribution of nitrogen among leaves of a vine (*Ipomoea tricolor* Cav.) grown horizontally to avoid mutual shading of leaves. Oecologia 97:451–457

Hikosaka K, Hanba YT, Hirose T, Terashima I (1998) Photosynthetic nitrogen-use efficiency in woody and herbaceous plants. Funct Ecol 12:896–905

Hikosaka K, Murakami A, Hirose T (1999a) Balancing carboxylation and regeneration of ribulose-1,5-bisphosphate in leaf photosynthesis in temperature acclimation of an evergreen tree, *Quercus myrsinaefolia*. Plant Cell Environ 22:841–849

Hikosaka K, Sudoh S, Hirose T (1999b) Light acquisition and use of individuals competing in a dense stand of an annual herb, *Xanthium canadense*. Oecologia 118:388–396

Hikosaka K, Ishikawa K, Borjigidai A, Muller O, Onoda Y (2006) Temperature acclimation of photosynthesis: mechanisms involved in the changes in temperature dependence of photosynthetic rate. J Exp Bot 57:291–302

Hikosaka K, Nabeshima E, Hiura T (2007) Seasonal changes in temperature response of photosynthesis in canopy leaves of *Quercus crispula* in a cool-temperate forest. Tree Physiol 27:1035–1041

Hirose T (1984) Nitrogen use efficiency in growth of *Polygonum cuspidatum* Sieb. et Zucc. Ann Bot 54:695–704

Hirose T, Werger MJA (1987a) Nitrogen use efficiency in instantaneous and daily photosynthesis of leaves in the canopy of a *Solidago altissima* stand. Physiol Plant 70:215–222

Hirose T, Werger MJA (1987b) Maximizing daily canopy photosynthesis with respect to the leaf nitrogen allocation pattern in the canopy. Oecologia 72:520–526

Hirose T, Werger MJA, Pons TL, van Rheenen JWA (1988) Canopy structure and leaf nitrogen distribution in a stand of *Lysimachia vulgaris* L. as influenced by stand density. Oecologia 77:145–150

Hogan KP, Smith AP, Ziska LH (1991) Potential effects of elevated CO_2 and changes in temperature on tropical plants. Plant Cell Environ 14:763–778

Hollinger DY (1989) Canopy organization and foliage photosynthetic capacity in a broad-leaved evergreen montane forest. Funct Ecol 3:53–62

Ito A (2010) Changing ecophysiological processes and carbon budget in East Asian ecosystems under near-future changes in climate: implications for long-term monitoring from a process-based model. J Plant Res 123:577–588

Jurik TW (1986) Temporal and spatial patterns of specific leaf weight in successional northern hardwood tree species. Am J Bot 73:1083–1092

Kikuzawa K (2003) Phenological and morphological adaptations to the light environment in two woody and two herbaceous plant species. Funct Ecol 17:29–38

Laisk A, Eichelmann H, Oja V, Rasulov B, Padu E, Bichele I, Pettai H, Kull O (2005) Adjustment of leaf parameters to shade in a natural canopy: rate parameters. Plant Cell Environ 28:375–388

Long SP, Ainsworth EA, Rogers A, Ort DR (2004) Rising atmospheric carbon dioxide: plants FACE the future. Annu Rev Plant Biol 55:591–628

Makino A, Mae T, Ohara K (1983) Photosynthesis and ribulose 1,5-bisphosphate carboxylase in rice leaves. Plant Physiol 73:1002–1007

Makino A, Shimada T, Takumi S, Kaneko K, Matsuoka M, Shimamoto K, Nakano H, Miyao-Tokutomi M, Mae T, Yamamoto N (1997) Does decrease in ribulose-1,5-bisphosphate carboxylase by 'antisense' RbcS lead to a higher nitrogen-use efficiency of photosynthesis under conditions of saturating CO_2 and light in rice plants? Plant Physiol 114:483–491

McClendon JH (1962) The relationship between the thickness of deciduous leaves and their maximum photosynthetic rate. Am J Bot 49:320–322

Medlyn BE (1996) The optimal allocation of nitrogen within the C_3 photosynthetic system at elevated CO_2. Aust J Plant Physiol 23:593–603

Medlyn BE, Badeck FW, de Pury DGG, Barton CVM, Broadmeadow M, Ceulemans R, de Angelis P, Forstreuter M, Jach ME, Kellomaki S, Laitat E, Marek M, Philippot S, Rey A, Strassemeyer J, Laitinen K, Liozon R, Portier B, Roberntz P, Wang K, Jarvis PG (1999) Effects of elevated CO_2 on photosynthesis in European forest species: a meta-analysis of model parameters. Plant Cell Environ 22:1475–1495

Migita C, Chiba Y, Tange T (2007) Seasonal and spatial variations in leaf nitrogen content and resorption in a *Quercus serrata* canopy. Tree Physiol 27:63–67

Milthorpe FL, Newton P (1963) Studies on the expansion of the leaf surface. III. The influence of radiation on cell division and leaf expansion. J Exp Bot 14:483–495

Monsi M, Saeki T (1953) Über den Lichtfaktor in den Pflanzengesellschaften und seine Bedeutung für die Stoffproduktion. Jap J Bot 14:22–52

Muller O, Hikosaka K, Hirose T (2005) Seasonal changes in light and temperature affect the balance between light harvesting and light utilisation components of photosynthesis in an evergreen understory shrub. Oecologia 143:501–508

Muller O, Oguchi R, Hirose T, Werger MJA, Hikosaka K (2009) The leaf anatomy of a broad-leaved evergreen allows an increase in leaf nitrogen content in winter. Physiol Plant 136:299–309

Muller O, Hirose T, Werger MJA, Hikosaka K (2011) Optimal use of leaf nitrogen explains seasonal changes in leaf nitrogen content of an understorey evergreen shrub. Ann Bot 108:529–536

Naidu SL, Delucia EH (1997) Growth, allocation and water relations of shade-grown *Quercus rubra* L. saplings exposed to a late-season canopy gap. Ann Bot 80:335–344

Naidu SL, Delucia EH (1998) Physiological and morphological acclimation of shade-grown tree seedlings to late-season canopy gap formation. Plant Ecol 138:27–40

Niinemets Ü (1997) Role of foliar nitrogen in light harvesting and shade tolerance of four temperate deciduous woody species. Funct Ecol 11:518–531

Niinemets Ü, Kull O, Tenhunen JD (1998) An analysis of light effects on foliar morphology, physiology, and light interception in temperate deciduous woody species of contrasting shade tolerance. Tree Physiol 18:681–696

Niinemets Ü, Ellsworth DS, Lukjanova A, Tobias M (2001) Site fertility and the morphological and photosynthetic acclimation of *Pinus sylvestris* needles to light. Tree Physiol 21:1231–1244

Nobel PS, Zaragoza LJ, Smith WK (1975) Relation between mesophyll surface area, photosynthetic rate, and illumination level during development for leaves of *Plectranthus parviflorus* Henckel. Plant Physiol 55:1067–1070

Oguchi R, Hikosaka K, Hirose T (2003) Does the photosynthetic light-acclimation need change in leaf anatomy? Plant Cell Environ 26:505–512

Oguchi R, Hikosaka K, Hirose T (2005) Leaf anatomy as a constraint for photosynthetic acclimation: differential responses in leaf anatomy to increasing growth irradiance among three deciduous trees. Plant Cell Environ 28:916–927

Oguchi R, Hikosaka K, Hiura T, Hirose T (2006) Leaf anatomy and light acclimation in woody seedlings after gap formation in a cool-temperate deciduous forest. Oecologia 149:571–582

Oguchi R, Hikosaka K, Hiura T, Hirose T (2008) Costs and benefits of photosynthetic light acclimation by tree seedlings in response to gap formation. Oecologia 155:665–675

Onoda Y, Hikosaka K, Hirose T (2005) The balance between RuBP carboxylation and RuBP regeneration: a mechanism underlying the interspecific variation in acclimation of photosynthesis to seasonal change in temperature. Funct Plant Biol 32:903–910

Osada N, Onoda Y, Hikosaka K (2010) Effects of atmospheric CO_2 concentration, irradiance and soil nitrogen availability on leaf photosynthetic traits on *Polygonum sachalinense* around the natural CO_2 springs in northern Japan. Oecologia 164:41–52

Röhrig E (1991) Climatic conditions. In: Röhrig E, Ulrich B (eds) Temperate deciduous forests. Elsevier, Amsterdam

Romme WH, Martin WH (1982) Natural disturbance by tree falls in old-growth mixed mesophytic forest: Lilley Cornett woods, Kentucky. In: Proceedings from the Fourth Central Hardwood Forest conference, University of Kentucky, Lexington

Ryel RJ, Beyschlag W (2000) Gap dynamics. In: Marshall B, Roberts JA (eds) Leaf development and canopy growth. Sheffield Academic Press, Sheffield

Sage RF (1994) Acclimation of photosynthesis to increasing atmospheric CO_2: the gas exchange perspective. Photosyn Res 39:351–368

Sims DA, Pearcy RW (1992) Response of leaf anatomy and photosynthetic capacity in *Alocasia-macrorrhiza araceae* to a transfer from low to high light. Am J Bot 79:449–455

Terashima I, Evans JR (1988) Effects of light and nitrogen nutrition on the organization of the photosynthetic apparatus in spinach. Plant Cell Physiol 29:143–155

Terashima I, Hikosaka K (1995) Comparative ecophysiology of leaf and canopy photosynthesis. Plant Cell Environ 18:1111–1128

Terashima I, Miyazawa SI, Hanba YT (2001) Why are sun leaves thicker than shade leaves? Consideration based on analyses of CO_2 diffusion in the leaf. J Plant Res 114:93–105

Turnbull MH, Doley D, Yates DJ (1993) The dynamics of photosynthetic acclimation to changes in light quantity and quality in three Australian rainforest tree species. Oecologia 94:218–228

Turnbull TL, Kelly N, Adams MA, Warren CR (2007) Within-canopy nitrogen and photosynthetic gradients are unaffected by soil fertility in field-grown *Eucalyptus globulus*. Tree Physiol 27:1607–1617

Verbelen JP, De Greef JA (1979) Leaf development of *Phaseolus vulgaris* L. in light and darkness. Am J Bot 66:970–976

von Caemmerer S, Evans JR (1991) Determination of the average partial pressure of carbon dioxide in chloroplasts from leaves of several C-3 plants. Aust J Plant Physiol 18:287–306

von Caemmerer S, Farquhar GD (1981) Some relationships between the biochemistry of photosynthesis and the gas exchange of leaves. Planta 153:376–387

Warren CR, Adams MA (2001) Distribution of N, Rubisco, and photosynthesis in *Pinus pinaster* and acclimation to light. Plant Cell Environ 24:597–609

Webber AN, Nie GY, Long SP (1994) Acclimation of photosynthetic proteins to rising atmospheric CO_2. Photosyn Res 39:413–425

Wilson KB, Baldocchi DD, Hanson PJ (2000) Spatial and seasonal variability of photosynthetic parameters and their relationship to leaf nitrogen in a deciduous forest. Tree Physiol 20:565–578

Wilson KB, Baldocchi DD, Hanson PJ (2001) Leaf age affects the seasonal pattern of photosynthetic capacity and net ecosystem exchange of carbon in a deciduous forest. Plant Cell Environ 24:571–583

Wright IJ, Leishman MR, Read C, Westoby M (2006) Gradients of light availability and leaf traits with leaf age and canopy position in 28 Australian shrubs and trees. Funct Plant Biol 33:407–419

Yamashita N, Ishida A, Kushima H, Tanaka N (2000) Acclimation to sudden increase in light favoring an invasive over native trees in subtropical islands, Japan. Oecologia 125:412–419

Yamori W, Noguchi K, Hikosaka K, Terashima I (2009) Cold-tolerant crop species have greater temperature homeostasis of leaf respiration and photosynthesis than cold-sensitive species. Plant Cell Physiol 50:203–215

Yamori W, Noguchi K, Hikosaka K, Terashima I (2010) Phenotypic plasticity in photosynthetic temperature acclimation among crop species with different cold tolerance. Plant Physiol 152:388–399

Yasumura Y, Onoda Y, Hikosaka K, Hirose T (2005) Nitrogen resorption from leaves under different growth irradiance in three deciduous woody species. Plant Ecol 178:29–37

Chapter 2
Ecophysiological Aspects of Phloem Transport in Trees

Teemu Hölttä, Maurizio Mencuccini, and Eero Nikinmaa

Abstract The primary function of the phloem is the transport of assimilate products from mature leaves to other tissues. Here we examine this function from a whole tree perspective and relate it to assimilate production, tree water relations, and tree structure. We argue that the turgor and osmotic pressures driving flow in the phloem are determined by these factors. An example calculation of these interactions is presented. The generalizations and possible shortcomings of the Münch flow hypothesis, the simplest theoretical framework used in describing phloem transport, are also discussed.

2.1 The Münch Flow Hypothesis

Long distance transport of water from soil through the xylem to the leaves is a fairly well understood process (see Chap. 6), but the long distance transport of assimilate products in the adjacent phloem tissue is a much less understood, and a more complicated process. The formulation of the theory of phloem transport is acknowledged to date to the work by Edward Münch in the 1930s, although very similar ideas were already developed in the nineteenth century (reviewed in Knoblauch and Peters 2010). According to the Münch flow hypothesis, sugars produced in leaf mesophyll cells are loaded into the sieve tubes in the phloem. Sugar loading can be an active process, or it can happen passively along a concentration gradient, depending on the species (Turgeon 2010). This loading decreases the osmotic

T. Hölttä (✉) • E. Nikinmaa
Department of Forest Sciences, University of Helsinki, PO Box 24, 00014 Helsinki, Finland
e-mail: teemu.holtta@helsinki.fi

M. Mencuccini
School of GeoSciences, University of Edinburgh, Crew Building, West Mains Road, Edinburgh EH9 3JN, UK

M. Tausz and N. Grulke (eds.), *Trees in a Changing Environment*, Plant Ecophysiology 9, DOI 10.1007/978-94-017-9100-7_2,

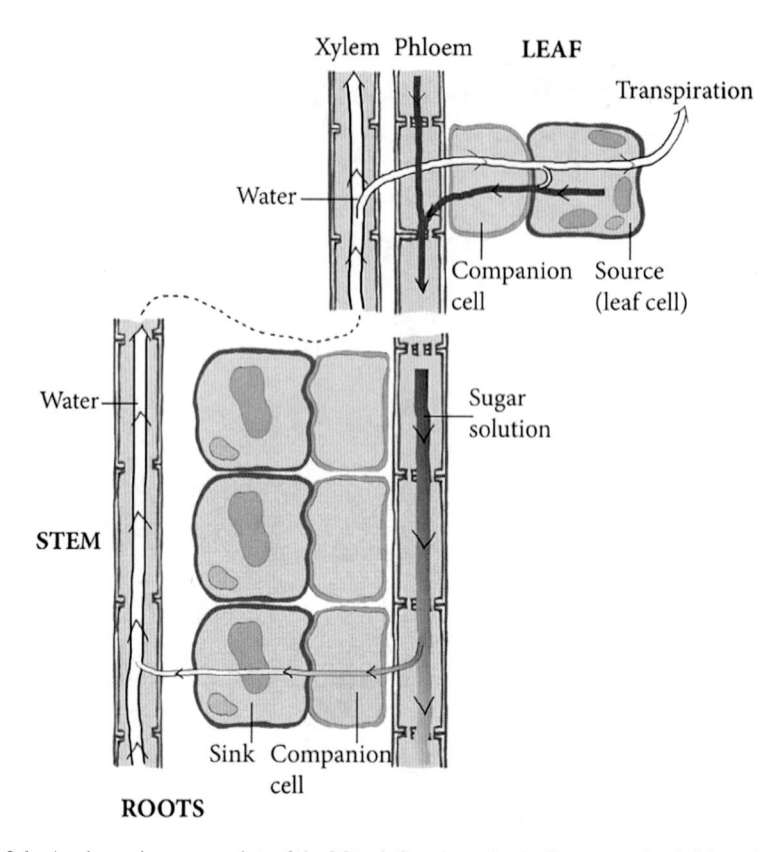

Fig. 2.1 A schematic presentation of the Münch flow hypothesis. Sugars are loaded into the sieve tubes of the phloem at the sugar source (leaf). The osmotic pressure of the sieve tubes is decreased; water is taken up from the surrounding tissue, which results in an increase in turgor pressure at the source. Unloading of sugars from the sieve tubes at the sinks (stem and roots) increases the sieve tube osmotic pressure, water escapes to surrounding tissue, and turgor pressure is reduced. The turgor pressure gradient between the source and the sink is the driving force for phloem transport. Phloem function is linked to xylem transport as water circulates between the two tissues. Phloem takes up water from the xylem along the whole transport pathway unless sugar unloading is locally very large. The transpiration stream in the leaves can also pass apoplastically along the cell walls without crossing the membrane of cells (Figure drawn by Kari Heliövaara, University of Helsinki)

potential of the sieve tubes, which results in water flow from the surrounding tissue, and a subsequent increase in turgor pressure. The active unloading of sugars from the sieve tubes at the sinks decreases the sieve tube osmotic concentration, resulting in water efflux and a reduction in turgor pressure. This turgor pressure gradient between the source and the sink is the driving force for phloem transport. According to the Münch flow hypothesis, sieve tubes are thought to be symplastically continuous along the whole plant axis, and the phloem sap flow totally passive outside the source and sink areas. The Münch flow hypothesis is depicted schematically in Fig. 2.1.

2.1.1 Possible Shortcomings of the Münch Flow Hypothesis

There are doubts over whether the Münch flow hypothesis alone is sufficient to explain phloem transport. According to the hypothesis, active processes are only required at the sources and sinks, and the transport pathway is assumed to be passive (non-energy consuming). Many experiments have demonstrated that this is not fully the case, as for example metabolic impairment of the transport phloem by cold blocking or metabolic poisons has been shown to affect phloem transport dramatically (e.g., Peuke et al. 2006). Continuous leakage and reloading of solutes has been observed along the phloem translocation pathway (Minchin and Thorpe 1987; McQueen et al. 2005).

Doubts over the role and significance of passive phloem transport have been expressed, especially for trees due to long transport distances. In the case of tall trees, the turgor pressure difference required to drive phloem flow between the sources and sinks, and the speed of information transmission between the sources and sinks, could become quite large based on standard theory (Thompson 2006). Experimental tests of the basic tenet that the turgor pressure gradient will drive solution flow in a compartment bounded by semipermeable membranes are difficult to conduct at realistic micrometer scales. Jensen et al. (2010, 2011) used lab-on-a-chip technology to develop a system with dimensions approximating those of real sieve elements (50–200 μm) and experimentally verified that osmotic gradients can generate large enough pressure gradients to move the solution at speeds predicted by the theory.

2.2 Relationships Between Whole-Tree Transport of Assimilated Sugars, Phloem Turgor Pressure, and Osmotic Concentration

A turgor pressure gradient in the phloem sieve tubes is required to overcome the frictional resistance between the phloem sap and sieve tubes walls and the sieve plate pores between adjoining sieve tube elements. An important issue in reconciling phloem transport with the Münch flow hypothesis is the magnitude of these turgor gradients and the flow rates maintained in the sieve tubes (e.g., Thompson 2006). The pressure difference required for the transport of photosynthates from the leaves can be approximated from simple equations, provided that the structural characteristics of the phloem are known. The transport rate of sugars in the phloem sap is proportional to the turgor pressure gradient times the concentration of the phloem sap. The relation can be expressed mathematically as follows:

$$J_s = \frac{\Delta P}{l} c \frac{kA}{\mu} \qquad (2.1)$$

ΔP is the turgor pressure difference between the sugar source and sink, l is the distance between the sugar source and sink, c is sugar concentration, k is phloem specific conductivity, A is phloem cross-sectional area, and μ is phloem sap viscosity.

Unfortunately there are no direct measurements of phloem specific conductivity k, but from anatomical measurements of sieve tube diameter and the size and number of the sieve pore plates adjoining the tubes, it has been estimated to vary in the range of 0.22–56 μm^2 among different species (Thompson and Holbrook 2003; Mullendore et al. 2010). The specific conductivity ranged between 4 and 12 μm^2 among the tree species studied. Confirmations of these theoretical calculations are missing since there are very few simultaneous measurements of flow rates and turgor pressure gradients (but see Gould et al. 2005). For example, it is uncertain how open the sieve plate pores are in their natural state.

There is another restriction to phloem transport, which relates to the water exchange with the adjacent tissues. Sieve tube water potential has to be nearly in equilibrium with the water potential of the surrounding tissue, which is determined mainly by xylem water transport, at least in the case of trees. Neglecting radial water potential losses (Thompson and Holbrook 2003), a relationship between phloem turgor pressure, osmotic concentration, and xylem water potential is established:

$$\Psi = P - \sigma cRT \qquad (2.2)$$

Ψ is the water potential of surrounding tissue (xylem), P is phloem turgor pressure, c is the osmotic concentration in the sieve tube, R is the gas constant, T is temperature and σ is the reflection coefficient of the membrane separating the sieve tube from its surroundings. Often σ is assumed to have a value of 1, i.e. that of a semipermeable membrane. All of the variables in these equations can be dependent on the axial position along the phloem transport system, so that solving for the pressure and concentration as a function of height is a complex problem, especially as viscosity is highly concentration dependent. To demonstrate the calculation of the phloem turgor pressure gradient, a simplified approach can be taken in which the equations are solved only at one position, i.e. the axial variability in the variables is ignored. Adopting this approach, and assuming that all sugars produced by photosynthesis are transported from the leaves to the soil, a unique solution for the turgor pressure and sugar concentration can be obtained from Eqs. 2.1 and 2.2. This is done here numerically, based on values at the source, as the concentration dependency of viscosity makes the equations impossible to solve analytically.

The turgor pressure at the top of our model tree is approximately 1.0 MPa (Fig. 2.2a), leading to a turgor pressure difference of 0.5 MPa between the source and sink. The osmotic concentration at the source is approximately 800 mol m^{-3} (Fig. 2.2b). Both of these values increase with decreasing phloem permeability and

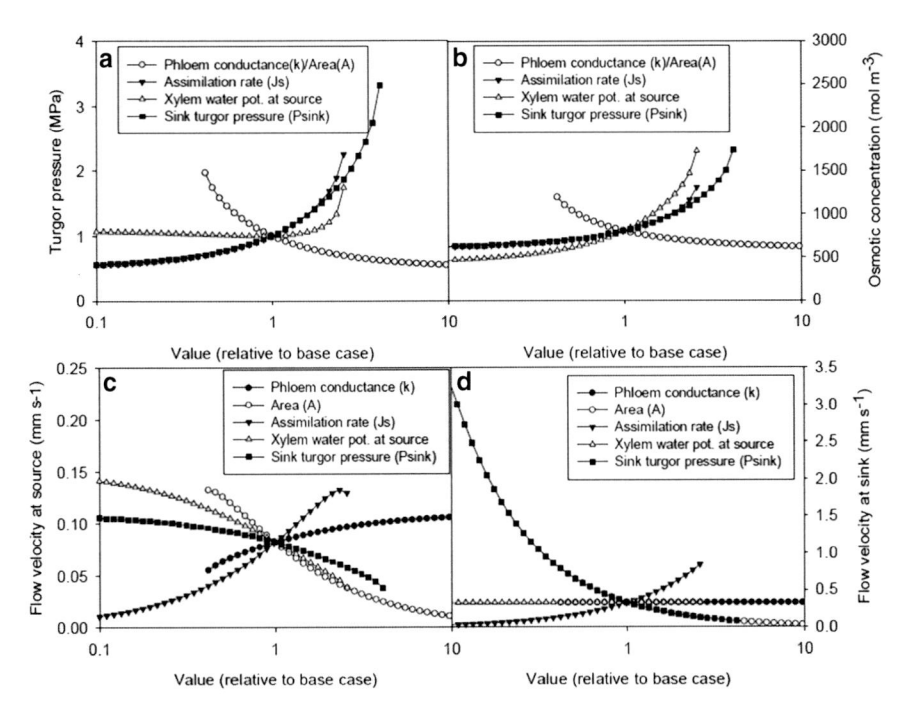

Fig. 2.2 The turgor pressure (**a**) and osmotic concentration (**b**) at the source phloem, and the phloem sap flow velocity at the source (**c**) and sink (**d**), and their changes when phloem parameterization (in Eqs. 2.1 and 2.2) is altered. The parameter values used in the base case calculation are as follows: tree height (h) 10 m, phloem cross-sectional area (A) 0.25×10^{-3} m^2 (which equates to 1 mm thick phloem on a 0.16 m diameter stem), phloem permeability (k) 4 μm^2, whole tree assimilation rate (J_s) 17 μmol s^{-1} of sucrose, xylem water potential (ψ) at source and sink − 1 and 0 MPa, respectively, and a sink turgor pressure (P) 0.5 MPa. Phloem sap viscosity μ was made a function of sugar concentration. The parameter values for which no values can be found in the graph (e.g. low conductance/phloem cross-sectional area) represent cases where the phloem cannot transport all assimilated sugars. Tree height was not varied because a change in tree height is likely to lead to simultaneous changes in almost all of the parameters. In (**a**) and (**b**), changes in phloem conductance and phloem cross-sectional area both produce exactly the same changes, and are therefore not drawn separately

phloem cross-sectional area, and with increasing whole tree photosynthesis rate and sink turgor pressure. The turgor pressure and osmotic concentration at the source increase with decreasing xylem water potentials. The turgor pressure also slightly increases at very high xylem water potentials as the phloem sap gets diluted with regard to sugar concentration. The values for the turgor pressure gradient and osmotic concentration are in agreement with measured values (Turgeon 2010). The resulting phloem sap flow velocity is approximately 0.1 mm s^{-1} at the source (Fig. 2.2c), but it increases considerably towards the sink (Fig. 2.2d) to compensate for the decreasing sugar concentration towards the sugar sink, as the sugar flux rate is the same everywhere. In our example base case calculation, the osmotic concentration at the sink is only one fourth of the osmotic concentration at the source, and the flow velocity therefore is five times higher, i.e. 0.3 mm s^{-1}, which is very similar to values found in experimental NMR studies (e.g., Windt et al. 2006). The sap flow velocity is also sensitive to the parameter choices: it increases with increasing phloem conductance and assimilation rate, and with decreasing phloem cross-sectional area, xylem water potential and sink turgor pressure (Fig. 2.2c, d).

Note that the solution presented is the only possible (stable) solution to Eqs. 2.1 and 2.2. Therefore, no other combination of values of turgor pressure (or its gradient) and osmotic concentration is possible in steady state. For example, a change in turgor pressure at any point along the pathway would lead to a perturbation from steady state, and the turgor pressure gradient would eventually (although not necessarily very quickly), have to return back to the steady state solution. This could also indicate that it is not the pressure difference that controls the flow rate, but rather the flow rate is determined by the transport needs of the plant, which determines the whole tree pressure gradient, as hypothesized by Knoblauch and Peters (2010). This could be compared to the situation in the xylem. It can be argued that the xylem water potential gradient does not control the xylem sap flux rate. Instead, the xylem water potential gradient is determined by xylem conductance and the evaporative demand of water, which is tightly controlled by the stomata in the leaves.

However, it seems possible that plants could exert control over phloem conductivity, and therefore be able to "tune" the values of plant turgor pressure and its gradient. Phloem conductivity is likely to be very dependent on the effective radii of the sieve pores and other obstructions to flow, which could be modified very quickly (Thorpe et al. 2010). It also seems possible that the phloem conducting area varies over time (Windt et al. 2006). Theoretically, a modification in phloem conductivity and flow conducting cross-sectional area would change the turgor and osmotic pressures and could therefore act as a mechanism to control the turgor pressure and osmotic concentrations in the phloem. In addition, there could be many other ways for a plant to regulate turgor pressure in the short term, such as changes in the reflection coefficient of the sieve tubes (see Eq. 2.2) or the active loading and unloading of sugars along the phloem transport pathway.

2.2.1   Speed of Link Between Phloem Sources and Sinks

One peculiar feature of phloem transport is that the propagation of changes in sugar concentration is predicted to occur faster than the movement of the individual sugar molecules themselves. The situation arises since the driving force for the advection-driven phloem transport is created by the substance itself, i.e. sugar. This can be predicted from the mathematical formulation of the Münch flow hypothesis (e.g., Ferrier et al. 1975; Thompson 2006), but its demonstration is beyond the scope of this chapter. Increase in sink sugar concentration due to source action can occur before the actual sugar molecules have traveled from the source to the sink. The experimental evidence for sugar concentrations propagating faster than individual sugar molecules is scarce, probably since the measurement of phloem sugar concentration dynamics is very difficult. However, a meta-analysis by Mencuccini and Höltta (2010) revealed that the time lag between a change in photosynthesis rate and a change in soil or ecosystem respiration rate was on average shorter, and much less dependent on transport distance, than the transit time of individual, isotopically marked molecules. Faster propagation of changes in sugar concentration could be important as it allows sinks and sources at different parts of the plant to sense changes in other parts. A reader interested in the time scales associated with the speed of link should turn to Thompson (2006), for example.

2.2.2 Tree Size and Phloem Transport

Since large transport distances are hypothesized to lead to large pressure differences between the sources and sinks and to a slow propagation of molecules and information between the sources and sinks, how can tall trees cope with the challenges of transporting photoassimilates? An increasing turgor pressure difference with increasing transport distance could be inferred simply from Eq. 2.1: the turgor pressure difference necessary to transport a given amount of solution is linearly proportional to transport distance, provided that the cross-sectional area and permeability remain constant. Experimental evidence of turgor pressure gradients in trees and especially its scaling with tree size is very limited. Turgeon (2010) claimed that some indications exist that the turgor pressure differences between the sources and sinks do not scale with tree size. On the contrary, calculations by Mencuccini and Höltta (2010) from a meta-analysis of time lags between photosynthesis and soil respiration pointed towards phloem pressure gradients (calculated from the observed time-lags) increasing with plant size, although in the final statistical analysis the calculated rate of change in pressure gradients with tree size was highly correlated with the corresponding rate of change in specific conductivity, making their separation difficult.

Very little is known about the variation in phloem conduit properties and flow conducting area with changing tree size and axial position. In the neighboring xylem tissue, the water potential difference between the leaves and roots (which is equivalent to the turgor pressure difference in the phloem as the driving force for transport) normally does not differ much between small and tall trees. It is therefore evident that conduit properties must change with tree size so that sufficient water supply to the leaves is maintained with increasing transport distance. This is achieved by an increase in xylem conduit size according to a power law to prevent the loss of water transport capacity due to increased transport length (West et al. 1999). This general scaling theory also suggests that a similar principle must apply to phloem conduits to compensate for the increasing transport distance with increasing tree height, but very few studies have been made to address this topic directly. Perhaps the only systematic study was the one conducted by Mencuccini et al. (2010), who found that phloem conduits did increase in size with increasing tree size, therefore alleviating the problems of phloem transport over long distances.

Changes in the phloem cross-sectional area may also influence phloem transport. However, quantitative measurements of conducting area along the stem axis are rare. Along the stem height, the actual thickness of conducting phloem cells seems to be fairly constant (Quilho et al. 2000), while the cross-sectional area of water conducting wood increases clearly from the stem top towards the base. This means that along the conducting axis, the ratio of phloem to xylem tissue increases towards the leaves, being close to one in leaf veins. Hölttä et al. (2006) showed that such a structure can maintain higher phloem loading rates than if the proportions were equal along the entire axis.

An alternative mechanism that would allow trees to overcome the problems of long transport distances is the use of the so-called solute relays. If solute relays were present, the sieve tubes would not be symplastically connected over the whole transport distance, but there would be loading and unloading of sugars at specific points along the transport pathway (Lang 1979). The pressure difference and the speed of information transmission within a tree would then decrease in approximately linear proportion to the number of relays (Hölttä et al. 2009). Lang (1979) estimated that approximately 2 % of the transported sugars would be used up in each relay because of the metabolic cost of unloading and reloading. Hypothetically, this metabolic cost could be compared with the cost of building larger or more phloem sieve tubes without relays to achieve identical sugar translocation rates. However, there is no substantial experimental evidence either in favour or against the existence of solute relays.

2.3   Connections Between Phloem and Xylem Transport

While xylem transport must occur against gravity, phloem transport occurs downward and therefore can take advantage of gravity (particularly at night). However, during the day, phloem transport must occur against the tree water potential gradient, which is generated by the transpiration stream in the xylem. Phloem transport should therefore be strongly coupled to whole plant water relations, i.e., xylem water potential and its gradient. Detailed modeling studies (Ferrier et al. 1975; Thompson and Holbrook 2003; Hölttä et al. 2006) demonstrate this, as does inference from our example calculation by modifying the water potential term in Eq. 2.2. There is also accumulating experimental evidence showing decreased phloem turgor pressures and exudation rates in connection with more negative xylem water potentials (e.g., Cernusak et al. 2009).

The impairments in phloem transport have also recently been raised as one candidate that may affect tree function, and even survival, during drought (McDowell and Sevanto 2010; Sala et al. 2010). For example, drought has been shown to decrease the export of sugars from the leaves and to decouple photosynthetic production from below-ground processes (e.g., Ruehr et al. 2009). From a theoretical point of view, the turgor pressure has to decrease and/or the osmotic concentration to increase due to the decrease in xylem water potential during drought (see Eq. 2.2). A decrease in the turgor pressure gradient clearly slows down phloem transport. Sugars also need to accumulate in the phloem as water potential decreases to prevent an excessive loss of turgor in the phloem. In our earlier example calculation, the total phloem sugar pool consists of approximately 1 week's worth of photosynthesized sugars (total amount of sugars in the phloem divided by the photosynthesis rate). If whole plant water potential decreases by half, the amount of sugars in the phloem must double to maintain a constant turgor pressure (see Eq. 2.2). This requires 1 week's worth of photosynthesis, which is taken away from active metabolic processes and/or must be found in some storage compartment. Another issue with low xylem water potential could be the increase in phloem sap viscosity with increasing sugar concentrations. Sucrose is the main constituent of sugars transported in the phloem, and sap viscosity is an exponentially rising function of sucrose concentration. Highly concentrated solutions become too viscous to be transported efficiently (e.g., Hölttä et al. 2009). This problem could perhaps be partly alleviated by the use of other osmotic substances, such as potassium, to create the turgor pressure required for turgor maintenance and transport without an increase in viscosity. However, in contrast to sucrose, large stores of potassium may not be available.

Also xylem functionality has been found to be dependent on phloem function. Xylem embolism refilling has been shown to require the integrity of phloem transport. Embolism refilling does not occur below a point of phloem girdling (e.g., Salleo et al. 2004). Xylem transport capacity also seems to be dependent on the recycling of ions from the phloem. Xylem sap ionic concentration apparently modifies the hydration state, and therefore the conductivity, of the xylem inter-conduit pit membranes (Zwieniecki et al. 2001). In addition, phloem sap flow also creates a flow in the xylem, so called Münch counterflow, which could be important in driving the transport of, for example, nutrients in the xylem in the absence of transpiration.

2.4 Measuring Phloem Transport

In contrast to the dead, hollow water conducting tissue in the xylem, the phloem consists of living sieve cell complexes, made up of companion cells and sieve tubes (themselves made up of sieve elements). It is very difficult to measure phloem flow and pressure gradients non-intrusively. The living phloem tissue responds strongly to disturbances which has made its function difficult to study in vivo (Van Bell 2003). Phloem tubes are under high pressure and defensive mechanisms readily seal off any wounded sieve elements to prevent phloem leaking, for example, after a puncture or an injury. This makes the anatomical and physiological analysis of cut samples difficult and potentially unreliable.

Phloem transport has been measured using radioisotopes, stable isotopes and nuclear magnetic resonance imaging. Findings from isotopic studies suggest that sugars are transported quite directly from photosynthesis to the phloem as the isotopic ratio of phloem sugars has been found to lag behind the isotopic ratio of photosynthetic products by only a short time (Keitel et al. 2003; Gessler et al. 2008). Nuclear magnetic resonance imaging has provided the possibility to monitor the phloem sap flow velocities under laboratory conditions (Windt et al. 2006). Recently, direct measurements of the phloem turgor pressure and its gradient have been obtained by pressure probes glued to exuding stylets of aphids (Gould et al. 2005). Jensen et al. (2011) developed an ingenious system to detect phloem sap velocity based on the use of an aqueous solution of 5(6)-carboxyfluorescein, which is placed on the surface of a gently abraded leaf surface. Once absorbed, the movement of this substance into the plant is followed by shining a low-intensity laser beam on the leaf petiole and following the movement of the front of the photo-bleached fluorescing dye using coupled highly sensitive photo-diodes (Schulz 1992).

A potential candidate to measure phloem function non-intrusively under field conditions is stem diameter change measurements. The diameter of both the xylem and the inner bark vary in response to changes in pressure, as any elastic material does respond. Both the xylem and the inner bark diameter change mainly in response to changes in xylem water potential due to transpiration, but one would also expect that stem diameter changes also responded to changes in phloem sugar dynamics (Sevanto et al. 2003; DeSchepper and Steppe 2010). However, no theoretical framework has yet been presented to obtain phloem flow dynamics from stem diameter change measurements.

2.5 Assimilate Production and Phloem Transport

Assimilate transport from leaves and utilization at sugar sinks are necessary requirements to maintain continuous photosynthetic production in the leaves. Without sufficient transport capacity, the storage capacity of the leaves would eventually become exhausted. Sugar and starch accumulation in the leaves will cause stomatal gas exchange to be limited and photosynthesis to be down-regulated (e.g., Goldschmidt and Huber 1992; Paul and Foyer 2001). This has been shown to occur for example in response to elevated CO_2 concentrations and phloem girdling.

Similarly, as stomatal conductance efficiently controls the leaf gas exchange (Cowan and Farquhar 1977), especially to avoid dangerous levels of xylem embolism, it could also simultaneously control efficient assimilate transport in the leaf phloem. Due to xylem-phloem linkages, insufficient stomatal control could lead not only to xylem cavitation but also to a reverse flow in the phloem, sugar accumulation in the leaves, and to a loss of turgor pressure elsewhere in the phloem transport pathway (Höltä et al. 2006). The water potential gradient controlling xylem flow and affecting phloem flow is created at the leaf tissue where transpiration creates the hydrostatic pull lifting water up from soil through xylem. Similarly, photosynthesis and sugar phloem loading in leaves facilitate moving assimilates down the phloem. Depending on the degree of stomatal opening and the prevailing environmental conditions, the relationships between these driving gradients, xylem water transport and assimilate transport from the leaves change.

Acknowledgments T. Hölttä received funding from Academy of Finland project #1132561. M. Mencuccini received funding from NERC grant number NE/I011749/1.

References

Cernusak LA, Arthur DJ, Pate JS, Farquhar GD (2009) Water relations link carbon and oxygen isotope discrimination to phloem sap sugar concentration in *Eucalyptus globulus*. Plant Physiol 131:1544–1554

Cowan IR, Farquhar GD (1977) Stomatal function in relation to leaf metabolism and environment. Symp Soc Exp Biol 31:471–505

DeSchepper V, Steppe K (2010) Development and verification of a water and sugar transport model using measured stem diameter variations. J Exp Bot 61:2083–2099

Ferrier JM, Tyree MT, Christy AL (1975) The theoretical time-dependent behavior of a Münch pressure-flow system: the effect of sinusoidal time variation in sucrose loading and water potential. Can J Bot 53:1120–1127

Gessler A, Tcherkez G, Peuke AD, Ghashghaie J, Farquhar GD (2008) Experimental evidence for diel variations of the carbon isotope composition in leaf, stem and phloem sap organic matter in *Ricinus communis*. Plant Cell Environ 31:941–953

Goldschmidt EE, Huber SC (1992) Regulation of photosynthesis by end-product accumulation in leaves of plants storing starch, sucrose and hexose sugars. Plant Physiol 99:1443–1448

Gould N, Thorpe MR, Koroleva O, Minchin PEH (2005) Phloem hydrostatic pressure relates to solute loading rate: a direct test of the Münch hypothesis. Funct Plant Biol 32:1019–1026

Hölttä T, Vesala T, Sevanto S, Perämäki M, Nikinmaa E (2006) Modeling xylem and phloem water flows in trees according to cohesion theory and Münch hypothesis. Trees 20:67–78

Hölttä T, Mencuccini M, Nikinmaa E (2009) Linking phloem function to structure: analysis with a coupled xylem–phloem transport model. J Theor Biol 259:325–337

Jensen KH, Lee J, Bohr T, Bruus H (2010) Osmotically driven flows in microchannels separated by a semipermeable membrane. J Fluid Mech 636:371–396

Jensen KH, Lee J, Bohr T, Bruus H, Holbrook NM, Zwieniecki MA (2011) Optimality of the Münch mechanism for translocation of sugars in plants. J R Soc Interface. doi:10.1098/rsif.2010.0578

Keitel C, Adams MA, Holst T, Matzerakis A, Mayer H, Rennenberg H, Gessler A (2003) Carbon and oxygen isotope composition of organic compounds in the phloem sap provides a short term measure for stomatal conductance of European beech (*Fagus sylvatica* L.). Plant Cell Environ 26:931–936

Knoblauch M, Peters WS (2010) Münch, morphology, microfluidics – our structural problem with the phloem. Plant Cell Environ 33:439–1452

Lang A (1979) A relay mechanism for phloem translocation. Ann Bot 44:141–145

McDowell NG, Sevanto S (2010) The mechanisms of carbon starvation: how, when, or does it even occur at all? New Phytol 186(2):264–266

McQueen JC, Minchin PEH, Thorpe MR, Silvester WB (2005) Short-term storage of carbohydrate in stem tissue of apple (*Malus domestica*), a woody perennial: evidence for involvement of the apoplast. Funct Plant Biol 32:1027–1031

Mencuccini M, Hölttä T (2010) The significance of phloem transport for the speed of link between canopy photosynthesis and belowground respiration. New Phytol 185:189–203

Mencuccini M, Hölttä T, Martinez-Vilalta J (2010) Design criteria for models of the transport systems of tall trees. In: Meinzer FC, Dawson T, Lachenbruch B (eds) Size- and age-related changes in tree structure and function. Springer tree physiology series. Springer-Verlag, New York (in press)

Minchin PEH, Thorpe MR (1987) Measurement of unloading and reloading of photoassimilate within the stem of bean. J Exp Bot 38:211–220

Mullendore DL, Windt CW, Van As H, Knoblauch M (2010) Sieve tube geometry in relation to phloem flow. Plant Cell 22:579–593

Paul MJ, Foyer CH (2001) Sink regulation of photosynthesis. J Exp Bot 52:1383–1400

Peuke AD, Windt CW, Van As H (2006) Effects of cold girdling on flows in the transport phloem in *Ricinus communis*: is mass flow inhibited? Plant Cell Environ 29:15–25

Quilhó T, Pereira H, Richter HG (2000) Within-tree variation in phloem cell dimensions and proportions in *Eucalyptus globulus*. IAWA J 21:31–40

Ruehr NK, Offermann CA, Gessler A, Winkler JB, Ferrio JP, Buchmann N, Barnard RL (2009) Drought effects on allocation of recent carbon: from beech leaves to soil CO_2 efflux. New Phytol 184:950–961

Sala A, Piper F, Hoch G (2010) Physiological mechanisms of drought-induced tree mortality are far from being resolved. New Phytol 186:274–281

Salleo S, LoGullo M, Trifilò P, Nardini A (2004) New evidence for a role of vessel-associated cells and phloem in the rapid xylem refilling of cavitated stems of *Laurus nobilis*. Plant Cell Environ 27:1065–1076

Schulz A (1992) Living sieve cells of conifers as visualized by confocal, laser-scanning fluorescence microscopy. Protoplasma 166:153–164

Sevanto S, Vesala T, Perämäki M, Nikinmaa E (2003) Sugar transport together with environmental conditions controls time lags between xylem and stem diameter changes. Plant Cell Environ 26:1257–1265

Thompson MV (2006) Phloem: the long and the short of it. Trends Plant Sci 11:26–32

Thompson MV, Holbrook NM (2003) Scaling phloem transport: water potential equilibrium and osmoregulatory flow. Plant Cell Environ 26:1561–1577

Thorpe MR, Furch ACU, Minchin PEH, Föller J, Van Bell AJE, Hafke JB (2010) Rapid cooling triggers forisome dispersion just before phloem transport stops. Plant Cell Environ 33:259–271

Turgeon R (2010) The puzzle of phloem pressure. Plant Physiol 154:578–581

Van Bell AJE (2003) The phloem, a miracle of ingenuity. Plant Cell Environ 26:125–149

West GB, Brown JH, Enquist BJ (1999) A general model for the structure and allometry of plant vascular systems. Nature 400:664–667

Windt CW, Vergeldt FJ, de Jager PA, Van As H (2006) MRI of long-distance water transport: a comparison of the phloem and xylem flow characteristics and dynamics in poplar, castor bean, tomato, and tobacco. Plant Cell Environ 29:1715–1729

Zwieniecki MA, Melcher PJ, Holbrook NM (2001) Hydrogel control of xylem hydraulic resistance in plants. Science 291:1059–1062

Chapter 3
Mycorrhizae and Global Change

Michael F. Allen, Kuni Kitajima, and Rebecca R. Hernandez

Abstract Mycorrhizal symbioses are essential components of terrestrial ecosystems. These symbioses are intimate associations between plants and fungi where the plant fixes C, exchanging it for nutrients and water from fungal hyphae that permeate and explore surrounding soil. Perturbations, whether acute (such as disturbance or cutting) or chronic (global change, N deposition) alter mycorrhizal functioning and thereby forest dynamics. Among these dynamics are C sequestration and alleviating nutrient stresses to optimize C:nutrient ratios. We explore three areas whereby global change might alter mycorrhizae, which in turn, will affect forest dynamics. First, increasing temperatures associated with elevated atmospheric CO_2 will increase soil temperature, thereby potentially increasing respiration. However, that may depend upon lags and the variation inherent in diel and seasonal variation. Second, the increased temperature will increase soil drying, and subsequently reduce the length of the growing season for mycorrhizal fungal hyphae. However, elevated CO_2 will simultaneously increase water-use efficiency, thereby increasing the length of the growing season. Third, mycorrhizae increase activity and nutrient uptake with elevated CO_2, negating some of the C:nutrient stress. This activity is dictated by both changing amounts of mycorrhizal hyphal growth and by shifting mycorrhizal fungal taxa, altering the strategies whereby nutrients are acquired and C allocated. This includes spatial (breadth and depth) as well as enzymatic shifts. Finally, we examine the longer-term implications of how global change can alter plant communities and plant dynamics on both ecological and evolutionary time scales.

M.F. Allen (✉) • K. Kitajima • R.R. Hernandez
Center for Conservation Biology, University of California, Riverside, CA 92521, USA
e-mail: michael.allen@ucr.edu

M. Tausz and N. Grulke (eds.), *Trees in a Changing Environment*,
Plant Ecophysiology 9, DOI 10.1007/978-94-017-9100-7_3,
© Springer Science+Business Media Dordrecht 2014

3.1   Introduction

Mycorrhizae play critical roles in all terrestrial ecosystems globally. The vast majority of plants are dependent upon mycorrhizae for acquisition of soil resources, largely because plant roots, including root hairs, are too large to thoroughly explore soils for water and nutrients. Mycorrhizal fungi require carbon (C) obtained from the plant in exchange for the nutrients the fungus provides (e.g., Allen 1991). In addition, because the fungi have the ability to cross soil air gaps and penetrate pores, down to large ultra-micropores, and because water flux occurs in response to potential gradients, the fungi can serve to exchange water between water pockets in soil and the host plants (Allen 2007). Plants that do not form mycorrhizae, largely annuals in the Caryophyllales, live in nutrient rich environments and rapidly exploit open patches of soil resources. For these plants, mycorrhizae are a net C drain (Allen and Allen 1990). A small group of plants form cluster roots, extremely fine roots that simulate mycorrhizal fungi, largely in extremely nutrient deficient soils, such as South Africa and Australia (e.g., Pate et al. 2001). For all other plants, mycorrhizae are the normal condition.

It is also important to remember that the mycorrhizal condition for fungi evolved multiple times, in multiple, independent lineages of fungi. The most common mycorrhiza is the arbuscular mycorrhiza (AM) formed between most plants and members of Glomales, a monophyletic group arising sometime in the Silurian, and expanding with land plants. The second most common mycorrhiza is an ectomycorrhiza (EM), which independently evolved many times among different plant groups, and among many fungi, including the Endogonales, Ascomycota, and Basidiomycota. A third mycorrhizal type, the ericoid mycorrhiza, merits consideration because it forms in extremely nutrient-deficient conditions, which may be altered by changing global conditions. There are specialized mycorrhizae, including orchid mycorrhizae that we will not address in this review. Thus, associations hinge upon diverse groups of fungi but only in the orchids are there interesting directional shifts within life stages, being important for biodiversity, but not global carbon management.

The impacts of global change are largely focused on increasing global temperatures (and resulting shifts in precipitation patterns) or on CO_2 directly. Increasing atmospheric temperature (T_a) increases fungal respiration (R_f) but it also increases evaporation, thereby reducing soil moisture (θ). Furthermore, changes in soil temperature (T_s) and θ could have dramatic effects on mycorrhizal composition and thereby mycorrhizal functioning. Changing CO_2 also has important consequences to mycorrhizal functioning. Increased atmospheric CO_2 increases water-use efficiency, delaying soil drying and increasing θ. Delays in soil drying are especially important in arid and semiarid environments. Shifting precipitation regimes also alter the depth distribution of soil moisture (Thomey et al. 2011), again a factor that is related to mycorrhizal dynamics (Querejeta et al. 2009). Finally, increasing atmospheric CO_2 alters the C:N and C:P ratios in soils, with dramatic impacts on mycorrhizae. During the history of terrestrial vegetation, abrupt changes in atmospheric CO_2 have had important consequences to the evolution of mycorrhizae (Allen et al. 1995).

Global change is a complex suite of changes resulting from chronic anthropogenic perturbation. In the past, most human impacts have been acute: heavy metal deposition, tillage, and soil disturbance. Global change is different in that it proceeds across generations of organisms, including many fungi and plants. This means that there are ecological interactions between organisms (e.g., Klironomos et al. 2005), and likely evolutionary changes incurred (altered gene frequencies within a population). This time scale, coupled with the global nature of the perturbation, means that virtually every interaction between plant-fungus-soil-fungus-plant is altered. These changes include direct CO_2 effects on fixation and C allocation, global-regional-local temperature regimes (with impacts on respiration, water use, and growth), water-use efficiency (WUE), and nutrient-use efficiency (NUE). In addition, not only are humans increasing atmospheric CO_2 through fossil fuel burning and increasing deforestation and desertification, we are also increasing N deposition, at a more local scale, again through fossil fuel consumption and food production. Ecosystems are extremely sensitive to C:N ratios, so that it is the interactions between elements of global change that pose the greatest challenge to understanding and managing forest resources. Similarly, at the scale of an individual plant: fungus mycorrhiza, as either CO_2 increases, or a soil resource increases (e.g., N, P), and the C:N or C:P ratio is altered; the mycorrhizal activity is increased or decreased depending upon the relative ratio (Treseder and Allen 2002).

A final caveat is that most of the results we will present are from our field sites in the southwestern US, particularly the James Reserve, a UC NRS site located near Idyllwild, CA (http://www.jamesreserve.edu) and the Sky Oaks Reserve, (http://fs.sdsu.edu/kf/reserves/sofs/), a San Diego State University reserve near Warner Springs, CA. Other programs are also relevant, but these provide a perspective from a system that appears to be very sensitive to the changing global environment and for which sites have been studied for almost two decades.

3.2   Resource Acquisition and CO_2 Allocation

Water acquisition and nutrient acquisition are fundamentally different processes, and should be so thought of in the context of global change and mycorrhizae. However, both directly affect CO_2 acquisition and nutrient use, so both processes need to be understood as separate parts of the mycorrhizal system. We will first deal with the two individually, then examine how they interact later.

3.2.1   Temperature, Water and Global Change

Elevated global temperatures result from actually a very small increment in long-wave re-radiation (a global average of 3 Wm^{-2}). Given that the peak radiation for a site such as the James Reserve is as high as 1,400 Wm^{-2}, this value appears quite small. But, the change occurs primarily at night (re-radiation), thereby slightly increasing soil temperature (T_s). Every GCM (Global Circulation Model) run shows increased T_a over the next half-century to century (IPCC 2007; Hayhoe et al. 2004) including ecosystems within southern California. This change alters both evaporation, thereby reducing θ, and microbial and root respiration because of increased T_s. But concomitant increases in atmospheric CO_2 will likely negate some of that response by increasing water use efficiency (WUE) and thereby increasing θ. Both of these directional changes are subtle and will require careful new modeling efforts. Nevertheless, some interesting hypotheses can be generated (Fig. 3.1).

In EM ecosystems, net ecosystem exchange (NEE) and soil respiration (R_s) appear to be sensitive to changes in T (Vargas et al. 2010). AM systems may be less sensitive to changing T_a, but that may be because AM plants occupy such a broad range in T_a (from alpine to hot deserts). In any case, increased T, a consistent response among GCM's, will have complex impacts on mycorrhizae. In an experimental AM system, Staddon et al. (2003a) reported that % AM root length increased with summer drought and winter warming. But, we do not know if this was due to decreasing fine root production relative to AM inoculum density, or increased fungal activity. Their observation of reduced hyphal density suggests that we need additional information on production and mortality of both roots and fungi. In theory, increasing T_a should increase microbial activity and R_s (e.g., Pritchard 2011). Thus, increasing T_a should result in increased T_s, thereby increasing mycorrhizal growth, respiration, and mortality. Many studies support the notion that higher T_s increases activity of soil organisms, including mycorrhizae (see Pritchard 2011). At the James Reserve, our modeling of R_s (Vargas and Allen 2008; Hasselquist et al. 2010) suggests that shifting T_s upward by a few degrees could have a measurable indirect impact on mycorrhizae and soil dynamics. However, we do not know if the changing T_a will reasonably alter T_s enough to directly alter soil microbial or mycorrhizal functioning. Applying DayCent modeling (http://www.nrel.colostate.edu/projects/daycent/) to our system, we increased T_a by 2 °C (Kitajima and Allen, unpublished data). There are some interesting problems with DayCent modeling, mostly dealing with the consequences of soil depth and lags in response times that are not captured by the model. Nevertheless, the overall effect is instructive. Two particular responses are relevant here. First, heterotrophic respiration (R_h) increased by approximately 6 % with increasing T_a (and subsequently T_s). Secondly, NEP and NPP increased by approximately 8 %.

In looking at changing T_s with seasonal change, the response of variables such as fine root production and leaf production is similar given the projected change in T_s. Is this enough (6–8 %) to cause a detectable ecosystem response? That is an issue requiring additional research. T_s varies greatly at this site on both a diurnal and on a

Fig. 3.1 Daily energy balance of James Reserve observed between May 1 and May 7, 2010

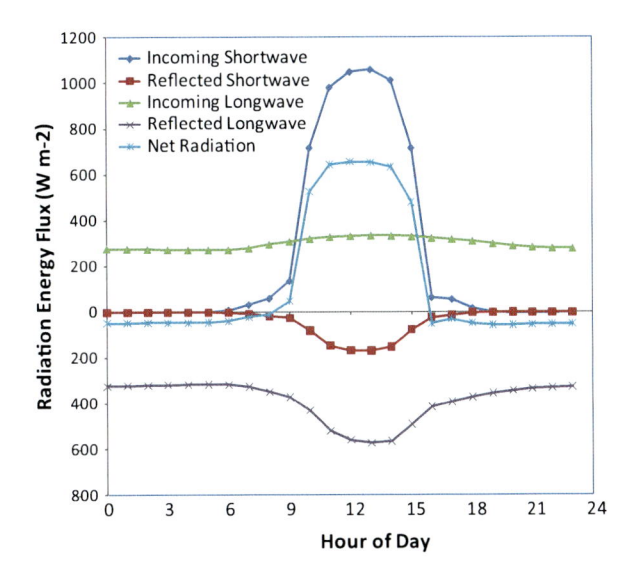

seasonal basis. Q_{10} analyses show a highly variable response within these ranges (Kitajima and Allen, unpublished data), in part associated with the hysteresis actually measured in the field (Vargas and Allen 2008). These differences can be greater than the 6–8 % responses measured.

This interpretation is also supported by the small variation in net radiation exhibited by elevated atmospheric CO_2 to 450 or 550 ppm at the scale of a single site (such as beyond 2,050 or 2,060). For values above these levels, variance estimates increase and predictability declines dramatically. However, we do not really know how this plays out with θ, especially as θ both deceases (higher evaporation) and increases (higher WUE with elevated CO_2). Hernandez and Allen (unpublished data) found that T_a will likely increase with all GCMs, scaled down to the James Reserve (Fig. 3.2). But precipitation varied greatly, depending upon both the model, and on the time element. The clearest signal was a T_a response that will likely modulate θ—presumably occurring via increased evaporation and longer dry seasons under a warming climate—in concert with a dynamic precipitation regime.

We ran two types of models. First, we looked at the ecosystem-scale responses to changing θ using both increasing and decreasing levels using DayCent, in comparison with direct measurements of NEP using eddy covariance (provided by M. Goulden, University of California-Irvine), sapflow, and fine-root dynamics. Second, we modeled R_s based on the Fick's second law of diffusion and measured CO_2 concentrations at three different soil depths (Tang et al. 2005). These relate to mycorrhizae as approximately 50 % of the R_s is newly fixed C based on $\delta^{14}C$ of respired soil CO_2 (Trumbore, personal communications). Further, in the case of this site, every root tip is mycorrhizal, so that we equate increasing fine root NPP with increasing mycorrhizal numbers (see Allen et al. 2010 for more discussion). The

Fig. 3.2 Projected air temperature (°C) and total growing season precipitation (cm) based on individual global climate models (n = 20) and mean projected temperature (black points) averaged over each decade at UC James Reserve (Idyllwild, CA). Models include: (*1*) CCMA (cgcm3 1 t63); (*2*) CNRM (cm3); (*3*) CSIRO (mk3 0); (*4*) GFDL (cm2 0); (*5*) GFDL (cm2 1); (*6*) GISS (aom); (*7*) GISS (model e h); (*8*) GISS (model e r); (*9*) IAP (fgoals1 0 g); (*10*) INM (cm3 0); (*11*) IPSL (cm4); (*12*) MIROC3 (2 hires); (*13*) MIROC3 (2 medres); (*14*) MIUB (echo g); (*15*) MPI (echam5); (*16*) MRI (cgcm2); (*17*) NCAR (ccsm3); (*18*) NCAR (pcm1); (*19*) UKMO (hadcm3); (*20*) UKMO (hadgem1)

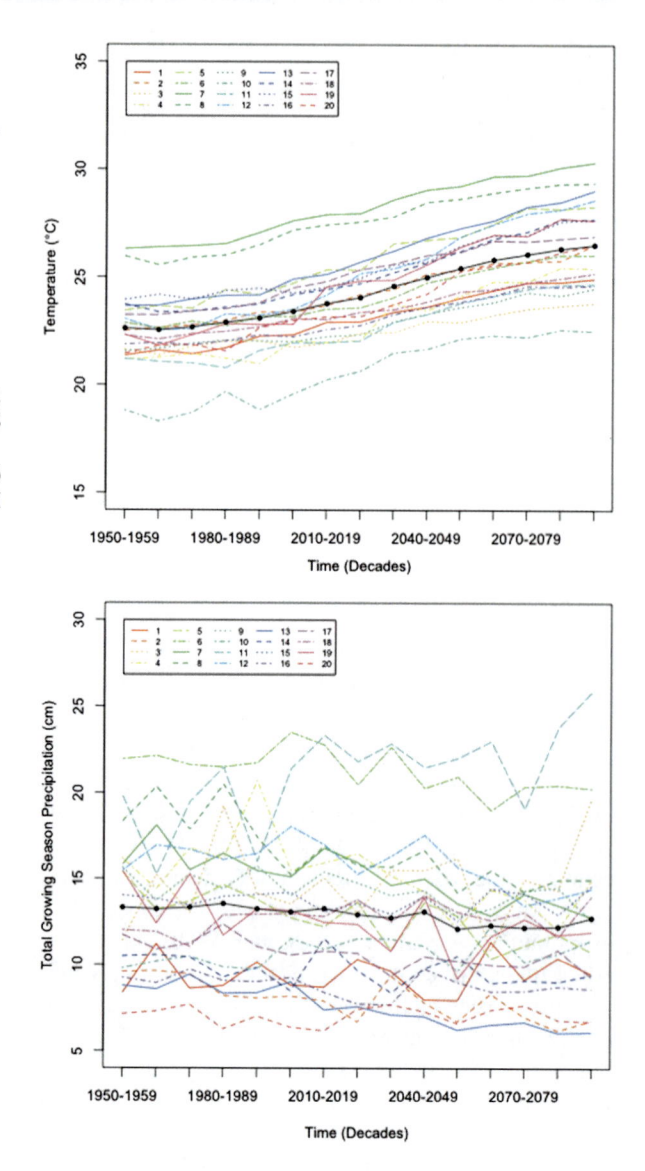

DayCent response of increasing precipitation (precipitation) is dramatic. Virtually all ecosystem parameters increased between 5 % and 10 %. These included NEP, NPP, fine and coarse root NPP, and R_s. As T_a increases, if precipitation also increases due to warming ocean temperatures (one of the GCM scenarios), then we can expect a dramatic increase in C allocation to roots and mycorrhizae, and subsequent sequestration in soils from fine root and mycorrhizal organic matter inputs (see Treseder and Allen 2000; Treseder et al. 2005b). Just as important, if winter precipitation drops, and the recharge of soil moisture declines, all relevant activity declines 10–15 %.

Beyond simply the predictability issues, a key question will be whether these changes are enough to affect other ecosystem processes, including accumulation of fine fuels, drying of soil water with increasing leaf production, or tree mortality due to periodic droughts. As EM fungi consists of chitin and other complex C compounds, rising T could both increase C sequestration by facilitating production, however, it could also decrease soil C due to increased R_s.

AM-dominated ecosystems are sensitive to changes in θ (Vargas et al. 2010) but across a far broader range (deserts to tropical forests). But under global change scenarios, accounting for θ is complicated. Treseder et al. (2003) found small increases in θ because of the increased WUE with elevated CO_2 in chaparral. AM activity also increased, but separating the effects of slight increases in θ from the direct increases in AM due to increased C allocation to AM was not possible. A change in T_a could increase evaporation and thereby reduce θ, and the subsequent length of the growing season. However, in that study, we did not change T (Fig. 3.3).

In that vein, changes in T and θ may also play out in altering the duration of the growing season. Specifically, if T_s increases evaporation, or if θ declines or increases directly due to changing precipitation, then the dry season will be longer or shorter, respectively. If we can establish life spans of fine roots and mycorrhizal fungal hyphae, then we could model how changing conditions would affect total annual production and C inputs to soils.

Estimates in the literature based on laboratory studies appear wildly inaccurate for field studies. For example, Staddon et al. (2003b) estimated that hyphal turned over in 3 days. If all AM hyphae in a field site had this type of lifespan, then the AM hyphae alone, over a growing season, would require an order of magnitude more C than is allocated to root systems. Thus, field observations of hyphal dynamics and growth characteristics are needed.

Allen et al. (2003) developed a hyphal expansion model based on the growth of new roots (from Allen 2001). In this model, as a root tip grows, a new infection forms. Specifically, the extramatrical hyphae branch out from the infection point, transferring nutrients to the plant in exchange for C. In laboratory observations, the hyphae grew at approximately one branching unit per day and up to eight branching orders. Subsequently, the hyphae died back from the tips. The hyphal tips lived only a few days, but the base of the network would survive for 40–50 days, and the runner (or arterial) hyphae could survive for a growing season. Using this model we could better estimate hyphal growth and mortality if we could observe hyphae directly in the field. Such observations would also confer the ability to measure standing crop of AM and EM hyphae. For tracking, we observe the hyphae radiating from a root or EM for identification, especially for EM, where there are no morphological differences.

Using an automated minirhizotron (Allen et al. 2007), which can resolve hyphal dynamics on a daily basis, we found that the mean life span for AM hyphae in the field was 46 days in a meadow at the James Reserve (unpublished data, Hernandez

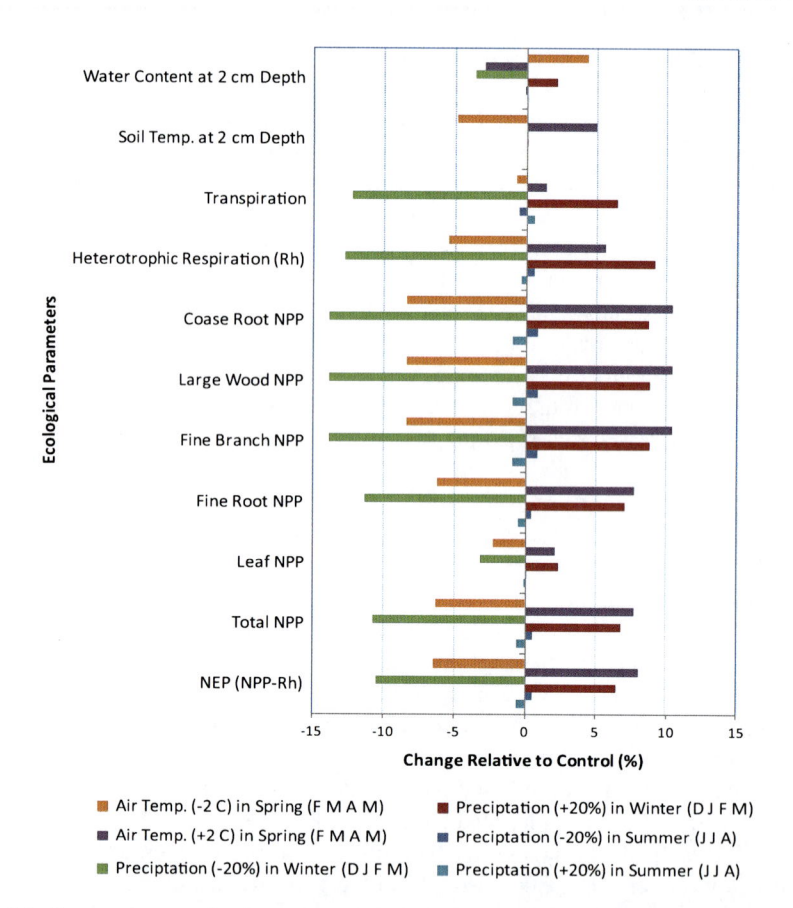

Fig. 3.3 Percent changes of various ecosystem components modeled with DayCent relative to the values estimated from the historical weather condition at the James Reserve (between 1943 and 2010). Six hypothetical weather conditions were simulated: increased/decreased air temperature in spring (Feb–May), precipitation in winter (Dec–Mar), and precipitation in summer (Jun–Aug)

and Allen). This is quite different from the estimate of 3 days from a grassland microcosm study (Staddon et al. 2003b). In the direct observations of individual hyphae, some lived only hours whereas others persisted for a full growing season, generally supporting the model of Allen et al. (2003). In addition, those observations throughout a growing season showed that the AM hyphae rapidly grow with the onset of winter rains. The average remains high throughout the growing season, although both production and mortality occur in response to weather events (Fig. 3.4). At the end of the growing season, in late spring, as T increases and θ declines, the AM hyphae rapidly disappear. In general, this pattern conforms to analyses from soil cores (Allen et al. 2005b) of AM dynamics through growing seasons in a shrubland. With these results, we can model the production of AM fungal hyphal C based on the standing crop and turnover, using traditional ecosystem calculations. This value could also be contrasted with fine root production and

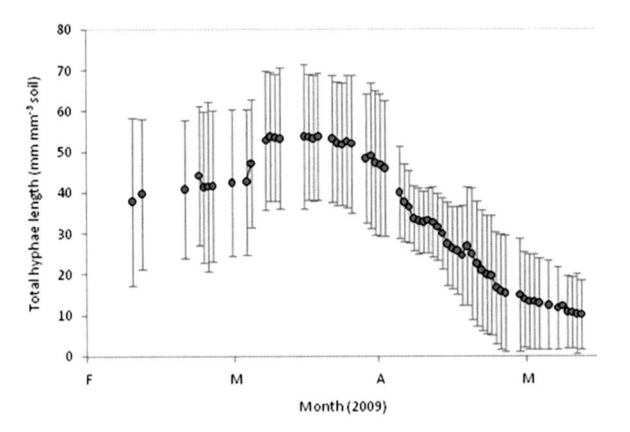

Fig. 3.4 Total hyphae length (mm) per mm^3 of soil as observed using an automated minirhizotron (n = 6) in a mixed conifer forest (UC James Reserve, Idyllwild, CA). Error bars are 95 % CI

mortality at the same site. Using this approach, we estimate that the mean standing crop of AM fungi was 20 g m^{-2} during the growing season, (compared with a peak standing crop of 21 g m^{-2} measured from soil cores, as per Allen et al. 2005). Assuming an average growing season length of 8 months, then the total allocation to AM fungi would be approximately 100 g m^{-2}, a value similar to that of fine root production, where the life span was slightly less than a single growing season (Kitajima et al. 2010). Based on this analysis, then, the annual allocation of C to mycorrhizal fungi may approach that of fine roots. Changes in T or θ that alter the growing season of AM fungi, or the total number of fine root new tips (that support the mycorrhizal fungi, Allen 2001) could dramatically affect C balance estimates, not to mention ecosystem dynamics.

In a final analysis, in modeled projections of environmental change, precipitation was extremely variable both among models and within each model through time (unpublished data, Hernandez and Allen). This is an extremely critical outcome. Increasing variability in precipitation even under decreased total precipitation may mean more biologically effective θ (Thomey et al. 2011). This result occurs because larger (even less frequent) storms increase the depth to which water infiltrates. Soils at the James Reserve are quite shallow, to the point where they are unable to store enough moisture to support the current growth of trees, including both oaks and conifers, through the dry season. Egerton-Warburton et al. (2003) found that trees, such as live oaks, utilized water in granite bedrock during the dry season. Egerton-Warburton et al. (2003) and Bornyasz et al. (2005) observed mycorrhizal roots in bedrock fractures and individual hyphae extending into the granite bedrock matrix. We analyzed the isotopic signatures of the water in the plant stems at the end of the growing season (Allen 2006). Mature plants were using water from the bedrock that had been deposited during winter rains. It is unlikely that the water diffused from the granite matrix to the roots, because of the extreme tortuosity of the material. However, the hyphae grow across these gaps, and provide a rapid pathway for water flow through that matrix (Allen 2007). Thus, if water increases infiltration with depth in response to greater storm variability (Thomey et al. 2011) or in response to greater precipitation, then mycorrhizae will play a critical role in acquiring that water for plant production and survival.

Just as important, both AM and EM fungi benefit from the water provided by hydraulic lift (Querejeta et al. 2003, 2007, 2009). Thus, the seasonal life spans of the mycorrhizal fungi can be extended beyond the growing season predicted by precipitation and T alone. Indeed, in our sapflow and R_s measurements, this was illustrated by an extension of activity into the summer drought, likely using water provided by hydraulic lift, and a result not predicted by the DayCent model.

3.2.2 Nutrients and Global Change

CO_2 fixation is linearly related to N concentration in the leaves, and curve-linearly related to P. Yet, CO_2 is taken up by the plant from the atmosphere, while water, N, and P (as well as other elements) from the soil. Uptake of CO_2 by the plant is direct, and depends upon its physiological status. The plant must either acquire soil resources through rather inefficient structures (roots) or "bargain" with the fungus for N or P. The exception being the formation of cluster roots by the plant, in which root systems are so fine as to co-opt mycorrhizal functioning (e.g., Adams et al. 2002) in surface area, and rhizosphere chemistry.

One clear fact emerges, however, and that is as atmospheric CO_2 increases, the need for N and, to a lesser extent P, increases. This creates a greater dependency by the plant on mycorrhizal fungi to acquire nutrients (e.g., Allen et al. 2003). We found that more C went through the mycorrhizal fungal energy channel under elevated CO_2 than in control model ecosystems in both sagebrush and *Populus tremuloides* (Klironomos et al. 1996, 1997). Further, elevated CO_2 altered the allocation of C within the mycorrhizal system (Rillig et al. 1998a, b, 1999). Alberton et al. (2007) also postulated that the increased C flow to mycorrhizal fungi requires increased N in the increased fungal mass, setting up a negative feedback to the host from the mycorrhizal fungi. However, whether mycorrhizal fungi increase soil C sequestration under elevated CO_2 is surprisingly controversial. Treseder and Allen (2000) and Treseder et al. (2005) proposed that as plants require more N and P, they allocate more C to mycorrhizal fungi, and the slower-decomposing compounds in the fungi remain, increasing soil C sequestration. Alternatively, Heinemeyer et al. (2007) argued that simply more C was respired by the mycorrhizal symbiosis; i.e., there was simply greater C throughput from plant to atmosphere. We argue that while throughput might increase, it is inconceivable that the increased mycorrhizal fungal C allocation would not add soil C based on the fungal chemical composition. Compounds like glomalin and chitin will remain in the soil well after the active life of the hypha. While it is likely that R_s increases with elevated CO_2, it is likely less than the net fixed. Indeed, studies of N additions, that reduce mycorrhizal activity, reduce soil C (Allen et al. 2010), evidenced by declining soil C age (measured by $\delta^{14}C$ analysis, Trumbore, personal communications) probably from increased saprotrophic activity, and decomposition of older carbon.

In a chaparral shrubland, we increased atmospheric CO_2 in 100 ppm increments (Allen et al. 2005) over a four growing season period. As atmospheric CO_2 increased, N became more limited in the host plant. N fixation in *Ceanothus greggii* (an actinorhizal-associated plant) increased, and uptake and transfer of N by EM, as estimated by isotopic fractionation increased. Root biomass and AM fungal mass increased. The amount of new C fixed into soil aggregates and glomalin also increased indicating that more C was being deposited in the soil through mycorrhizal fungi (Rillig et al. 1999; Treseder et al. 2003) in both the chaparral shrubland and in experimental annual forb/grasslands. In AM plants, the dominant fungi shifted, particularly from *Glomus* spp. to taxa in the Gigasporeaceae that form extensive mycelia networks. Production and soil functioning crashed under severe N limitation somewhere between 600 and 750 ppm atmospheric CO_2, a level projected (without CO_2 limits) to occur sometime in the next century.

These responses suggest that there is a response curve between soil resources and atmospheric CO_2 that will regulate mycorrhizal functioning along a changing global environment. Treseder and Allen (2002) described such a response surface in an experimental analysis of mycorrhizal response to and N and P gradient. Applying that curve to a CO_2 by N by P (or other limiting nutrient) response provides useful information on mycorrhizal responses to a variety of global change parameters (Fig. 3.5) but especially the impacts of CO_2. In one perspective, adding N deposition in an elevated CO_2 atmosphere, drives many processes back to pre-industrial relationships, albeit towards a more eutrophic environment overall.

Nitrogen is the next critical plant resource after T and θ in most ecosystems. Leaf N is linearly related to photosynthesis, such that as N increases, photosynthesis increases. But, if N is added, particularly through N fertilization or through anthropogenic N deposition, the plant does not need to exchange N with the mycorrhizal fungus to obtain the critical resource. Alternatively, if it declines (in a high CO_2 environment) the plant will be more dependent upon mycorrhizae. In most AM dominated ecosystems (Egerton-Warburton and Allen 2000, 2001; Egerton-Warburton et al. 2007; Johnson et al. 2008), as soil N increases through anthropogenic additions, AM activity of many plants declined (e.g., *Andropogon gerardii*, *Panicum virgatum*, *Bouteloua gracilis*, *B. eriopoda*). However, AM do not always decline. Some AM plants (*Juniperus monospermum*, *Elymus elymoides*, *Agropyron repens*) simply respond by increasing total production (Johnson et al. 2008; Corkidi et al. 2008; Allen et al. 2010). EM systems are less well measured for soil activity, but are likely very sensitive. In an N fertilization experiment, EM functioning declined in *Pinus edulis*, as measured both by total mycorrhizal root tips and by N fractionation between hyphae and leaf, although AM juniper showed increased production (Allen et al. 2010), as did mortality in red pine (Johnson et al. 2008; Corkidi et al. 2008; Allen et al. 2010). Mortality in red pine at the Harvard Forest was also observable with high levels of N fertilization, although AM angiosperms increased in production (Allen, unpublished observations). Theoretically, as CO_2 increases, and the ratio again shifts toward higher C:N ratios, mycorrhizal activity

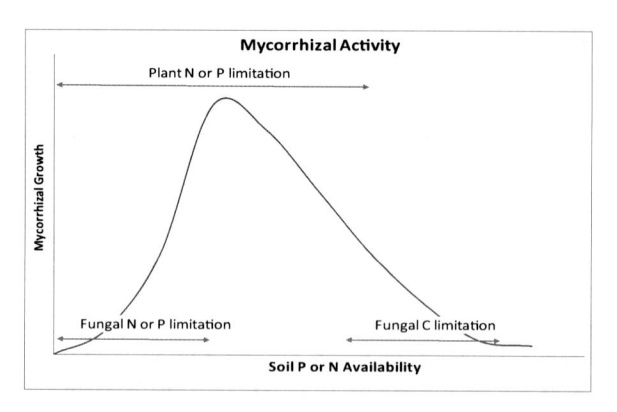

Fig. 3.5 The interaction of plant and fungal nutrient limitation on the biomass of mycorrhizal fungi. At high nutrient levels, fungi will receive little carbon from plants and will be C-limited. At lower nutrient levels, plants will be N- or P-limited and will allocate C to mycorrhizal fungi. At the same time, if N or P concentrations are sufficient for fungal growth, mycorrhizal fungi will proliferate. At the lowest nutrient levels, both fungi and plants should be nutrient limited, and fungal biomass will be low regardless of C allocation to the fungi by plants (Redrawn from Treseder and Allen 2002)

should again increase, and as N deposition increases, mycorrhizal functioning declines. If N deposition is controlled, then mycorrhizae should again become more important.

In addition, as atmospheric CO_2 increases, P becomes more limiting. Treseder and Allen (2002) found that shifting N:P ratios shifted AM fungal composition. In other studies, the Konza prairie has been noted as a model ecosystem for studying AM functioning. However, N fertilization actually increased AM activity (Johnson et al. 2008). This site has calcareous soil, which makes soil nutrients N and P extremely limiting, resembling high CO_2 in the case of both soil nutrients. Based on the Treseder and Allen (2002) model, the Konza prairie may exist nearer the extremely low nutrient condition end of the spectrum, and nutrient additions stimulate mycorrhizal activity, in a similar mechanism to increasing atmospheric CO_2. Mycorrhizae can also shift activity in response to altered P conditions. In a high P, but clayey soil, AM fungi produced oxalate crystals that enhanced P uptake by mycorrhizal fungi (Jurinak et al. 1986) but in sandy soils with high available P, AM fungi apparently did not produce oxalate crystals, although EM fungi did (Allen et al. 1996). Sites with serious nutrient deficiencies may make important test models for environments being altered by increasing CO_2 levels.

Mediterranean climate forest and shrubland ecosystems may be a useful test case system. Above ground, there is a large literature on convergent evolution because the climate has driven similarities in leaf structure and physiognomy between very different groups of plants in the different regions (the Mediterranean Basin, South Africa, Australia, Chilean coast, California coast). However, belowground, these ecosystems radically diverge in mycorrhizal types largely because soil nutrient conditions also differ. Australia and South Africa contain many plants forming cluster roots, a mechanism to acquire nutrients under extreme nutrient deficiencies when even mycorrhizae are limiting. There are a few legumes forming cluster roots in P deficient calcareous soil in the Mediterranean Basin. Many of these sites also contain plants forming ericoid mycorrhizae. These associations are especially effective at acquiring organic N in bogs and other highly organic environments. Interestingly, in Australia, South Africa, and (a few) in the Mediterranean Basin, these associations exist in arid soils with little organic matter, but where that organic matter is still critical for N cycling. In all regions, ectomycorrhizal and arbuscular mycorrhizal plants abound. In California and Chile, there is a special abundance of nonmycotrophic annual plants (e.g., annual Chenopodiaceae, Brassicaceae- although perennial taxa in these families are mycorrhizal, e.g., Allen and Allen 1990, unpublished observations) and many other invasive plants in particular that show minimal response to mycorrhizae such as annual grasses (*Bromus* spp., *Avena* spp.). There are no comparative studies among ecosystems to explore these dynamics, an area ripe for comparative research.

3.3 Global Change and Biodiversity

3.3.1 Shifting Fungal Composition

Measuring ecosystem response is a complex task, particularly teasing apart the multiple impacts of global change in highly variable environments, where daily values may exceed the change in response. However, organisms tend to track extremes and variation, often better than instrumentation and models. Fungi tolerate incredible variation in T, θ and nutrients. However, in a competitive environment, they may be readily overtaken by fungi less sensitive to altered conditions. In general, individual hyphae are presumed to have relatively short lives, although a mycelial network may be long-lived. This interpretation has also been applied to mycorrhizal fungi, as most studies of mycorrhizae are limited to short-lived pot culture studies (e.g., Staddon et al. 2003b). An outcome resulting from this assumption is that hyphae turn over rapidly and composition can likely change rapidly in time as one fungus replaces another following environmental change. Another problem is that assessments of mycorrhizal activity tend to be made from relatively infrequent coring in which space cannot be distinguished from time. Soils are remarkably heterogeneous, and even neighboring cores a few centimeters apart can result in very different communities (Allen and MacMahon 1985; Klironomos et al. 1999).

Allen et al. (2003) developed a simple stochastic model to study shifting the relative production under shifting ratios of C, N and P. This approach was built around known variation in physiological dynamics of different fungal taxa. Interestingly, this model showed a complex array of outcomes in nutrient allocation, plant growth, and fungal growth dependent upon the fungal physiology. Although increased diversity of mycorrhizal fungi is often related to increased plant performance (e.g., Van der Heijden et al. 1998), that is not the only outcome. In multiple stochastic runs of our stoichiometry model, in some cases, increasing fungal richness increased productivity. In others, increasing richness caused no change, or even reduced plant productivity, depending upon the characteristics of the individual fungi (Allen et al. 2003). These multiple outcomes can be observed in other published studies of plant growth responses to fungal diversity (Allen et al. 2003). We proposed that understanding the individual physiological characteristics of the participants under shifting environments is crucial to understanding the outcomes of global change.

A few experimental studies have demonstrated a change in mycorrhizal fungi in response to elevated CO_2. In a plant $\times$ AM fungal experiment, Wolfe et al. (2003) reported complex changes in AM fungal communities to increasing CO_2. In annual communities with a mix of invasive and native species, complex changes in the fungal communities emerged (Rillig et al. 1998a, 1999). In chaparral, Treseder et al. (2003) found that AM fungi shifted from a predominance of *Glomus* spp. in low to ambient CO_2 levels, to a predominance in Acaulosporaceae and Gigasporaceae in high atmospheric CO_2 levels. Apparently, the increased N and P deficiencies associated with elevated CO_2 increased the dependency upon AM fungi known to form an extensive mycelial network (Allen et al. 2005). However, only recently have molecular sequencing techniques been developed for AM fungi allowing for more extensive species-level community analyses. These have not yet been applied to AM fungal communities altered by elevated CO_2 experiments.

EM communities are far more complex and difficult to assess. Parent et al. (2006) found changes in EM fungal communities in response to elevated CO_2. However, there was a shift toward EM fungi located deeper in the soil profile (Pritchard et al. 2008). Alberton and Kuyper (2009) showed that two fungi with different N strategies (one nitrophilous, one not so) resulted in very different outcomes in N allocation and immobilization in response to elevated CO_2. A site like the James Reserve probably has somewhere between 40 and 200 species in a stand of plants (Allen et al. 2002). However, we have very little information on how the community composition changes in response to elevated CO_2, increasing T, or altered θ. Alternatively, as a test system, we might look at the impacts of N deposition on EM communities with the hypothesis of hindcasting back to higher C:N ratios (Hoeksema et al. 2010). If this model is appropriate, we have a few studies in which to examine mycorrhizal composition and even functional change.

In an early study, Karen et al. (1997) found that although there was a large decline in sporocarp production (see also Arnolds 1991), an analysis of root tips using RFLP analyses showed that a high diversity actually remained at the site. This result has been duplicated in other ecosystems (e.g., Lilleskov et al. 2001). In a cross-continent study, Lansing (2003), using an RFLP analysis of the ITSF region, found that richness declined slightly with high N fertilization, showing a shift in the species increment curve, but that species overlap between N-fertilized and control plots was low.

One hypothesis is that the turnover of EM would be increased under high N, such that individual mycorrhizal tips last shorter, returning the C back to the atmosphere more rapidly. We were unable to demonstrate a consistent response, however (Treseder et al. 2004). In response to fertilization, some fungi increased their C accumulation lifespan and others decreased. Another approach is needed to tease this community level dynamic apart.

Allen et al. (2010) found that with N fertilization, the total numbers and richness of EM declined. In the control plots, the pines obtained approximately 35 % of their N from mycorrhizal fungi, while allocating 20 % of the leaf NPP to the fungi in exchange for that N, based on Hobbie and Hobbie (2006). However, with fertilization, the needle biomass dramatically increased, making the ratio of leaf:root tip increase dramatically. N isotopic fractionation data showed that the trees could obtain all their N from the added fertilizers, and the EM were reduced to improving P uptake, or simply existing as commensalists or even parasites. The fungi might be C starved, as no fruiting of EM was found during the duration of the study in the N fertilized plots, although they continued fruiting in control plots and in the surrounding forest. Alternatively, in *Adenostoma fasciculatum*, under elevated CO_2, $\delta^{15}N$ fractionation between the leaf tissue and soils indicated that mycorrhizal fungi increased the fraction of N available for aboveground productivity (Allen et al. 2005).

Examination of high N deposition-fertilization studies in southern California also show interesting patterns that we can use to understand the implications of N deposition and, indirectly, develop testable hypotheses to elevated CO_2., using the Hobbie and Hobbie (2006) model of N and C allocation based on isotopic fractionation. Pinaceae appear to be particularly sensitive to N deposition, in part because aboveground productivity tends to increase beyond the sustainability of the root/mycorrhiza system to sustain water and P uptake (see Allen et al. 2010). In the San Bernardino Mountains, where N deposition has been studied extensively (Sirajuddin 2009, Allen, Sirajuddin and Fenn in preparation), we developed a regional phylogeny for the Transverse Ranges in southern California, focusing on a low N input site (Camp Osceola), a high N site (Camp Pavika) and with additional N fertilization at both sites (Fenn and Bytnerowicz 1993) along with the James Reserve, an intermediate depositional site. In the low N control site, we estimated that the trees obtain approximately 35 % of their N from mycorrhizal fungi (based on the Hobbie and Hobbie fractionation model). To obtain that N, the trees allocated approximately 21 % of the total NPP. In turn, the fungi allocated 41 % of their N to obtain that C. This value is within the expected allocation range for the allocation of C to mycorrhizal fungi. In the high N treatments, we could not generate an allocation. The dominant shift was from fungi such as *Rhizopogon* sp., *Russula acrifolia*, and *Cortinarius* sp. No fruiting was observed, although this may be compounded by the severe 2001–2002 drought. But, in both the high N deposition site, and following long-term N fertilization, the dominant fungi were *Cenococcum* sp. and unknown Thelephoraceae species found on the root tips. Sporocarps of *Cenococcum* have not been found and we were unable to find sporocarps of *Thelephora* spp. Our preliminary analysis suggests that the total allocation may remain similar, but the NPP of the sites increased dramatically with added N (Fengming Yuan, personal communication). Trees in the region did suffer high rates of mortality during the 2001–2002 severe drought.

Angiosperms, including oak (*Quercus agrifolia*), may simply increase production in response to added N, but alter their C allocation patterns. Again, using the Hobbie and Hobbie (2006) model and examining isotopic data from sporocarps collected from high and low N deposition sites (Allen et al. 2005) and the isotopic signatures in oak leaves (Cario 2005), we were able to determine changing allocation of C and N patterns. In a low N area (the Sky Oaks reserve), the trees received 35 % of their N from EM fungi, while allocating 20 % of the aboveground NPP (estimated as per Arbaugh et al. 1998) to the EM fungi in exchange. With N deposition (at the San Dimas Experimental Range), the plants only needed 23 % of their N from EM fungi and allocated only 10 % of their aboveground NPP.

Importantly, few of the fungi found in the control plots actually disappeared from the N added areas, although they did shift relative dominance. Those EM fungi lost could have simply been missed in the sampling. Species increment curves demonstrated that not all EM fungi present at any of the sites were accounted for. As CO_2 continues to increase, and potentially N deposition is controlled, will the species' relative abundances shift back or adjust to a new composition remains an interesting question.

One aspect that could be undertaken is to study individual fungi and use these to parameterize currently hypothetical stoichiometric models (Allen et al. 2003). In examining AMR images, we identified two EM morphotypes that correspond to morphotypes isolated by coring and identified by sequencing. Both were from our James Reserve site, a moderate N deposition location. We have analyzed the fungi, plants and soils for $\delta^{15}N$ and $\delta^{13}C$, along with soils and plants to evaluate dynamics and change. Hoeksema and Kummel (2003) found that turnover rates of fungi increased with elevated CO_2, although Treseder and colleagues (2004) did not find a consistent change with N fertilization in either sporocarp and root tip age or rhizomorph lifespans, respectively. Hobbie and Agerer (2010) developed a means to evaluate comparative strategies among fungi based on $\delta^{15}N$ fractionation. Two fungi in particular were identified at the James Reserve, located near each other. The first was *Russula acrifolia*. This fungus has only a small isotopic fractionation, suggesting that it is more exploratory, probably utilizing the same N sources as the plant (NO_3^-, NH_4^+, and amino acids). The other fungus is an unknown taxon of *Cortinarius*. This fungus has a high fractionation, suggesting that it is taking up organic N, fractionating, and transporting proteins to the plant. Based on the stoichiometric model, these two fungi together increase the ability of the plant to acquire multiple forms of N. If one or more of these strategies are lost, then we could expect important shifts in plant nutrient relationships to emerge. By coupling observations of individual tip and hyphal dynamics from AMR units, with sequencing for both identification and specific gene activities, and with isotopic analyses, hypothetical models such as the stoichiometric approaches could be accurately parameterized. With time and with experimental treatment, patterns of individual fungal behaviors might emerge that would provide the critical answers to this difficult question.

3.3.2 Shifting Plant Composition

There are numerous studies on the impacts of increased N and CO_2 on plant communities. We will not deal with this specific issue here, except as they relate to mycorrhizal functioning. Wolf et al. (2003) reported that elevated CO_2 altered the diversity of plants, which in turn changed the composition of AM fungi. Unfortunately, the site had been sterilized with methyl bromide prior to the imposition of the elevated CO_2 compromising what could have been a very important study.

The interesting shift in vegetation in response to elevated CO_2 relating to mycorrhizal dynamics occurred in the Mojave Desert FACE (Free Air CO_2 Enrichment) study. There, Smith et al. (2000) found that for a wet period, there was a dramatic increase in *Bromus tectorum*. This pattern was not consistent through time, but *Bromus* dominance appears to cycle (Salo 2004) with precipitation, and has important implications for mycorrhizal functioning. We would have expected to see an increase in AM fungal activity and even a potential shift in the species composition in response to elevated CO_2. But no change in percent infection, or glomalin (an indication of long term AM fungal activity) was found (Clark et al. 2009). Importantly, all species of *Bromus* are also considered nitrophilous, and often replace native shrubs and forbs under N deposition or fertilization (e.g., Allen 2004). *Bromus tectorum* seems to have little response to AM, although the relationships are formed (Wolfe et al. 2003). In other species of *Bromus*, the AM fungi shift in response to elevated CO_2 and to N (Rillig et al. 1998a, b, 1999), but the plant appears relatively unaffected. Clearly more work on the relationships between mycorrhizae, elevated CO_2, and plant community dynamics is needed.

3.3.3 Adaptation: A Missing Element from Global Change Studies

The rate of change in atmospheric is very rapid. There are many calls that although CO_2 levels are nowhere near highest levels geologically, the rate of change is unprecedented. As we do not know past rates of change (such as the elevated CO_2 shift from the Paleocene to Eocene with high CO_2 inputs), it is difficult to know if ecological or evolutionary adaptations can occur within the timescales of the current rate of change.

Across the long history of mycorrhizal symbioses, major changes in new mycorrhizal types occur at times of elevated CO_2 (Allen 1996). When the land was first invaded, atmospheric CO_2 was high, and nutrients in short supply. AM fungi, which were already present, possibly living symbiotically with algal communities as *Geosiphon* does today (Schuessler and Kluge 2001). During the Carboniferous Period, globally high levels of CO_2 were found (as high as 2,000 ppm). This is also the time period when there was a major change in the Pinales (i.e., order comprising extant conifers). Specifically, the EM Pinaceae separated from the Cupressaceae. The Cupressaceae and all other northern hemisphere gymnosperms are AM whereas the Pinaceae forms EM. The fungi forming AM are Glomeromycota, a distinct monophyletic of ancient fungi, whereas EM are Ascomycota and Basidiomycota (plus one advanced Zygomycota, the Endogonales). The timing of this event corresponds to a spike in atmospheric CO_2. On one final note, nonmycotrophic annual plants in the Caryophyllales are modern plants. They rapidly spread in high fertility soils during the Pleistocene, when atmospheric CO_2 levels were particularly low (180 ppm).

Klironomos et al. (2005) found, by looking at mycorrhizal infection over several generations of short-lived plants, that if the imposition of elevated CO_2 occurred abruptly, then a marked change in mycorrhizal activity and composition occurred. However, if the imposition of elevated CO_2 occurred gradually across multiple generations, then neither parameter changed significantly. This experiment suggests that we need to think about our systems and interpretations of rates of change. The advantage of this study was that it went across generations of plants. For many plants, changes in CO_2 levels, and warming conditions will occur across generations, unlike the immediate imposition of FACE and growth chamber experiments. However, for many tree species, the changes humans are imposing are well within a generation. Many old growth forests were initiated under atmospheric CO_2 at approximately 300 ppm, and individual trees will be alive as we exceed 450 ppm and beyond.

Whether the fungi are adapting across generations, or even evolutionary is an important question that deserves consideration. The Glomeromycota do not have cross walls, such that the mycelial network has unbounded, large numbers of nuclei scattered through the hyphae. There is still debate as to whether these are identical or vary within an "individual" but even within an individual there must be some mutation of individual haploid nuclei. In EM in the higher fungi, the dikaryons have variation upon which there can be changing gene frequencies under environmental change condition. Thus there is a potential for both ecological and evolutionary adjustment that cannot be studied in growth chamber studies or short term field enrichments.

3.4 Altered Environments: A Global View

Mycorrhizae contribute a large fraction of the sequestered C in soils. Further, soils contain approximately three times as much C as in the atmosphere; imparting small changes in soil processes with potential to confer dramatic impacts on the atmospheric CO_2 concentrations, and feedback effects such as climate change and C:nutrient dynamics. However, soil ecological research, and mycorrhizal research in particular suffers from a strong aversion to getting in the field and determining what is actually occurring. It is far easier to continue to run pot culture studies than to investigate belowground processes as they happen. There is a further notion that the tools of the 1970s, coring and laboratory measurement, are adequate to study soil processes. Unfortunately, these perspectives are hindering our understanding of C dynamics at ecosystem to global scales.

Measuring aboveground processes and subtracting to extract soil "differences" are extremely inaccurate. Even the best eddy co-variance system dumps much of the phenomena data (using best Ameriflux standards), much of which we believe is tied to soil dynamic processes. Coring is completely inadequate, manifesting as almost a classic Heisenberg uncertainty principle. As soon as the soil is extracted, it is changed. One can never take two cores in time, because the second core changes space as well. We know enough about soil communities to know that two cores samples differ in organisms, compositions, and structure (Allen and MacMahon 1985; Klironomos et al. 1999; Allen et al. 2007). At the very least, coring should be coupled with sensor technology to actually determine what is happening in the field, not what might happen based on theory.

There is a large research base with background information on the impacts of acute anthropogenic perturbation on mycorrhizae, from disturbance, to N and P fertilization, as well as immediately enriched CO_2 environments. The impacts of chronic change, in particular N deposition and increasing atmospheric CO_2 that changes at decadal to century scale adjustments are much more difficult. We often try to relate changes occurring today with those in the geological record, but we do not know if those occurred at the century or millennial scale. We can model (e.g., DayCent) and we can hindcast (e.g., Egerton-Warburton et al. 2001), but we need a combination of real long-term monitoring and perturbation studies focused on the soil environment and soil environmental change. New tools from soil observatories (Allen et al. 2007) to next generation sequencing capacity (Nilsson et al. 2011), with the ability to resolve processes, not just composition, provide incredible opportunities to study the impacts of global change. Observatories such as NEON and Ameriflux (and their international collaborations) have the potential to incorporate such exciting new technologies and ideas, but those programs must not be allowed to continue to flounder on past approaches, because it is easier for the entrenched bureaucracy, agency personnel, and entitlements to particular groups of scientists to continue with business as usual.

Acknowledgments We thank Tim Mok and Renee Wong for tracing hyphae and Leela Rao for helping us to run DayCent model. This research was funded by the National Science Foundation (EF-0410408 and CRR-0120778).

References

Adams MA, Bell TL, Pate JS (2002) Phosphorus sources and availability modify growth and distribution of root clusters and nodules of native Australian legumes. Plant Cell Environ 25:837–850

Alberton O, Kuyper TW, Gorissen A (2007) Competition for nitrogen between *Pinus sylvestris* and ectomycorrhizal fungi generates potential for negative feedback under elevated CO_2. Plant Soil 296:159–172

Allen MF (1991) The ecology of mycorrhizae. Cambridge University Press, New York

Allen MF (1996) The ecology of arbuscular mycorrhizae: a look back into the 20th century and a peek into the 21st. Mycol Res 100:769–782

Allen MF (2001) Modeling arbuscular mycorrhizal infection: is % infection an appropriate variable? Mycorrhiza 10:255–258

Allen EB (2004) Restoration of artemisia shrublands invaded by exotic annual *Bromus*: a comparison between Southern California and the Intermountain Region. USDA Forest Service proceedings RMRS-P-31

Allen MF (2006) Water dynamics of mycorrhizas in arid soils. In: Gadd GM (ed) Fungi in biogeochemical cycles. Cambridge University Press, New York

Allen MF (2007) Mycorrhizal fungi: highways for water and nutrients in arid soils. Vadose Zone J 6:291–297

Allen MF, Allen EB (1990) Carbon source of VA mycorrhizal fungi associated with *Chenopodiaceae* from a semi-arid steppe. Ecology 71:2019–2021

Allen MF, MacMahon JA (1985) Impact of disturbance on cold desert fungi: comparative microscale dispersion patterns. Pedobiologia 28:215–224

Allen MF, Morris SJ, Edwards F, Allen EB (1995) Microbe-plant interactions in Mediterranean-type habitats: shifts in fungal symbiotic and saprophytic functioning in response to global change. In: Moreno JM, Oechel WC (eds) Global change and Mediterranean-type ecosystems, vol 117, Ecological studies. Springer, New York

Allen MF, Figueroa C, Weinbaum BS, Barlow SB, Allen EB (1996) Differential production of oxalates by mycorrhizal fungi in arid ecosystems. Biol Fertil Soils 22:287–292

Allen MF, Lansing J, Allen EB (2002) The role of mycorrhizal fungi in composition and dynamics of plant communities: a scaling issue. Prog Bot 63:344–367

Allen MF, Swenson W, Querejeta JI, Egerton-Warburton LM, Treseder KK (2003) Ecology of mycorrhizae: a conceptual framework for complex interactions among plants and fungi. Ann Rev Phytopathol 41:271–303

Allen MF, Egerton-Warburton L, Treseder K, Cario C, Lindahl A, Lansing J, Querejeta I, Karen O, Harney S, Zink T (2005a) Biodiversity and mycorrhizal fungi in southern California. In: Kus B, Beyers JL (eds) Planning for biodiversity: bringing research and management together: proceedings of a symposium for the South Coast Ecoregion, March 2000, Pomona. USDA Forest Service Pacific Southwest Research Station general technical report PSW-GTR-195:43–56

Allen MF, Klironomos JN, Treseder KK, Oechel WC (2005b) Responses of soil biota to elevated CO_2 in a chaparral ecosystem. Ecol Appl 15:1701–1711

Allen MF, Vargas R, Graham E, Swenson W, Hamilton M, Taggart M, Harmon TC, Ratko A, Rundel P, Fulkerson B, Estrin D (2007) Soil sensor technology: life within a pixel. Bioscience 57:859–867

Allen MF, Allen EB, Lansing JL, Pregitzer KS, Hendrick RL, Ruess RW, Collins SL (2010) Responses to chronic N fertilization of ectomycorrhizal pinon but not arbuscular mycorrhizal juniper in a pinon-juniper woodland. J Arid Environ 74:1170–1176

Arbaugh MJ, Johnson DW, Pulliam WM (1998) Simulated effects of N deposition, ozone injury, and climate change on a forest stand in the San Bernardino Mountains. In: Miller PR, McBride JR (eds) Oxidant air pollution impacts in the montane forests of Southern California: a case study of the San Bernardino Mountains. Springer, New York

Arnolds E (1991) Decline of ectomycorrhizal fungi in Europe. Agric Ecosyst Environ 35:209–244

Bornyasz MA, Graham R, Allen MF (2005) Ectomycorrhizae in a soil-weathered granitic bedrock regolith: linking matrix resources to plants. Geoderma 126:141–160

Cario CH (2005) Elevated atmospheric carbon dioxide and chronic atmospheric nitrogen deposition change nitrogen dynamics associated with two Mediterranean climate evergreen oaks. Dissertation, University of California, Davis

Clark NM, Rillig MC, Nowaka RS (2009) Arbuscular mycorrhizal fungal abundance in the Mojave Desert: seasonal dynamics and impacts of elevated CO_2. J Arid Environ 73:834–843

Corkidi L, Evans M, Bohn J (2008) An introduction to propagation of arbuscular mycorrhizal fungi in pot cultures for inoculation of native plant nursery stock. Native Plants J 9:29–38

Egerton-Warburton LM, Allen EB (2000) Shifts in arbuscular mycorrhizal communities along an anthropogenic nitrogen deposition gradient. Ecol Appl 10:484–496

Egerton-Warburton LM, Allen MF (2001) Endo- and ectomycorrhizae in *Quercus agrifolia* Nee. (*Fagaceae*): patterns of root colonization and effects on seedling growth. Mycorrhiza 11:283–290

Egerton-Warburton LM, Graham RC, Allen EB, Allen MF (2001) Reconstruction of historical changes in mycorrhizal fungal communities under anthropogenic nitrogen deposition. Proc Roy Soc Lond B Biol Sci 1484:2479–2848

Egerton-Warburton LM, Graham RC, Hubbert KR (2003) Spatial variability in mycorrhizal hyphae and nutrient and water availability in a soil-weathered bedrock profile. Plant Soil 249:331–342

Egerton-Warburton LM, Johnson NC, Allen EB (2007) Mycorrhizal community dynamics following nitrogen fertilization: a cross-site test in five grasslands. Ecol Monogr 77:527–544

Fenn ME, Bytnerowicz A (1993) Dry deposition of nitrogen and sulfur to ponderosa and Jeffrey pine in the San Bernardino National Forest in southern California. Environ Pollut 81:277–285

Hasselquist NJ, Vargas R, Allen MF (2010) Using soil sensing technology to examine interactions and controls between ectomycorrhizal growth and environmental factors on soil CO_2 dynamics. Plant Soil 331:17–29

Hayhoe K, Cayan D, Field CB, Frumhoff PC, Maurer EP, Miller NL, Moser SC, Schneider SH, Cahill KN, Cleland EE, Dale L, Drapek R, Hanemann RM, Kalkstein LS, Lenihan J, Lunch CK, Neilson RP, Sheridan SC, Verville JH (2004) Emissions pathways, climate change, and impacts on California. Proc Natl Acad Sci 101:12422–12427

Heinemeyer A, Hartley IP, Evans SP, Carreira De La Fuente JA, Ineson P (2007) Forest soil CO_2 flux: uncovering the contribution and environmental responses of ectomycorrhizas. Glob Chang Biol 13:1786–1797

Hobbie EA, Agerer R (2010) Nitrogen isotopes in ectomycorrhizal sporocarps correspond to belowground exploration types. Plant Soil 327:71–83

Hobbie JE, Hobbie EA (2006) ^{15}N in symbiotic fungi and plants estimates nitrogen and carbon flux rates in Arctic tundra. Ecology 87:816–822

Hoeksema JD, Kummel M (2003) Ecological persistence of the plant-mycorrhizal mutualism: a hypothesis from species coexistence theory. Am Nat 162:S40–S50

Hoeksema JD, Chaudhary VB, Gehring CA, Johnson NC, Karst J, Koide RT, Pringle A, Zabinski C, Bever JD, Moore JC, Wilson GWT, Klironomos JN, Umbanhowar J (2010) A meta-analysis of context-dependency in plant response to inoculation with mycorrhizal fungi. Ecol Lett 13:394–407

Johnson NC, Rowland DL, Corkidi L, Allen EB (2008) Characteristics of plant winners and losers during grassland eutrophication – importance of biomass allocation and mycorrhizal function. Ecology 89:2868–2878

Jurinak JJ, Dudley LM, Allen MF, Knight WG (1986) The role of calcium oxalate in the availability of phosphorus in soils of semiarid regions: a thermodynamic study. Soil Sci 142:255–261

Karen O, Hogberg N, Dahlberg A, Jonsson L, Nylund JE (1997) Inter- and intraspecific variation in the ITS region of rDNA of ectomycorrhizal fungi in Fennoscandia as detected by endonuclease analysis. New Phytol 136:313–325

Kitajima K, Anderson KE, Allen MF (2010) Effect of soil temperature and soil water content on fine root turnover rate in a California mixed conifer ecosystem. J Geophys Res 115:G04032

Klironomos JN, Rillig MC, Allen MF (1996) Below-ground microbial and microfaunal responses to *Artemisia tridentata* grown under elevated atmospheric CO_2. Funct Ecol 10:527–534

Klironomos JN, Rillig MC, Allen MF, Zak DR, Kubiske M, Pregitzer KS (1997) Soil fungal-arthropod responses to *Populus tremuloides* grown under enriched atmospheric CO_2 under field conditions. Glob Chang Biol 3:473–478

Klironomos JN, Rillig MC, Allen MF (1999) Designing belowground field experiments with the help of semi-variance and power analyses. Appl Soil Ecol 12:227–238

Klironomos JN, Allen MF, Rillig MC, Piotrowski J, Makvandi-Nejad S, Wolfe BE, Powell JR (2005) Abrupt rise in atmospheric CO_2 overestimates community response in a model plant-soil system. Nature 433:621–624

Lansing J (2003) Comparing arbuscular and ectomycorrhizal fungal communities in seven North American Forests and their response to nitrogen fertilization. Dissertation, University of California/San Diego State University, Davis/San Diego

Lilleskov EA, Fahey TJ, Lovett GM (2001) Ectomycorrhizal fungal aboveground community change over an atmospheric nitrogen deposition gradient. Ecol Appl 11:397–410

Nilsson RH, Tederso L, Lindahl BD, Rasmus Kjøller R, Carlsen T, Quince C, Abarenkov K, Pennanen T, Stenlid J, Bruns T, Larsson K-H, Kõljalg U, Kauserud H (2011) Towards standardization of the description and publication of next-generation sequencing datasets of fungal communities. New Phytol 177:790–801

Parent JL, Morris WF, Vilgalys R (2006) CO_2 enrichment and nutrient availability alter ectomycorrhizal fungal communities. Ecology 87:2278–2287

Pate JS, Verboom WH, Galloway PD (2001) Turner review no. 4 co-occurrence of *Proteaceae*, laterite and related oligotrophic soils: coincidental associations or causative inter-relationships? Aust J Bot 49:529–560

Pritchard SG (2011) Soil organisms and global climate change. Plant Pathol 60:82–99

Pritchard SG, Strand AE, McCormack ML, Davis MA, Finzi AC, Jackson RB, Matamala R, Rogers HH, Oren R (2008) Fine root dynamics in a loblolly pine forest are influenced by free-air-CO_2-enrichment: a six-year-minirhizotron study. Glob Chang Biol 14:588–602

Querejeta JI, Egerton-Warburton L, Allen MF (2003) Direct nocturnal water transfer from oaks to their mycorrhizal symbionts during severe soil drying. Oecologia 134:55–64

Querejeta JI, Egerton-Warburton LM, Allen MF (2007) Hydraulic lift may buffer rhizosphere hyphae against the negative effects of severe soil drying in a California oak savanna. Soil Biol Biochem 39:409–417

Querejeta JI, Egerton-Warburton LM, Allen MF (2009) Topographic position modulates the mycorrhizal response of oak trees to inter-annual rainfall variability in a California woodland. Ecology 90:649–662

Rillig MC, Allen MF, Klironomos JN, Chiariello NR, Field CB (1998a) Plant-species specific changes in root-inhabiting fungi in a California annual grassland: responses to elevated CO_2 and nutrients. Oecologia 113:252–259

Rillig MC, Allen MF, Klironomos JN, Field CB (1998b) Arbuscular mycorrhizal percent root infection and infection intensity of *Bromus hordeaceus* grown in elevated atmospheric CO_2. Mycologia 90:199–205

Rillig MC, Field CB, Allen MF (1999) Fungal root colonization responses in natural grasslands after long-term exposure to elevated atmospheric CO_2. Glob Chang Biol 5:577–585

Salo LF (2004) Population dynamics of red brome (*Bromus madritensis* subsp. *rubens*): times for concern, opportunities for management. J Arid Environ 57:291–296

Schuessler A, Kluge M (2001) *Geosiphon pyriforme*, an endocytosymbiosis between fungus and cyanobacteria, and its meaning as a model system for AM research. In: Hock B (ed) The mycota, vol IX, Fungal associations. Springer, Berlin

Sirajuddin AT (2009) Impact of atmospheric nitrogen pollution on belowground mycorrhizal fungal community structure and composition in the San Bernardino Mountains. Dissertation, University of California, Riverside

Smith SD, Huxman TE, Zitzer SF, Charlet TN, Housman DC, Coleman JS, Fenstermaker LK, Seemann JR, Nowak RS (2000) Elevated CO_2 increases productivity and invasive species success in an arid ecosystem. Nature 408:79–82

Staddon PL, Thompson K, Jakobsen I, Grime JP, Askew AP, Fitter AH (2003a) Mycorrhizal fungal abundance is affected by long-term climatic manipulations in the field. Glob Chang Biol 9:186–194

Staddon PL, Ramsey CB, Ostle N, Ineson P, Fitter AH (2003b) Rapid turnover of hyphae of mycorrhizal fungi determined by AMS microanalysis of C-14. Science 300:1138–1140

Tang J, Misson L, Gershenson A, Cheng W, Goldstein AH (2005) Continuous measurements of soil respiration with and without roots in a ponderosa pine plantation in the Sierra Nevada Mountains. Agr Forest Meteorol 132:212–227

Thomey ML, Collins SL, Vargas R, Johnson JE, Brown RF, Natvig DO, Friggens MT (2011) Effect of precipitation variability on net primary production and soil respiration in a Chihuahuan Desert grassland. Glob Chang Biol 17:1505–1515

Treseder KK, Allen MF (2000) Mycorrhizal fungi have a potential role in soil carbon storage under elevated CO_2 and nitrogen deposition. New Phytol 147:189–200

Treseder KK, Allen MF (2002) Direct N and P limitation of arbuscular mycorrhizal fungi: a model and field test. New Phytol 155:507–515

Treseder KK, Egerton-Warburton LM, Allen MF, Cheng Y, Oechel WC (2003) Alteration of soil carbon pools and communities of mycorrhizal fungi in chaparral exposed to elevated CO_2. Ecosystems 6:786–796

Treseder KK, Masiello CA, Lansing JL, Allen MF (2004) Species-specific measurements of ectomycorrhizal turnover under N-fertilization: combining isotopic and genetic approaches. Oecologia 138:419–425

Treseder KK, Allen MF, Ruess RW, Pregitzer KS, Hendrick RL (2005a) Lifespans of fungal rhizomorphs under nitrogen fertilization in a pinyon-juniper woodland. Plant Soil 270:249–255

Treseder KK, Morris SJ, Allen MF (2005b) The contribution of root exudates, symbionts, and detritus to carbon sequestration in the soil. In: Wright F, Zobel R (eds) Roots and soil management– interactions between roots and soil, Agronomy monograph no 48. American Agronomy Society, Madison

Van der Heijden MGA, Klironomos JN, Ursic M, Moutoglis P, Streitwolf-Engel R, Boller T, Wiemken A, Sanders IR (1998) Mycorrhizal fungal diversity determines plant biodiversity, ecosystem variability and productivity. Nature 396:69–72

Vargas R, Allen MF (2008) Dynamics of fine root, fungal rhizomorphs and soil respiration in a mixed temperate forest: integrating sensors and observations. Vadose Zone J 7:1055–1064

Vargas R, Baldocchi DD, Allen MF, Bahn M, Black TA, Collins SL, Yuste JC, Hirano T, Jassal RS, Pumpanen J, Tang J (2010) Looking deeper into the soil: biophysical controls and seasonal lags of soil CO_2 production and efflux. Ecol Appl 20:1569–1582

Wolfe J, Johnson NC, Rowland DL, Reich PB (2003) Elevated CO_2 and plant species richness impact arbuscular mycorrhizal fungal spore communities. New Phytol 157:579–588

Chapter 4
Dynamic Stomatal Changes

Hartmut Kaiser and Elena Paoletti

Abstract Stomatal pores regulate CO_2 uptake and water loss from leaves. Stomatal responses are dynamic by nature and often lag behind the faster changing environmental conditions as is common in tree canopies. Even under constant conditions, gas exchange of angiosperms occasionally shows cycling fluctuations, called stomatal oscillations. They are interpreted as an effect of feedback control failing to achieve stable regulation and thus demonstrate that stomata not only respond to external factors, but also to the environment inside the leaf. The processes which translate transpiration into turgor are called the physiological gain. The physical processes and environmental conditions which control stomatal aperture, stomatal conductance and transpiration are called the physical gain. More research on the physiological gain is needed in order to understand these processes. In order to overcome the epidermal backpressure, guard cell turgor has to reach a certain threshold level, although guard cell swelling anticipates the opening. When the pore opens, the relation between pore area and stomatal conductance determines the physical gain. In contrast to the Fick's first law of diffusion, this relation is not linear, but convex shaped, with a rapid increase of conductance just after opening and much less effect of aperture changes at large apertures. The high and abruptly changing gain at smallest pore openings can promote overshooting oscillatory responses, as supported by microscopic observations of stomatal apertures. A review of the literature suggests that stomatal movements are metabolically active responses of guard cells to local water status. A full understanding of the mechanisms, however, is complex because stomatal movements result from the interaction of two processes that are difficult to separate experimentally: hydraulic effects, and active osmotic adjustment of guard cells and epidermal cells. Hydropassive

H. Kaiser
Botanical Institute, Christian-Albrechts University, Kiel, Germany
e-mail: hkaiser@bot.uni-kiel.de

E. Paoletti (✉)
Institute of Plant Protection, National Research Council, Florence, Italy
e-mail: e.paoletti@ipp.cnr.it

M. Tausz and N. Grulke (eds.), *Trees in a Changing Environment*,
Plant Ecophysiology 9, DOI 10.1007/978-94-017-9100-7_4,
© Springer Science+Business Media Dordrecht 2014

movement, resulting from an unbalance of turgor pressure between guard cells and the surrounding epidermis, may also occur. An example of hydropassive movement is the so-called Iwanoff effect or Wrong Way Response (WWR), i.e. a fast opening response followed by a slow closure, that occurs as a response to a steep increase in the leaf to air difference in water vapor pressure and may last 2.5–38 min depending on the species and the experimental conditions. An additional 10–60 min may be required for completing the closing response. In contrast to the rather slow osmo-regulatory negative feedback, hydraulic responses act fast, starting within seconds and completing within minutes, and have been suggested as a key mechanism in stomatal oscillations. In a plant displaying oscillations, movements of individual stomata are more or less synchronized on a very small scale within a leaf (1–2 mm). The nature of the synchronizing mechanism is not clear. Synchronization can also occur among leaves, ultimately leading to concerted cycling of gas exchange of entire plants. Comprehensive models of stomatal behaviour based on the mecha-nisms operating in and around stomatal guard cells are still missing, and may help explaining gas exchange response to stressors. Studies with the air pollutant of most concern to forests, i.e. ground-level ozone, suggest that stomata show a transient decrease of stomatal conductance upon exposure and are sluggish in responding to further stimuli.

4.1 Introduction

Ever since vascular plants started the conquest of land, the interior of photosynthesizing organs had to be maintained in a state of sufficient hydration by evaporational barriers which limit transpiration, while at the same time putting minimal constraints on CO_2 supply for photosynthesis. The solution was stomatal pores which in response to a multitude of environmental factors perform the task of adjusting leaf conductance to an optimal compromise between the needs of water conservation and photosynthetic carbon gain. As the multi-factorial microclimatic conditions, which determine optimal leaf conductance, are usually in permanent fluctuation, stomatal responses are dynamic by nature. Due to the slow rates of movements, stomatal responses often lag behind the faster changing environmental conditions. All this makes the stomatal response in natural environments transitory and fugitive, most often far from the optimum and equilibrium state which at best can be observed under constant laboratory conditions.

   We now know a great deal about gas exchange of trees (see Thomas and Winner 2002), although most insight comes from young trees and steady-state measure-ments. Eddy-correlation measurements provide an integrated assessment of canopy-level gas exchange, but cannot untangle the leaf-level dynamics as a response to fluctuating environmental stimuli. Tree canopies are very dynamic environments where all physical parameters vary with space and time (Zhang and

Xu 2000; Wang and Jarvis 1990), for example, variable light may represent two thirds of the incident light in forest canopies (Pearcy 1990). The ability to adjust gas exchange to rapid changes in environmental stimuli is an index of successful adaptation of trees.

Stomatal opening is driven by the accumulation of K^+ salts and sugars in guard cells, which is mediated by electrogenic proton pumps in the plasma membrane and/or metabolic activity. Opening responses are achieved by coordination of light signalling, light-energy conversion, membrane ion transport, and metabolic activity in guard cells. Great progress has been made in elucidating the signal transduction pathways by which stomatal guard cells respond to changes in light intensity and CO_2 concentration (Assmann and Shimazaki 1999; McAinsh et al. 2000; Assmann and Wang 2001; Hetherington 2001; Vavasseur and Raghavendra 2005), and short-term changes in hydraulic variables such as humidity (Mott and Parkhurst 1991; Monteith 1995; Oren et al. 1999), xylem hydraulic conductance (Saliendra et al. 1995; Cochard et al. 2002; Brodribb and Holbrook 2004; Powles et al. 2006), and soil water status (Fuchs and Livingston 1996; Comstock and Mencuccini 1998). Substantial progress has been made in elucidating the mechanisms leading to stomatal closure (Pei and Kuchitsu 2005; Schroeder et al. 2001). Briefly, as a response to a sudden exposure to a stressor, production of reactive oxygen species (ROS) in guard cells increases. This leads to suppression of plasma membrane H^+ and Ca^{+2}, adenosine $5'$-triphosphatases, and perturbations in membrane polarization and ion permeability, particularly to Ca^{+2}. The ultimate result is loss of osmotic substances and a decrease in stomatal pore width. The whole cascade of events may be completed within 5–10 min (Pei and Kuchitsu 2005), although 10–60 min may be required for complete stomatal closure.

As soon as methods for continuous observations of stomatal responses were available, scientists found that even under constant conditions, gas exchange occasionally showed cycling fluctuations, a baffling observation as it does not reconcile well with the idea of an optimal stomatal aperture. Stomatal oscillations were soon interpreted as an effect of feedback control failing to achieve stable regulation and as such, demonstrated that stomata not only respond to external factors, but also to the environment inside the leaf which is affected by the diffusional streams through stomatal pores, thus forming a negative feedback loop.

The aim of this chapter is to summarize the present state-of-knowledge and future prospects about dynamic stomatal changes, with a focus on stomatal oscillations.

4.2 Stomatal Oscillations

The interest in these peculiar responses has led to a large number of published observations from a diverse range of species (Barrs 1971) both from monocotyledonous and dicotyledonous angiosperms. In gymnosperms, there is only one casual observation (Stålfelt 1928) of uncertain quality. Other authors only found damped oscillations in conifers (Phillips et al. 2004). Apparently stomatal oscillations have never been observed in *Pteridophyta*.

Stomatal oscillations were observed in plants with various stomatal anatomies ranging from the simple anomocytic type (without specialized subsidiary cells, Barrs 1968; Kaiser and Kappen 2001; Marenco et al. 2006) to plants with more complicated stomatal complexes with an apparatus of several subsidiary cells (Nikolic 1925; Brun 1961). Oscillations were also observed in *Gramineae* type stomata (Raschke 1965; Brogårdh and Johnsson 1973; Prytz et al. 2003). Stomatal oscillations may occur in herbaceous plants (Ehrler et al. 1965; Shaner and Lyon 1979; Santrucek et al. 2003; Yang et al. 2003; Wallach et al. 2010), grasses (Florell and Rufelt 1960; Raschke 1965; Johnsson 1973; Johnsson et al. 1979), shrubs (Ehrler et al. 1965; Shirazi and Stone 1976a; Rose et al. 1994; Kaiser and Kappen 2001; Marenco et al. 2006) and trees (Levy and Kaufmann 1976; Elias 1979; Reich 1984; Naidoo and von Willert 1994; Herppich and von Willert 1995; Zipperlen and Press 1997; Steppe et al. 2006). Therefore it can be concluded that stomatal oscillations may occur in any clade of angiosperms, irrespective of life form and stomatal anatomy.

Stomatal oscillations were most often observed under laboratory conditions for the simple reason that under fluctuating outdoor conditions oscillatory responses cannot be easily discerned from responses to environmental fluctuations. Nonetheless, a number of observations in the field (Elias 1979; Hirose et al. 1994; Dzikiti et al. 2007) show that stomatal oscillations do not only occur under artificial lab conditions but can be of relevance for real life situations of plants.

Oscillations can be observed on different spatial scales, ranging from movements of individual stomata to whole tree fluctuations of gas-exchange and stem-flux. The temporal and spatial resolution of the applied methods is determined by the degree to which responses of individual stomata are integrated. Most observations were made at leaf level by measuring gas-exchange of leaves or parts of leaves. Leaf patches, often separated by veins, may however show independent dynamics, with phase shifted oscillations (Cardon et al. 1994). This variation can be detected by chlorophyll fluorescence imaging which visualizes effects of different CO_2 supply on the photosynthesizing tissue (Cardon et al. 1994; Siebke and Weis 1995; West et al. 2005). Using chlorophyll fluorescence parameters as a proxy for stomatal apertures, however, only allows qualitative inferences as long as the causal chain stomatal aperture $\rightarrow$ conductance $\rightarrow$ C_i $\rightarrow$ fluorescence yield is not quantified. Another method to determine spatial differences in transpiration is thermography of leaf temperature, which responds to transpirational cooling (Prytz et al. 2003; West et al. 2005). The spatial resolution of these imaging methods is not so much restricted by pixel resolution as by thermal conduction and CO_2-diffusion blurring the image. Nonetheless they may approach sub-millimeter resolution and thus offer the most spatially inclusive and comprehensive measurement of stomatal actions. However, even the smallest discernible area contains many stomata which may include significant variation. The degree of variation among stomata is little known as only a few reports of directly observed aperture oscillations exist (Kaiser and Kappen 2001).

Integrating measurements hide the variation in amplitude and frequency between individual stomata, leaf patches or different leaves. Therefore they cannot answer central questions of the mechanism of stomatal oscillations: How do individual stomata get synchronized to a degree that periodic oscillations become observable on a higher scale? At what level does variation among stomata prevent coordination and has a damping effect on oscillations?

4.2.1 The Mechanism of Stomatal Oscillations

Stomatal oscillations disclose the action of negative feedback loops in aperture regulation, which under certain conditions produce an unstable response (Cowan 1972). Feedback controlled systems have one or more inputs from signal sources conveying information on the state of the parameter to be controlled. Dependent on the magnitude of this input, an output is produced, which has an effect on the state of the controlled parameter. In negative feedback loops, the effect of this output is inverse to the deviation of the controlled parameter, thus having a stabilizing effect. In positive feedback the output is positively related to input, thus enforcing deviations and having a destabilizing effect. The dynamics of feedback control is determined by some basic properties. The degree of regulation, called feedback gain, is the amplification a signal receives when being translated into regulatory output. A high gain promotes oscillations. All processes in the feedback loop, the sensing of the system state, the generation of an output and the response of the controlled parameter to this output usually do not occur instantaneously but with a certain lag, which introduces delays and possibly overshooting responses and oscillatory cycling. Feedback loops may also consist of a mixture of negative and positive feedback acting with separate kinetics.

In leaves, two separate negative feedback loops could be involved in regulation of stomatal aperture (Fig. 4.1). The first one is the feedback loop which keeps intercellular CO_2 (C_i) concentration at sufficient levels for photosynthesis. This loop is formed via guard cell sensitivity to CO_2 and aperture response to photosynthetically decreased C_i, thus allowing higher diffusional influx of CO_2 into the leaf, which then increases C_i. The other feedback loop balances leaf hydration: guard cells respond to transpirational water loss with stomatal closure and thus decrease transpiration.

The relative contribution of each of these interacting feedback loops has been analyzed experimentally only in a few cases. Reducing C_i towards the CO_2 compensation point and thus preventing feedback related to C_i fluctuations did not prevent oscillations (Bravdo 1977) nor the period of oscillations (Marenco et al. 2006). The (difficult) inverse experiment, keeping transpiration constant and allowing C_i fluctuations, apparently has not been performed, therefore it is not known if oscillations based on CO_2-feedback alone can develop. The prominent role of hydraulic relations in oscillations is obvious in most studies on stomatal oscillations where multiple parameters were measured. Hydraulic fluctuations

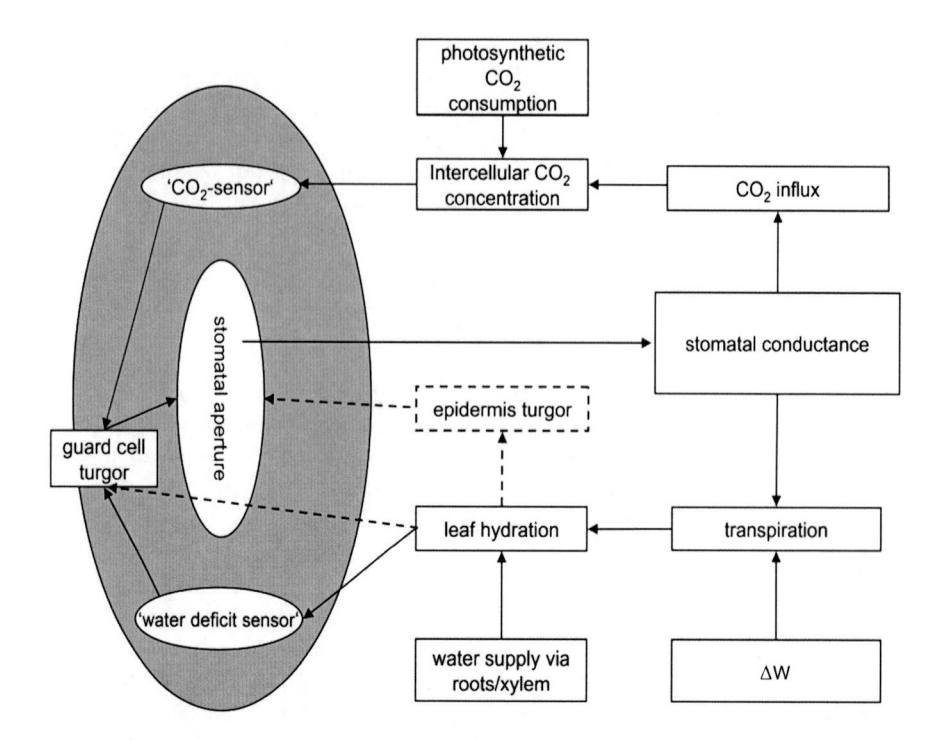

Fig. 4.1 CO_2 and water-related feed-back loops in stomatal regulation of gas exchange. Stomatal aperture governs stomatal conductance, which controls both CO_2 influx and transpiration. CO_2 influx, together with photosynthetic CO_2 uptake, affect intercellular CO_2-concentration which is sensed by guard cells. This feedback loop keeps CO_2-concentration in the mesophyll at sufficient levels for photosynthesis. Transpiration, driven by leaf to air concentration gradient of water vapour (ΔW) and controlled by stomatal conductance affects leaf hydration, directly impacts epidermal and guard cell turgor (*dashed lines*) and results in 'hydropassive' movements. This rapid effect leads to increased stomatal opening upon increased transpiration (and vice versa), thus forming a positive feedback-loop. A putative 'water deficit sensor' perceives leaf hydration and elicits active stomatal osmoregulation leading to changes in guard cell turgor and aperture. This active response of guard cells constitutes a negative feedback loop, keeping transpiration below a threshold level

affect any parameter related to water status like transpiration, leaf thickness, trunk diameter and sap flow. In contrast, as fluctuations in photosynthesis remain comparably small, it can be concluded that stomatal oscillations are mainly caused by instabilities in the water-related feedback loop with a possible additional contribution of CO_2-related feedback. The feedback mechanism involved in oscillations therefore appears to be identical to the mechanisms involved in regulation of leaf water loss. Unfortunately, these mechanisms are not well understood although examined and heatedly debated over decades (Buckley 2005). One of the points of dispute was whether air humidity could also be sensed directly, i.e., without transpiration through the stomata necessarily being involved. The claim for this so-called feed-forward response was bolstered by observations of a disproportionally strong closing response to dry air (Schulze et al. 1972; Farquhar 1978). If only negative feedback via transpiration sensing was involved, transpiration should gradually approach a maximum with increasing leaf to air difference in mole fraction of water vapor (ΔW). Contrary to this expectation, in some cases, transpiration at highest ΔW decreased again, which could not be explained by feedback regulation alone. These observations led to the proposition of a feed-forward response and to a search for mechanisms providing the claimed direct sensitivity to air humidity outside the leaf without proportional transpirational water loss. Many of the proposed mechanisms involved water loss of stomata through cuticles (Farquhar 1978; Maier-Maercker 1983; Grantz 1990), but experimentally such a mechanism could not be confirmed (Meidner 1986; Kerstiens 1997).

The idea of feed-forward lost momentum after it was shown that stomata are sensitive to changes in transpiration when ΔW was kept constant, but not to changes in ΔW under constant transpiration (Mott and Parkhurst 1991). A reanalysis of existing data (Monteith 1995) and further experiments (Franks et al. 1997) questioned the general existence of a true feed-forward response (e.g., direct sensitivity to external humidity) and offered alternative explanations for the earlier observations. The current evidence supports feedback-response of stomata to effects of transpiration on leaf water status (Buckley 2005).

The mechanism by which changes in transpiration translate into stomatal responses, however, is still not identified. A number of possible mechanisms have been proposed, either involving localized transduction processes, which are confined to the guard or adjacent cells, or spatially distributed mechanisms involving other leaf tissues. A locally confined sensing mechanism could be developed through transpiration-dependent accumulation of substances in the apoplast of guard cells (Lu et al. 1997; Zhang and Outlaw 2001). Evaporation from the guard cell apoplast should induce a local accumulation of apoplastic solutes. This effect has been confirmed for sucrose (Outlaw and De Vlieghere He 2001). Evidence against such strictly local sensing mechanisms in or at individual guard cells comes from the observation that blocking the transpiration of a single stoma has no effect on its humidity response (Kaiser and Legner 2007). Only after additionally blocking adjacent stomata could closure in dry air be observed. Sensing of transpiration therefore is not located at individual guard cells but appears to be a function of the local tissue, integrating the transpiration of several stomatal pores on a sub-millimeter spatial scale. These results support the general notion, derived from gas exchange and water status measurements that stomatal responses are controlled by local tissue leaf water potential. For a discussion see Buckley (2005).

Understanding the involved mechanisms is difficult due to the fact that stomatal movements are a result of an interaction of hydraulic effects, and active osmotic adjustment of guard cells and epidermal cells. Both processes, acting simultaneously but with different kinetics, are hard to separate experimentally. The simplest conceivable mechanism would be a direct drawdown of guard cell turgor by an increase in transpiration, without active osmotic adjustment. This does not reconcile, however, with the mechanical relations between guard and epidermal cells. Stomatal aperture is regulated by the balance of turgor pressures between guard cells and the surrounding epidermis. One crucial feature of this counterbalance is the so called *mechanical advantage* of epidermal cells over guard cells (DeMichele and Sharpe 1973; Sharpe et al. 1987; Franks et al. 1998). This means that a change in epidermal turgor has a larger effect on aperture than a similar change in guard cell turgor. A shift in water status similarly affecting epidermal and guard cell turgor will therefore cause a so-called hydropassive movement. Such hydropassive movements have been detected for any possible perturbation of the balance between water supply and loss, irrespective if the cause is a change in leaf water supply (Iwanoff 1928; Powles et al. 2006) or altered transpiration rate due to an increase in ΔW (Kappen et al. 1987; Kaiser and Legner 2007). The typical stomatal response to a steep increase in ΔW is a fast hydropassive opening response (also called Iwanoff effect or Wrong Way Response, WWR), starting almost immediately and finished within a few minutes, followed by a more or less delayed closing response (Fig. 4.2a). The hydropassive opening response further increases transpiration and acts as positive feedback within the control loop. These hydraulic processes therefore have a tendency to destabilize the regulatory loop, making the concept of a negative feedback regulation of transpiration based on purely hydraulic processes implausible.

Nonetheless, there were some attempts to develop hydraulic models explaining the observed stomatal responses on the basis of transpiration-induced micro-gradients between mesophyll, epidermis and guard cells (e.g. Farquhar 1978; Dewar 1995; Eamus and Shanahan 2002). These models require intricate additional assumptions, like variable flow resistance in the hydraulic continuum, to describe the biphasic stomatal response to a step change in humidity (Buckley and Mott 2002) which lack experimental support. Therefore, the most parsimonious hypothesis for stomatal response to transpiration is a metabolically active response of guard cells to local water status (for a detailed discussion see Buckley 2005).

The sensing mechanisms leading to the 'physiological' response to transpiration are not well understood. They could involve osmo-sensing (Yoshida et al. 2006) or mechano-sensitive channels (Zhang et al. 2007), which are triggered by hydraulic disturbances in guard cells or the adjacent tissues. Another possibility could be a local perturbation of the chemical composition of the apoplastic solution (Harris et al. 1988; Zhang and Outlaw 2001), possibly involving pH and its interaction with partitioning and redistribution of abscisic acid (Wilkinson and Davies 2008). As it is not yet known if abscisic acid (ABA), pH, mechanical stresses or other proximal effectors transduce transpiration-related changes into leaf water relations into active stomatal responses, there is no reason to delve into intracellular details of signal transduction. In this field, much more recent progress has occurred than in the question of transpiration sensing by guard cells. How little this research field is settled is demonstrated by recently proposed mechanisms for transpiration sensing, which are fundamentally different from the existing models (Peak and Mott 2011; Pieruschka et al. 2010).

We will now try to identify properties of the stomatal feedback system which favor oscillations. In feedback-controlled systems, oscillations are promoted by a

Fig. 4.2 Time course of *Phaseolus vulgaris* stomatal conductance (g_s) after severing a leaf at time 0. *Graph A* shows an example of Wrong Way Response (*WWR*), a transient increase of g_s after severing, followed by a decrease with increasing leaf water stress. *Graph B* shows that water stress and ozone (O_3) exposure resulted in stomatal sluggishness: longer WWR duration and lower degree of g_s decrease (Δg_s) relative to the controls (Hoshika et al. 2012)

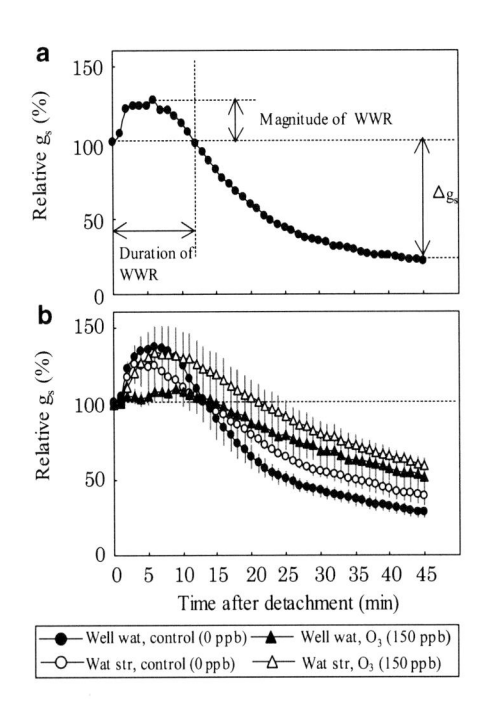

high feedback gain, and delays in the response to changed input with involvement of positive feedback. Feedback gain in negative regulation of transpiration is the degree of regulation of the causal chain: transpiration → physiological (osmotic) activity of stomata → turgor → aperture → stomatal conductance (g_s) → transpiration. The processes which translate transpiration into turgor determine the physiological gain, whereas the following translation into stomatal aperture, g_s and transpiration is governed by physical processes and environmental conditions, which can be summarized as the physical gain. The overall gain is the product of physiological and physical gain. The physiological gain is somewhat obscure, as the underlying physiological events are hardly understood and there is a lot of variability induced by plant species, acclimation responses, diurnal variations and large stoma to stoma variability. Therefore, the physiological gain at the current state of knowledge can only be addressed in a "black box" approach without much prospect to better understand its influence on oscillations. The physical gain, however, is better understood and has some clear effects on the susceptibility to oscillations. First, it depends on the relation between guard cell turgor and aperture, which has a sigmoidal or convex shaped relation (Franks et al. 1998). This relation is strongly dependent on the epidermal backpressure. Notably, in order to overcome the given epidermal backpressure, guard cell turgor has to reach a certain threshold level. As a consequence, at the lowest range of the physiologically possible turgor pressures, the pore is simply closed and any osmotic activity below the opening threshold has no effect on leaf diffusion resistance, and the total gain of the feedback loop is zero. The opening threshold may also lead to a delay in opening, as demonstrated in *Sambucus nigra*, where guard cell swelling was observed for up to 30 min before the pore initially opened (Kaiser and Kappen 2001). As soon as the pore has opened, the relation between pore area and gs determined the physical gain. In contrast to widespread notions, based on a simple application of Fick's first law of diffusion, this relation is not linear, but convex shaped, with a rapid increase of conductance just after opening of the pore and much less effect of aperture changes at large apertures (Kaiser 2009). The reasons for this non-linearity can be found in the three dimensional shape of the pore and additional mesophyll resistances to water vapor diffusion (Kaiser 2009). Consequently physical gain is variable and changes discontinuously within the available physiological range of turgor pressures. It is zero at small pressures, leaps to maximum gain at initial opening and gradually decreasing again, as the pore opens further (Fig. 4.3). The high and abruptly changing gain at smallest pore openings should promote overshooting oscillatory responses.

This view is supported by microscopic observations of stomatal apertures of *Sambucus nigra* during oscillations (Kaiser and Kappen 2001), which revealed that most stomata were closed completely in the troughs of the oscillations and opened only slightly during their respective maxima. Similar observations of oscillations in another four species (Fig. 4.4 and Kaiser, unpublished) confirmed that stomata always cycled between the completely closed and slightly opened state. During the troughs of oscillations, Marenco et al. (2006) estimated that 22 % of the stomata were open. Gas-exchange measurements of oscillation often show very small, minimal conductance, also indicating temporary complete closure (e.g. Rose and Rose 1994; Steppe et al. 2006). Some measurements, on the other hand, appear to contradict this view as oscillations occur at a rather high g_s, indicating on average significantly opened pores (Hirose et al. 1994; Santrucek et al. 2003). The integrating gas exchange signal, however, may hide a lot of variation between individual stomata and asynchronously oscillating leaf patches (Cardon et al. 1994). Each individual pore may completely close in the troughs, but at any time there are enough open pores to maintain a high leaf conductance (Kaiser and Kappen 2001). In summary, experimental evidence indicates that intermittent, complete closure is the typical mode of stomatal oscillations, which is in accordance with the idea that the high and discontinuously changing gain at small apertures, promote oscillations.

Delays in feedback loops contribute to oscillating behavior. For the case of stomata, the time required to perceive transpiration and produce osmotic activity of guard cells introduces a significantly lagging response. The lag is difficult to determine, as physiological responses of stomata to changes in humidity are always intermixed with the hydropassive wrong way opening response. Typical lag times, defined as the time span between switching to high ΔW and the reversal of the initial transient opening response range between 2.5–4 min in *Phaseolus vulgaris* (Meidner 1987), 5–8 min in *Xanthium strumarium* (Mott 2007), 8–20 min in *Sambucus nigra* (Kaiser and Legner 2007) and 8–38 min in *Vicia faba* (Kappen et al. 1987; Assmann and Gershenson 1991; Kaiser and Legner 2007). An additional 10–60 min may be required for completing the closing response. The lag times for opening are similar, as opening speed as an energy-requiring process is

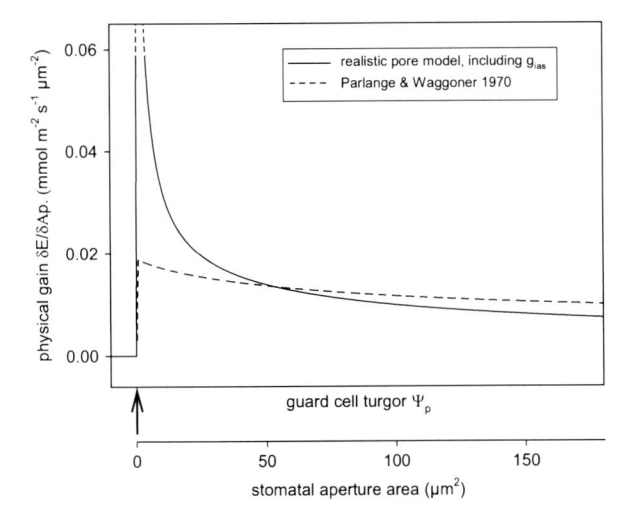

Fig. 4.3 Schematic representation of the dependence of physical gain of the stomatal feed-back loop on aperture area. Physical gain here is defined as $\delta E/\delta$Aperture, which is the unit change of transpiration (E) per unit change in aperture. Gain was calculated using the widely used formula describing the relationship between aperture and stomatal conductance given by Parlange and Waggoner (1970) or a realistic pore model including a detailed pore geometry and mesophyll resistances (Kaiser 2009). As aperture area is, in first approximation, linearly related to guard cell turgor pressure (ψ_p), the graph can also be read as $\delta E/\delta\psi_p$ vs. ψ_p. At the ψ_p value indicated by the arrow, guard cell turgor is sufficient to overcome the backpressure of the epidermal cells, leading to rapid opening from zero to maximum gain (The figure is redrawn from Kaiser 2009)

often slower than the closing response. These delays in signal transduction can delay the response to changed transpiration to the extent that it coincides with the opposite phase of the cycle and this feedback becomes positive.

A destabilizing component of positive feedback is also introduced by the hydropassive response, which tends to open pores further upon increasing transpiration and vice versa. The gain of this feedback, like the gain of negative feedback, also depends on the relation between aperture and g_s, and is largest at small apertures. This effect should further increase instability of response of nearly closed stomata. In contrast to the rather slow osmoregulatory negative feedback, hydraulic responses act fast, starting within seconds and completing within minutes. This effect is known to speed up any response in dry air (Assmann and Grantz 1990; Kaiser and Kappen 2000), and has been suggested as a key mechanism in stomatal oscillations (Cox 1968).

Low air humidity, therefore, has a dual effect on the development of oscillations: an increased water vapor gradient proportionally increases transpiration and thus the physical gain of the feedback loop and, second, the aperture at which the target transpiration is attained is shifted to smaller apertures, where a larger effect of aperture changes on g_s further increases the gain.

In a plant displaying oscillations of gas exchange, movements of individual stomata are more or less synchronized. Obviously mechanisms exist which

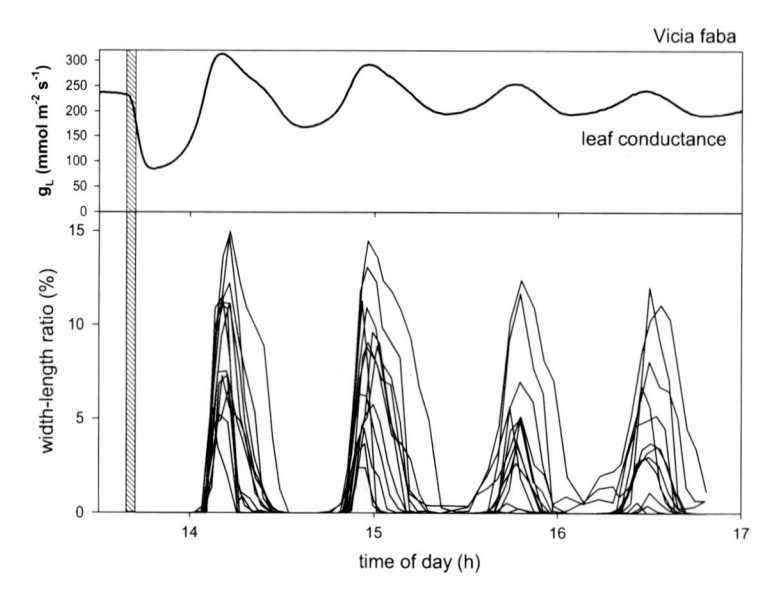

Fig. 4.4 Typical stomatal oscillations in *Vicia faba*. Simultaneous recordings of leaf conductance and stomatal opening (length:width ratio) of 12 randomly selected stomata. Oscillations were elicited by darkening the leaf for 3 min (*hatched bar*). No apertures were recorded before 14 h. The leaf was kept at 25 °C, a photon flux density of 145 μmol mol m^{-2} s^{-1} and ΔW of c. 25 mmol mol^{-1} (Data author and holder: H. Kaiser)

synchronize the individual oscillators. Synchronization has been observed on many different scales, between neighboring stomata (Kaiser and Kappen 2001), and on leaf patches measuring a few mm (Siebke and Weis 1995; West et al. 2005) within leafy organs (Teoh and Palmer 1971). Even entire trees can oscillate synchronously (Herppich and von Willert 1995; Dzikiti et al. 2007).

Due to a lack of detailed microscopic observations, the minimum number of synchronized stomata needed in order to sustain oscillations is unknown. In *Sambucus nigra*, out-of-phase oscillations were observed in stomata with a distance of less than 2 mm (Kaiser and Kappen 2001). More recent experiments revealed a tight synchronization when the distance between stomata was less than 1 mm, gradually decreasing with increasing distance (Kaiser, unpublished), which indicates synchronizing mechanisms acting on a very small scale. This view is supported by observing the effect of blocked transpiration of selected stomata on their humidity response (Kaiser and Legner 2007): the transpiration of a single pore affected active closure of other stomata within an area of 0.5 mm^2. In *Helianthus annuus*, oscillations on only one face of the leaf were observed, without transmitting to the other surface (Nagarajah 1978). Similar evidence comes from Mott et al. (1993) who found different patchy patterns of stomatal opening on the two faces of leaves of *Xanthium strumarium*. Additionally, Mott (2007) found no response of stomata on one face of the leaf, if only the other face was subjected to dry air, despite a substantial decrease in epidermal turgor on the treated face of the leaf. This lack of synchronization between the two leaf surfaces points to a localization of feedback mechanisms in the epidermis rather than in the mesophyll.

The nature of the synchronizing mechanism acting on this small scale is not clear. Based on the assumption that stomata respond to variations in local water potential in the epidermis or the adjoining layers of mesophyll (Buckley 2005), local transpiration could affect larger areas by gaseous diffusion within the intercellular spaces. Additionally, gradients of water potential could be equalized by symplastic and apoplastic flows of water. Another possible mechanism providing lateral synchronization could be the generation of a chemical signal by epidermal or mesophyll cells which is spread by diffusion or mass flow within the tissue. Only one of these possibilities, the hydraulic coupling of the tissue, has solid experimental support. Streaming dry air to a small region of a leaf, which was otherwise kept humid, led to hydropassive opening in a distance of up to 0.4 mm (Mott and Franks 2001), which demonstrates that positive hydraulic feedback of pore transpiration also affects the responses of adjacent stomata. A spatial model of hydraulically connected stomata (Haefner et al. 1997) showed agreement with observed patch formation and dynamics.

Lateral transmission of hydraulic disturbance within this small scale network of hydraulically coupled stomata relies on cell to cell water transport. The hydraulic interaction of different regions of the leaf (Buckley and Mott 2000) most likely involves water transport in xylem vessels, which are able to transmit water potential changes to distant regions of the leaf due to their low resistance compared to extravascular pathways (Sack and Holbrook 2006). Interactions between hydraulically coupled leaf patches, forming a higher level network, may allow for pattern formation and synchronization (Johnsson 2007). Hydraulic resistance in itself is highly dynamic and its fluctuation appears to play a role in the development of stomatal oscillations. In *Helianthus annuus*, Marenco et al. (2006) found periodic xylem embolism and refilling corresponding with the fluctuations in transpiration, with highest percentage of embolised vessels at peak transpiration. Embolism occurring under increasing transpiration further impairs leaf water status, and amplifies hydropassive opening. This not only boosts positive feedback, but also synchronizes responses within the area supplied by the affected vessel.

Synchronization can also occur between leaf organs, ultimately leading to concerted cycling of gas exchange of entire crop plants (Cox 1968; Marenco et al. 2006) or trees (Steppe et al. 2006; Dzikiti et al. 2007). The responses in *Citrus sinensis* (Dzikiti et al. 2007) are in good agreement with a water balance model considering water reservoirs and hydraulic resistances within the entire plant. The role of cavitations in the generation of whole plant oscillations, however, is still hypothetical and needs further research (Marenco et al. 2006).

Stomatal oscillations promise insight into the stomatal control system; therefore many attempts have been made to construct models that will allow testing their assumptions (Johnsson 2007). Any modeling of complex systems faces the dilemma of choosing between a simple and manageable but possibly nonrealistic model, and the futile attempt to comprehensively describe all sub-processes. Earlier attempts were optimistic in that they focused on the hydraulic processes which are comparably easy to formalize. These first modeling approaches described the leaf in terms of hydraulic capacitors connected by flows with corresponding hydraulic resistances (Cowan 1972; Shirazi et al. 1976b; Delwiche and Cooke 1977), assuming no short-term, active adjustment of guard cell osmotic potential in response to transpiration. These models were valuable in that they enhanced the understanding of physical processes within the stomatal hydro-mechanic apparatus. They treated the leaf as a "lumped model" consisting of one guard cell and one of each of the interacting components, epidermal cells, xylem, etc. As a spatio-temporal dynamic was obvious from observations of patchy oscillatory behavior, models of hydraulically interacting stomata were developed (Rand et al. 1982; Haefner et al. 1997), which were based on known hydraulic interactions, and included stomatal variability and its influence on pattern formation (Laisk et al. 1980). These models successfully simulated patchy stomatal coordination and dynamics similar to those occurring in real leaves. However, the active regulation of guard cell osmotic pressure in response to local leaf water status, and the spatial and temporal dynamics of leaf hydraulic resistances involved in stomatal oscillations were not satisfactorily accounted for.

4.3 Rapid Transient Variation of Stomatal Conductance Under Ozone Exposure

A very interesting example of rapid stomatal responses to environmental stimuli is the rapid transient decline of g_s (RTD, Fig. 4.5) induced by ozone (Vahisalu et al. 2010). Ground-level or tropospheric ozone (O_3) is the gaseous pollutant at present of most concern for forest health (Serengil et al. 2011). Ozone is also used as a tool to induce ROS production and investigate their effects. RTD coincided with a burst of ROS in guard cells of 11 Arabidopsis ecotypes (Vahisalu et al. 2010). Mutants deficient in various aspects of stomatal function revealed that the SLAC1 protein, essential for guard cell plasma membrane S-type anion channel function, and the protein kinase OST1 were required for the ROS-induced fast stomatal closure. The recovery of g_s occurred even during O_3 exposure (Fig. 4.5) and stomata did not respond to additional O_3 pulses until a resting period for the guard cells allowed them to sense and respond to O_3 again (Vahisalu et al. 2010).

The temporary desensitization of stomata may be a cause of the sluggish responses to environmental stimuli observed after O_3 exposure (Paoletti and Grulke 2010). Sluggishness is defined as a delay in stomatal response to changing environmental factors relative to controls (Fig. 4.2), and has been demonstrated in different plant physiognomic classes (Paoletti and Grulke 2010). Sluggishness results from a longer time to respond to the closing signal and slower rate of closing. Sluggish stomatal responses to light variation with O_3 exposure were first postulated in Norway spruce using a transpirational assay, i.e. by measuring water

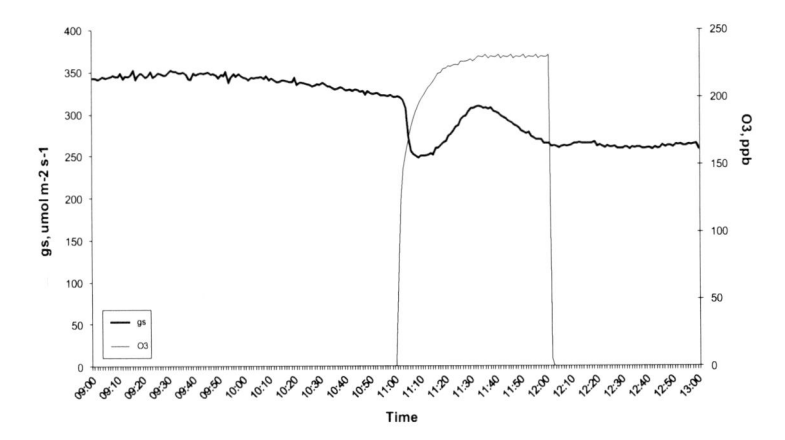

Fig. 4.5 Transient decline of stomatal conductance (g_s) in a *Helianthus annuus* leaf exposed to 230 ppb ozone (O_3). The recovery of gs occurred even during the O_3 exposure, suggesting a desensitization of stomata (Data authors and holders: H. Kaiser and E. Paoletti)

losses over time in detached needles (Keller and Häsler 1984). Delayed stomatal response following O_3 exposure has since been reported with changes in leaf to air vapour pressure deficits (Tjoelker et al. 1995; Kellomaki and Wang 1997; Grulke et al. 2007b), fluctuating photosynthetic photon flux density (PPFD) (Reich and Lassoie 1984; Reiling and Davison 1995; Paoletti 2005; Grulke et al. 2007b; Paoletti and Grulke 2010), and water stress (Reich and Lassoie 1984; Paoletti 2005; Mills et al. 2009; Grulke et al. 2007b; Paoletti et al. 2009; Hoshika et al. 2013). Drought stress itself, however, is able to induce stomatal sluggishness (Hoshika et al. 2013). Sluggish stomatal control over transpiration may increase water loss at the leaf level. At the crown-level, however, O_3 exposure reduced gas exchange and accelerated leaf shedding, thus compensating for sluggishness-increased water loss (Hoshika et al. 2012). Several mechanisms by which O_3 may induce sluggishness can be found in the published literature. Omasa (1990) reported a slight increase in permeability of epidermal cell membranes and alteration of the osmotic pressure after O_3 exposure that may modulate a balance in turgor between guard and subsidiary cells. Vahisalu et al. (2010) found that Ca^{2+}-dependent signalling and O_3-induced stomatal movements were independent, and noted that the temporary desensitization of the guard cells was due to blocked K^+ channels. Ozone may also delay stomatal responses by stimulating ethylene production and reducing stomatal sensitivity to ABA (Wilkinson and Davies 2010). Another cause of sluggishness may be O_3-induced lower rates of transpiration, which permit leaves to take longer to perceive the same change in water status or light variation.

4.4 Concluding Remarks

Stomatal regulation is the primary function for balancing the efficiency of water expenditure in relation to carbon gains (Cowan 1977). Optimization theory states that for each set of environmental conditions, an optimal stomatal conductance exists. It is immediately evident that g_s during stomatal oscillations is not at its optimum most of the time, either expending too much water in relation to carbon gain at peak conductance, or unnecessarily limiting carbon gain in the troughs. Nonetheless, modeling the effect of stomatal oscillations on time-averaged water use efficiency, Upadhyaya et al. (1988) identified conditions where oscillations slightly improved water use efficiency at reduced transpiration when compared to constant conductance. However, there is no experimental support for these observations. The marginally positive effect of a relatively rare phenomenon is unlikely to provide sufficient selection for this resulting complex feature.

Considering the prominent role of hydraulic positive feedback in stomatal oscillations, another hypothesis can be suggested. Hydropassive positive feedback in stomatal mechanics is a property only existing in seed plants and has not been found in ferns and mosses which appear to respond with hydropassive closure to increased transpiration (Brodribb and McAdam 2011). Positive hydraulic feedback evolved in seed plants along with a more sophisticated control of stomata through leaf water relations (McAdam and Brodribb 2012). Acceleration of stomatal opening as well as closing by hydropassive positive feedback enables larger and faster responses with the same metabolic effort. This allows a faster tracking of the dynamic environmental conditions resulting in an on average smaller deviation from the floating optimum. The metabolic costs necessary for dynamic stomatal movements (Vico et al. 2011) could be reduced due to hydraulic amplification of osmotic activity. Stomatal oscillations therefore may not in itself enhance efficiency of water use, but could be seen as a side effect of an aggressive tuning of feedback-regulation, which has evolved because it allows a faster response to environmental fluctuations.

The control of gas exchange by leaf stomata has broad implications for the response of terrestrial vegetation to changes in environmental conditions, including climate change (Hetherington and Woodward 2003). The feedback mechanism involved in oscillations appears to be identical to the mechanisms involved in regulation of leaf water loss. Unfortunately, there is still no consensus regarding the identity of the effectors involved in stomatal responses to hydraulic perturbations, nor regarding the biophysical mechanisms by which those effectors induce changes in stomatal conductance (Buckley and Mott 2002b; Meinzer 2002; Franks 2004; Buckley 2005). Although a vast amount of knowledge has been gathered on the intracellular events of guard cell signal transduction, these processes are both too complex and still too poorly understood to be described other than in a 'black box' approach. Moreover, the mechanism providing sensorial input of local leaf water relations into guard cell signaling is still obscure. Integrating these signaling events and metabolic actions merely as empirical functions into the models is difficult as the response is highly variable depending on – among others – species, previous treatment, and circadian effects.

Similar to cellular processes, the higher levels of hydraulic interaction between stomata, leaf regions or different leaves or branches require a better understanding before these pivotal processes can be integrated into models of stomatal dynamics at the leaf or whole plant level. A prerequisite is to monitor rapid changes in plant gs by means of gas-exchange measurement devices with high-time resolution (Grulke et al. 2007a), ideally coupled with microscopical observation of individual stomata (Kappen et al. 1987; Kaiser and Kappen 2000, 2001).

References

Assmann S, Gershenson A (1991) The kinetics of stomatal responses to VPD in *Vicia faba*: electrophysiological and water relations models. Plant Cell Environ 14:455–465

Assmann SM, Grantz DA (1990) The magnitude of the stomatal response to blue light – modulation by atmospheric humidity. Plant Physiol 93:701–709

Assmann SM, Shimazaki K (1999) The multisensory guard cell. Stomatal responses to blue light and abscisic acid. Plant Physiol 119:337–361

Assmann SM, Wang XQ (2001) From milliseconds to millions of years: guard cells and environmental responses. Curr Opin Plant Biol 4:421–428

Barrs HD (1968) Effect of cyclic variations in gas exchange under constant environmental conditions on ratio of transpiration to net photosynthesis. Physiol Plant 21:918–920

Barrs HD (1971) Cyclic variation in stomatal aperture, transpiration, and leaf water potential under constant environmental conditions. Ann Rev Plant Physiol 22:223–236

Bravdo BA (1977) Oscillatory transpiration and CO_2 exchange of *Citrus* leaves at the CO_2 compensation concentration. Physiol Plantarum 41:36–41

Brodribb TJ, Holbrook NM (2004) Stomatal protection against hydraulic failure: a comparison of coexisting ferns and angiosperms. New Phytol 162:663–670

Brodribb TJ, McAdam SAM (2011) Passive origins of stomatal control in vascular plants. Science 331:582–585

Brogårdh T, Johnsson A (1973) Oscillatory transpiration and water uptake of *Avena* plants II. Effects of deformation of xylem vessels. Physiol Plantarum 28:341–345

Brun WA (1961) Photosynthesis and transpiration of banana leaves as affected by severing the vascular system. Plant Physiol 36:577–580

Buckley TN (2005) The control of stomata by water balance. New Phytol 168:275–292

Buckley TN, Mott KA (2000) Stomatal responses to non-local changes in PFD: evidence for long-distance hydraulic interactions. Plant Cell Environ 23:301–309

Buckley T, Mott K (2002) Dynamics of stomatal water relations during the humidity response: implications of two hypothetical mechanisms. Plant Cell Environ 25:407–419

Cardon ZG, Mott KA, Berry JA (1994) Dynamics of patchy stomatal movements, and their contribution to steady-state and oscillating stomatal conductance calculated using gas-exchange techniques. Plant Cell Environ 17:995–1007

Cochard H, Coll L, Le Roux X, Améglio T (2002) Unraveling the effects of plant hydraulics on stomatal closure during water stress in walnut. Plant Physiol 128:282–290

Comstock JP, Mencuccini M (1998) Control of stomatal conductance by leaf water potential in *Hymenoclea salsola* (T. & G.), a desert subshrub. Plant Cell Environ 21:1029–1038

Cowan IR (1972) Oscillations in stomatal conductance and plant functioning associated with stomatal conductance: observations and a model. Planta 106:185–219

Cowan IR (1977) Stomatal behaviour and environment. Adv Bot Res 4:117–228

Cox EF (1968) Cyclic changes in transpiration of sunflower leaves in a steady environment. J Exp Bot 19:167–175

Delwiche MJ, Cooke JR (1977) An analytical model of the hydraulic aspects of stomatal mechanics. J Theor Biol 69:113–141

DeMichele DW, Sharpe PJH (1973) An analysis of the mechanics of guard cell motion. J Theor Biol 41:77–96

Dewar RC (1995) Interpretation of an empirical model for stomatal conductance in terms of guard cell function. Plant Cell Environ 18:365–372

Dzikiti S, Steppe K, Lemeur R, Milford JR (2007) Whole-tree level water balance and its implications on stomatal oscillations in orange trees *Citrus sinensis* (L.) Osbeck under natural climatic conditions. J Exp Bot 58:1893–1901

Eamus D, Shanahan S (2002) A rate equation model of stomatal responses to vapour pressure deficit and drought. BMC Ecol 2:8

Ehrler WL, Nakayama FS, Vanbavel CH (1965) Cyclic changes in water balance and transpiration of cotton leaves in a steady environment. Physiol Plant 18:766–775

Elias P (1979) Stomatal oscillations in adult forest trees in natural-environment. Biol Plant 21:71–74

Farquhar G (1978) Feedforward responses of stomata to humidity. Aust J Plant Physiol 5:787–900

Florell C, Rufelt H (1960) Transpiration of wheat plants cultivated under different environmental conditions. Physiol Plant 13:482–486

Franks PJ (2004) Stomatal control and hydraulic conductance, with special reference to tall trees. Tree Physiol 24:865–878

Franks P, Cowan I, Farquhar G (1997) The apparent feedforward response of stomata to air vapour pressure deficit: information revealed by different experimental procedures with two rainforest trees. Plant Cell Environ 20:142–145

Franks P, Cowan I, Farquhar G (1998) A study of stomatal mechanics using the cell pressure probe. Plant Cell Environ 21:94–100

Fuchs EE, Livingston NJ (1996) Hydraulic control of stomatal conductance in Douglas fir [*Pseudotsuga menziesii* (Mirb.) Franco] and alder [*Alnus rubra* (Bong)] seedlings. Plant Cell Environ 19:1091–1098

Grantz DA (1990) Plant response to atmospheric humidity. Plant Cell Environ 13:667–679

Grulke NE, Neufeld HS, Davison AW, Chappelka A (2007a) Stomatal behavior of O_3-sensitive and -tolerant cutleaf coneflower (*Rudbeckia laciniata* var. *digitata*) Great Smoky Mountain National Park. New Phytol 173:100–109

Grulke NE, Paoletti E, Heath RL (2007b) Comparison of calculated and measured foliar O_3 flux in crop and forest species. Environ Pollut 146:640–647

Haefner J, Buckley T, Mott K (1997) A spatially explicit model of patchy stomatal responses to humidity. Plant Cell Environ 20:1087–1097

Harris M, Outlaw WH Jr, Mertens R, Weiler EW (1988) Water-stress-induced changes in the abscisic acid content of guard cells and other cells of *Vicia faba* L. leaves as determined by enzyme-amplified immunoassay. Proc Nat Acad Sci 85:2584–2588

Herppich WB, von Willert DJ (1995) Dynamic changes in leaf bulk water relations during stomatal oscillations in mangrove species. Continuous analysis using a dewpoint hygrometer. Physiol Plant 94:479–485

Hetherington AM (2001) Guard cell signaling. Cell 107:711–714

Hetherington AM, Woodward FI (2003) The role of stomata in sensing and driving environmental change. Nature 424:901–908

Hirose T, Ikeda M, Izuta T, Miyake H, Totsuka T (1994) Stomatal oscillation in peanut leaves observed under field conditions. Jpn J Crop Sci 63:162–163

Hoshika Y, Omasa K, Paoletti E (2012) Whole-tree water use efficiency is decreased by ambient ozone and not affected by O_3-induced stomatal sluggishness. PLoS ONE 7:e39270

Hoshika Y, Omasa K, Paoletti E (2013) Both ozone exposure and soil water stress are able to induce stomatal sluggishness. Environ Exp Bot 88:19–23

Iwanoff L (1928) Zur Transpirationsbestimmung am Standort. Berichte der Deutschen Botanischen Gesellschaft 46:306–310

Johnsson A (1973) Oscillatory transpiration and water uptake of *Avena* plants I. Preliminary observations. Physiol Plant 28:40–50

Johnsson A (2007) Oscillations in plant transpiration. In: Mancuso S, Shabala S (eds) Rhythms in plants. Springer, Berlin

Johnsson M, Brogardh T, Holje Ø (1979) Oscillatory transpiration of *Avena* plants: perturbation experiments provide evidence for a stable point of singularity. Physiol Plant 45:393–398

Kaiser H (2009) The relation between stomatal aperture and gas exchange under consideration of pore geometry and diffusional resistance in the mesophyll. Plant Cell Environ 32:1091–1098

Kaiser H, Kappen L (2000) In-situ-observation of stomatal movements and gas exchange of *Aegopodium podagraria* L. in the understory. J Exp Bot 51:1741–1749

Kaiser H, Kappen L (2001) Stomatal oscillations at small apertures: indications for a fundamental insufficiency of stomatal feedback-control inherent in the stomatal turgor mechanism. J Exp Bot 52:1303–1313

Kaiser H, Legner N (2007) Localization of mechanisms involved in hydropassive and hydroactive stomatal responses of *Sambucus nigra* to dry air. Plant Physiol 143:1068–1077

Kappen L, Andresen G, Lösch R (1987) *In situ* observations of stomatal movements. J Exp Bot 38:126–141

Keller T, Häsler R (1984) The influence of a fall fumigation with ozone on the stomatal behavior of spruce and fir. Oecologia 64:284–286

Kellomäki S, Wang KY (1997) Effects of elevated O_3 and CO_2 concentrations on photosynthesis and stomatal conductance in Scots pine. Plant Cell Environ 20:995–1006

Kerstiens G (1997) Cuticular water permeance and its physiological significance. J Exp Bot 47:1813–1832

Laisk A, Oja V, Kull K (1980) Statistical distribution of stomatal apertures of *Vicia faba* and *Hordeum vulgare* and the Spannungs phase of stomatal opening. J Exp Bot 31:49–58

Levy Y, Kaufmann MR (1976) Cycling of leaf conductance in citrus exposed to natural and controlled environments. Can J Bot 54:2215–2218

Lu P, Outlaw WH Jr, Smith BG, Freed GA (1997) A new mechanism for the regulation of stomatal aperture size in intact leaves: accumulation of mesophyll-derived sucrose in the guard-cell wall of *Vicia faba*. Plant Physiol 114:109–118

Maier-Maercker U (1983) The role of peristomatal transpiration in the mechanism of stomatal movement. Plant Cell Environ 6:369–380

Marenco RA, Siebke K, Farquhar GD, Ball MC (2006) Hydraulically based stomatal oscillations and stomatal patchiness in *Gossypium hirsutum*. Funct Plant Biol 33:1103–1113

McAdam SAM, Brodribb TJ (2012) Stomatal innovation and the rise of seed plants. Ecol Lett 15:1–8

McAinsh MR, Gray JE, Hetherington AM, Leckie CP, Ng C (2000) Ca^{2+} signalling in stomatal guard cells. Biochem Soc Trans 28:476–481

Meidner H (1986) Cuticular conductance and the humidity response of stomata. J Exp Bot 37:517–525

Meidner H (1987) The humidity response of stomata and its measurement. J Exp Bot 38:877–882

Meinzer FC (2002) Coordination of vapour and liquid phase water transport properties in plants. Plant Cell Environ 25:265–274

Mills G, Hayes F, Wilkinson S, Davies WJ (2009) Chronic exposure to increasing background ozone impairs stomatal functioning in grassland species. Glob Change Biol 15:1522–1533

Monteith JL (1995) A reinterpretation of stomatal responses to humidity. Plant Cell Environ 18:357–364

Mott KA (2007) Leaf hydraulic conductivity and stomatal responses to humidity in amphistomatous leaves. Plant Cell Environ 30:1444–1449

Mott KA, Franks PJ (2001) The role of epidermal turgor in stomatal interactions following a local perturbation in humidity. Plant Cell Environ 24:657–662

Mott KA, Parkhurst DF (1991) Stomatal response to humidity in air and helox. Plant Cell Environ 14:509–515

Mott KA, Cardon ZG, Berry JA (1993) Asymmetric patchy stomatal closure for the 2 surfaces of *Xanthium strumarium* L leaves at low humidity. Plant Cell Environ 16:25–34

Nagarajah S (1978) Some differences in the responses of stomata of the two leaf surfaces in cotton. Ann Bot 42:1141–1147

Naidoo G, von Willert DJ (1994) Stomatal oscillations in the mangrove *Avicennia germinans*. Funct Ecol 8:651–657

Nikolic E (1925) Beiträge zur Physiologie der Spaltöffnungsbewegung. II. Über die Beziehung der Stomatarbewegung zur Lichtintensität. Beihefte zum Botanischen Centralblatt 41:309–346 (in German)

Omasa K (1990) Study on changes in stomata and their surroundings cells using a nondestructive light microscope system: responses to air pollutants. J Agric Meteorol 45:251–257 (in Japanese with English summary)

Oren R, Sperry JS, Katul GG, Pataki DE, Ewers BE, Phillips N, Schafer KVR (1999) Survey and synthesis of intra- and interspecific variation in stomatal sensitivity to vapour pressure deficit. Plant Cell Environ 22:1515–1526

Outlaw WH Jr, De Vlieghere-He X (2001) Transpiration rate. An important factor controlling the sucrose content of the guard cell apoplast of broad bean. Plant Physiol 126:1716–1724

Paoletti E (2005) Ozone slows stomatal response to light and leaf wounding in a Mediterranean evergreen broadleaf, *Arbutus unedo*. Environ Pollut 134:439–445

Paoletti E, Grulke NE (2010) Ozone exposure and stomatal sluggishness in different plant physiognomic classes. Environ Pollut 158:2664–2671

Paoletti E, Contran N, Bernasconi P, Günthardt-Goerg MS, Vollenweider P (2009) Structural and physiological responses to ozone in Manna ash (*Fraxinus ornus* L.) leaves of seedlings and mature trees under controlled and ambient conditions. Sci Total Environ 407:1631–1643

Parlange J-Y, Waggoner PE (1970) Stomatal dimensions and resistance to diffusion. Plant Physiol 46:337–342

Peak D, Mott KA (2011) A new, vapour-phase mechanism for stomatal responses to humidity and temperature. Plant Cell Environ 34:162–178

Pearcy RW (1990) Sunflecks and photosynthesis in plant canopies. Annu Rev Plant Physiol Plant Mol Biol 41:421–453

Pei Z-M, Kuchitsu K (2005) Early ABA signalling events in guard cells. J Plant Growth Regul 24:296–307

Phillips NG, Oren R, Licata J, Linder S (2004) Time series diagnosis of tree hydraulic characteristics. Tree Physiol 24:879–890

Pieruschka R, Huber G, Berry JA (2010) Control of transpiration by radiation. Proc Natl Acad Sci 107:13372–13377

Powles JE, Buckley TN, Nicotra AB, Farquhar GD (2006) Dynamics of stomatal water relations following leaf excision. Plant Cell Environ 29:981–992

Prytz G, Futsaether CM, Johnsson A (2003) Thermography studies of the spatial and temporal variability in stomatal conductance of *Avena* leaves during stable and oscillatory transpiration. New Phytol 158:249–258

Rand RH, Storti DW, Upadhyaya SK, Cooke JR (1982) Dynamics of coupled stomatal oscillators. J Math Biol 15:131–149

Raschke K (1965) Die Stomata als Glieder eines schwingungsfähigen CO_2-Regelsystems Experimenteller Nachweis an *Zea mays* L. Z Naturforsch 20b:1261–1270, in German

Reich PB (1984) Oscillations in stomatal conductance of hybrid poplar leaves in the light and dark. Physiol Plant 61:541–548

Reich PB, Lassoie JP (1984) Effects of low level O_3 exposure on leaf diffusive conductance and water-use efficiency in hybrid poplar. Plant Cell Environ 7:661–668

Reiling K, Davison AW (1995) Effects of ozone on stomatal conductance and photosynthesis in populations of *Plantago major* L. New Phytol 129:587–594

Rose MA, Rose MA (1994) Oscillatory transpiration may complicate stomatal conductance and gas-exchange measurements. HortSci 29:693–694

Rose MA, Beattie DJ, White JW (1994) Oscillations of whole-plant transpiration in 'Moonlight' rose. J Am Soc Hortic Sci 119:439–445

Sack L, Holbrook NM (2006) Leaf hydraulics. Ann Rev Plant Biol 57:361–381

Saliendra NZ, Sperry JS, Comstock JP (1995) Influence of leaf water status on stomatal response to humidity, hydraulic conductance, and soil drought in *Betula occidentalis*. Planta 196:357–366

Santrucek J, Hronkova M, Kveton J, Sage RF (2003) Photosynthesis inhibition during gas exchange oscillations in ABA-treated *Helianthus annuus*: relative role of stomatal patchiness and leaf carboxylation capacity. Photosynthetica 41:241–252

Schroeder JI, Allen GJ, Hugouvieux V, Kwak JM, Waner D (2001) Guard cell signal transduction. Annu Rev Plant Physiol Plant Mol Biol 52:627–658

Schulze ED, Lange OL, Buschbom U, Kappen L, Evenari M (1972) Stomatal responses to changes in humidity in plants growing in the desert. Planta 108:259–270

Serengil Y, Augustaitis A, Bytnerowicz A, Grulke N, Kozovitz AR, Matyssek R, Müller-Starck G, Schaub M, Wieser G, Coskun AA, Paoletti E (2011) Adaptation of forest ecosystems to air pollution and climate change: a global assessment on research priorities. iForest Biogeosci For 4:44–48

Shaner DL, Lyon JL (1979) Stomatal cycling in *Phaseolus vulgaris* L. in response to glyphosate. Plant Sci Lett 15:83–87

Sharpe PJH, Wu H, Spence RD (1987) Stomatal mechanics. In: Zeiger E, Farquhar GD, Cowan IR (eds) Stomatal function. Stanford University Press, Stanford

Shirazi GA, Stone JF (1976) Oscillatory transpiration in a cotton plant. I. Experimental characterization. J Exp Bot 27:608–618

Shirazi GA, Stone JF, Bacon CM (1976) Oscillatory transpiration in a cotton plant. II. A model. J Exp Bot 27:619–633

Siebke K, Weis E (1995) Assimilation images of leaves of *Glechoma*-Hederaceae – analysis of nonsynchronous stomata related oscillations. Planta 196:155–165

Stålfelt MG (1928) Die Abhängigkeit der photischen Spaltöffnungsreaktionen von der Temperatur. Planta 6:183–191

Steppe K, Dzikiti S, Lemeur R, Milford JR (2006) Stomatal oscillations in orange trees under natural climatic conditions. Ann Bot 97:831–835

Teoh CT, Palmer JH (1971) Nonsynchronized oscillations in stomatal resistance among sclerophylls of *Eucalyptus umbra*. Plant Physiol 47:409–411

Thomas SC, Winner WE (2002) Photosynthetic differences between saplings and adult trees: an integration of field results by meta-analysis. Tree Physiol 22:117–127

Tjoelker MG, Volin JC, Oleksyn J, Reich PB (1995) Interaction of ozone pollution and light effects on photosynthesis in a forest canopy experiment. Plant Cell Environ 18:895–905

Upadhyaya SK, Rand RH, Cooke JR (1988) Role of stomatal oscillations on transpiration, assimilation and water-use efficiency of plants. Ecol Model 41:27–40

Vahisalu T, Puzorjova I, Brosché M, Valk E, Lepiku M, Moldau H, Pechter P, Wang Y-S, Lindgren O, Salojarvi J, Loog M, Kangasjarvi J, Kollist H (2010) Ozone-triggered rapid stomatal response involves the production of reactive oxygen species, and is controlled by SLAC1 and OST1. Plant J 62:442–453

Vavasseur A, Raghavendra AS (2005) Guard cell metabolism and CO_2 sensing. New Phytol 165:665–682

Vico G, Manzoni S, Palmroth S, Katul G (2011) Effects of stomatal delays on the economics of leaf gas exchange under intermittent light regimes. New Phytol 192:640–652

Wallach R, Da-Costa N, Raviv M, Moshelion M (2010) Development of synchronized, autonomous, and self-regulated oscillations in transpiration rate of a whole tomato plant under water stress. J Exp Bot 61:3439–3449

Wang YP, Jarvis PG (1990) Influence of crown structural properties on PAR absorption, photosynthesis, and transpiration in Sitka spruce: application of a model (MAESTRO). Tree Physiol 7:297–316

West JD, Peak D, Peterson JQ, Mott KA (2005) Dynamics of stomatal patches for a single surface of *Xanthium strumarium* L. leaves observed with fluorescence and thermal images. Plant Cell Environ 28:633–641

Wilkinson S, Davies WJ (2008) Manipulation of the apoplastic pH of intact plants mimics stomatal and growth responses to water availability and microclimatic variation. J Exp Bot 59:619–631

Wilkinson S, Davies W (2010) Drought, ozone, ABA and ethylene: new insights from cell to plant community. Plant Cell Environ 33:510–525

Yang H-M, Zhang W-X, Wang G-X, Li Y, Wei X-P (2003) Cytosolic calcium oscillation may induce stomatal oscillation in *Vicia faba*. Plant Sci 165:1117–1122

Yoshida R, Umezawa T, Mizoguchi T, Takahashi S, Takahashi F, Shinozaki K (2006) The regulatory domain of SRK2E/OST1/SnRK2.6 interacts with ABI1 and integrates abscisic acid (ABA) and osmotic stress signals controlling stomatal closure in *Arabidopsis*. J Biol Chem 281:5310–5318

Zhang SQ, Outlaw WH Jr (2001) Abscisic acid introduced into the transpiration stream accumulates in the guard-cell apoplast and causes stomatal closure. Plant Cell Environ 24:1045–1054

Zhang X, Xu D (2000) Seasonal changes and daily courses of photosynthetic characteristics of 18-year-old Chinese fir shoots in relation to shoot ages and positions within tree crown. Sci Silvae Sin 19(03):36

Zhang W, Fan LM, Wu WH (2007) Osmo-sensitive and stretch-activated calcium-permeable channels in *Vicia faba* guard cells are regulated by actin dynamics. Plant Physiol 143:1140–1151

Zipperlen SW, Press MC (1997) Photosynthetic induction and stomatal oscillations in relation to the light environment of two Dipterocarp rain forest tree species. J Ecol 85:491–503

Chapter 5
The Regulation of Osmotic Potential in Trees

Andrew Merchant

Abstract The availability of water and the ability of plants to acquire it influence productivity among many ecosystems. For all plants, both terrestrial and marine, the modification of internal osmotic potential (ψ_π) is a commonly observed response to changes in water availability. This chapter highlights that modification of ψ_π and its effects on plant water potential (ψ) should be considered with regard to physiological relevance, particularly when discussing the physiology of trees. A limited range of solutes are suitable as cellular osmotica and we highlight the physiochemical properties that help to maintain physiological function at low ψ. Overall, differences in the capacity to regulate ψ_π are among many functional adaptations that enable growth and productivity of trees across a wide range of environments.

5.1 Principles of Monitoring Osmotic Regulation in Trees

The regulation of osmotic potential (ψ_π) is one of many adaptive traits that enable trees to obtain water from their environment. Combined with structural and physiological adaptations, the regulation of ψ_π in response to saline and/or dry conditions has been demonstrated across many tree species leading to the compilation of several reviews on the topic (e.g. Abrams 1988a; Hoch et al. 2003; Lemcoff et al. 1994; Popp et al. 1997). Variation in the capacity to modify ψ_π among tree species has been proposed to explain differing salt (e.g. Grieve and Shannon 1999; Martinez et al. 1991; Nikam and McComb 2000) and drought (e.g. Li 1998; Gebre et al. 1998) tolerances leading to suggestions of its use as a selection criterion for high performing individuals (Abrams 1988a, 1990; Lemcoff et al. 1994; van der Moezel et al. 1991).

A. Merchant (✉)
Faculty of Agriculture and Environment, University of Sydney, Sydney, NSW, Australia
e-mail: andrew.merchant@sydney.edu.au

M. Tausz and N. Grulke (eds.), *Trees in a Changing Environment*,
Plant Ecophysiology 9, DOI 10.1007/978-94-017-9100-7_5,

While little doubt remains that the regulation of ψ_π plays a significant role in tree survival and growth, it is also important to recognize that regulation can take several forms. To explain these phenomena, it is first useful to outline the components of water potential as it pertains to tree physiology and some common methods of measurement.

5.2   Calculating Osmotic and Water Potential

For the purposes of tree physiology, water potential (ψ) is a measure of the free energy contained within a solution or the 'chemical potential of water' (Jones 1992) and is generally expressed in mega-Pascals (MPa). In order to occupy a lower energy state, water will flow from regions of higher chemical potential to regions of lower chemical potential. To maintain a net influx of water into cells, trees may modify the chemical potential of cellular water through a number of mechanisms. Osmotic potential is a component of ψ and is expressed as:

$$\psi = \psi_\pi + \psi_g + \psi_m + \psi_p \tag{5.1}$$

Where ψ_g, ψ_m and ψ_p represent gravitational, matric and pressure forces. By reducing ψ_π inside cells, plants are able to obtain water from their environment as water becomes less available. Several methods for the assessment of osmotic potential have been developed including pressure chamber (Turner 1988), psychrometer (Martinez et al. 2011), and pressure probe (Tomos 2000) technologies. Investigations seeking to characterize changes in ψ_π generally encompass responses to imposed stress under controlled conditions with less focus on responses to environmental variation across seasonal cycles and the heritability of such traits among populations. Values of ψ_π are most commonly expressed at full turgor (ψ_{100}) in response to stress conditions (Abrams 1988a; Turner 1988; Clayton-Green 1983; Lenz et al. 2006; Merchant et al. 2007a; Schulte and Hinckley 1985) and are most commonly derived from analysis of pressure-volume curves (PV curves, Fig. 5.1; cf. Chap. 6, Fig. 6.1). PV curves are generated with the use of a pressure chamber (Turner 1981, 1988) and rely upon the repetitive measurement of ψ and relative water content (RWC). Plotting the inverse of ψ against the reciprocal of RWC produces a two phase relationship (Fig. 5.1). ψ_π at full turgor ($\psi_{\pi ft}$) of the tissue is represented by extrapolation of the linear phase of the PV curve (Fig. 5.1). A significant advantage of PV curve analysis is the ability to simultaneously quantify adjustments of ψ_π at the point of zero turgor ($\psi_{\pi tlp}$), relative water content at turgor loss point (RWC_{tlp}), and apoplastic water volume and cell bulk modulus of elasticity (ε). Adjustments of ε (hence cell volume) may be employed by plants to modify ψ via influences on ψ_π and ψ_p. Inter-specific variation in the capacity to adjust ε is observed among trees in response to water deficits (Fan et al. 1994) and correlates with environmental origin. Whilst ε is likely influenced by tissue age

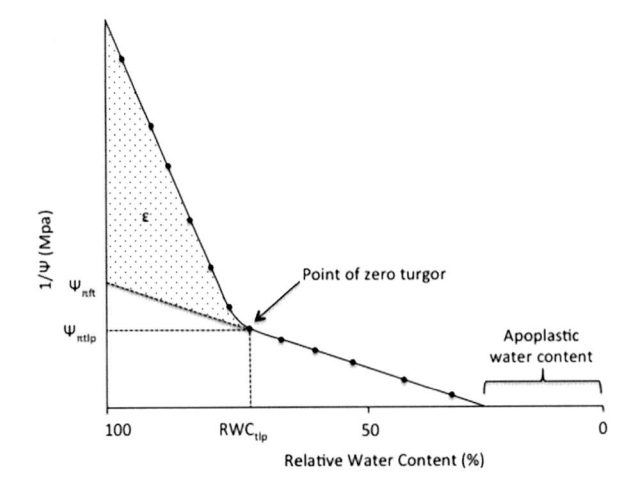

Fig. 5.1 Pressure volume (PV) curve depicting a typical two phase relationship between the relative water content (RWC) and the inverse of the water potential (ψ). Note the relationship becomes linear after the point of zero turgor is reached. Osmotic potential at turgor loss point ($\psi_{\pi tlp}$) is obtained from the y-axis at the point of zero turgor whilst the osmotic potential at full turgor ($\psi_{\pi ft}$) is obtained by extrapolation of the linear phase of the PV curve to the y-axis. Relative water content at turgor loss point (RWC_{tlp}) is obtained from the x-axis. Bulk modulus of elasticity (ϵ) is obtained via the integration of the area under the pressure volume curve bound by the extrapolation of the linear phase to the y-axis. Apoplastic water volume can be calculated as the residual of the linear phase at the point of intersection with the x-axis

(Fan et al. 1994), the capacity to adjust ϵ undoubtedly plays a role in regulating ψ among trees species.

An alternative to pressure volume analysis is the quantification of compounds from leaf material combined with measures of tissue water content. Using this approach, the ψ_π of the cell solution may be *approximated* by the van't Hoff equation (Eq. 5.2):

$$\psi_\pi = -RTc_s \tag{5.2}$$

Where R is the universal gas constant ($8.32\ J\ mol^{-1}\ K^{-1}$), T is the temperature in Kelvin and c_s is the concentration of solutes in solution ($mol\ m^{-3}$). This *approximation* of ψ_π is suitable for most biological systems despite the influence of 'non-ideal' (Cameron et al. 1997) behavior of solutions at high concentrations (see also Jones 1992 and references therein). Whilst both PV analysis and chemical analysis have important limitations, used in combination they provide information regarding the structural and chemical traits adopted by plant tissues to regulate water content. Information gleaned from such approaches is now beginning to characterise the mechanisms employed by plants to regulate ψ_π and the identity of the compounds used to perform this role.

5.3   Considerations for Physiological Relevance

Trees occupy a range of environments that encompass a wide range of edaphic and climatic conditions. Despite the seemingly central role of regulating ψ_π in maintaining plant function, several aspects are often not considered when characterizing the nature of the response. Changes in ψ_π can arise via several mechanisms working at different temporal and spatial scales. Consequently, clear definitions must be established to accurately describe often complex and co-occurring processes.

5.3.1   Inducible and Constitutive Responses

Many physiological and chemical studies characterize induced plant responses to the onset of stressful conditions. Plants may increase synthesis of compounds to reduce ψ_π in response to stress, a process termed 'osmotic adjustment' (Turner and Jones 1980). Alternatively, accumulation of solutes in plant tissues due to reductions in breakdown or changes in solute transport may also lead to increases in solute concentration that do not fall under the definition of osmotic adjustment. Constitutive presence of solutes (rather than changes in concentration upon exposure to altered conditions), offers an additional mechanism for the regulation of ψ_π and may provide a more valid reflection of a plants ability to withstand low external ψ. Constitutive accumulation may be considered a pre-emptive arrangement by plants to tolerate conditions of periodic stress and underpin an-isohydric behavior (e.g. Tardieu and Simonneau 1998) observed among many tree species.

A well characterized example of both inducible and constitutive mechanisms is that of the broadly distributed tree genus, *Eucalyptus*. The magnitude of interspecific species variation *changes* of ψ_π is consistent across the genus despite significant differences in environmental origin and growth habit (Table 5.1).

In contrast, the taxonomic pattern of osmotic potential at full turgor ($\psi_{\pi100}$) shows a clear delineation between those species growing in arid environments. These species exhibit values of ψ_{100} consistently < -1.5 MPa. Differences in ψ_π remain clear and significant despite considerable variation among tissues both in treatments and tissue age among studies (not shown). Whilst it appears likely that the capacity to lower osmotic potential as a means of maintaining turgor under mild osmotic stress is a widespread phenomenon – and the constitutive presence of osmotica lead to inherently low osmotic potentials – both inducible and constitutive effects appear independent from responses arising from the restriction of growth.

Table 5.1 Osmotic potential at full turgor among a range of *Eucalyptus* species grown under 'stressed' and 'unstressed' conditions

Species	Environmental origin	Location	Method	π_{FT} unstressed	π_{FT} stressed	Osmotic adjustment	Reference
Eucalyptus leucoxylon	*Xeric*	Field	Pressure volume	−3.20	−3.18	−0.02	White et al. (2000)
Eucalyptus platypus	*Xeric*	Field	Pressure volume	−2.48	−2.78	0.30	White et al. (2000)
Eucalyptus polyanthemos	*Xeric*	Field	Extracted	−1.63	−2.14	0.51	Myres and Neales (1986)
Eucalyptus behriana	*Xeric*	Field	Extracted	−1.63	−1.93	0.30	Myres and Neales (1986)
Eucalyptus microcarpa	*Xeric*	Field	Extracted	−1.72	−1.98	0.26	Myres and Neales (1986)
Eucalyptus miniata	*Xeric*	Field	Pressure volume	−1.81	−1.83	0.02	Myers et al. (1997)
Eucalyptus teterodonta	*Xeric*	Field	Pressure volume	−1.83	−1.93	0.10	Myers et al. (1997)
Eucalyptus melliodora	*Xeric*	Field	Pressure volume	−2.18			Clayton-Greene (1983)
Eucalyptus melliodora	*Xeric*	Field	Pressure volume	−1.57			Clayton-Greene (1983)
Eucalyptus microcarpa	*Xeric*	Field	Pressure volume	−2.60			Clayton-Greene (1983)
Eucalyptus tricarpa	*Xeric*	Controlled	Pressure volume	−1.62	−2.17	0.55	Merchant et al. (2007a)
Eucalyptus cladocalyx	*Xeric*	Controlled	Pressure volume	−1.65	−2.29	0.64	Merchant et al. (2007a)
Eucalyptus polyanthermos	*Xeric*	Controlled	Pressure volume	−1.40	−2.34	0.93	Merchant et al. (2007a)
Eucalyptus grandis	*Mesic*	Controlled	Pressure volume	−1.33	−1.53	0.20	Lemcoff et al. (1994)
Eucalyptus viminalis	*Mesic*	Controlled	Pressure volume	−1.21	−1.60	0.39	Lemcoff et al. (1994)
Eucalyptus teretecornis	*Mesic*	Controlled	Pressure volume	−1.20	−1.70	0.50	Lemcoff et al. (1994)
Eucalyptus grandis	*Mesic*	Controlled	Pressure volume	−1.10	−1.08	−0.02	Fan et al. (1994)
Eucalyptus viminalis	*Mesic*	Controlled	Pressure volume	−1.40			Ladiges (1975)
Eucalyptus viminalis	*Mesic*	Controlled	Pressure volume	−1.45			Ladiges (1975)
Eucalyptus viminalis	*Mesic*	Controlled	Pressure volume	−1.44			Ladiges (1975)
Eucalyptus marginata	*Mesic*	Controlled	Pressure volume	−1.14	−1.54	0.40	Stoneman et al. (1994)
Eucalyptus globulus	*Mesic*	Controlled	Pressure volume	−0.81	−1.03	0.23	Pita and Pardos (2001)
Eucalyptus globulus	*Mesic*	Field	Pressure volume	−1.33	−1.59	0.26	Correia et al. (1989)
Eucalyptus obliqua	*Mesic*	Controlled	Pressure volume	−1.15	−1.71	0.56	Merchant et al. (2007a)
Eucalyptus rubida	*Mesic*	Controlled	Pressure volume	−1.12	−1.93	0.80	Merchant et al. (2007a)
Eucalyptus camaldulensis	*Phreatophytic*	Controlled	Pressure volume	−0.95	−2.02	1.07	Merchant et al. (2007a)

5.3.2 Acute and Adaptive Responses

When distinguishing the chemical processes leading to solute accumulation in plant tissues, one must also consider the context of regulatory control. Plant responses may take the form of either acute or acclimatory response mechanisms (Smith and Stitt 2007) with major implications for plant function. Imposition of short term, intense stress conditions (as is the case for many investigations) often leads to acute responses as a result of the temporary repression of physiological processes such as photosynthesis, metabolism and transport. Delineation of acute and adaptive responses is a major challenge to understanding plant responses to stress and serves to highlight the importance of field based and long term monitoring campaigns. Relatively slow imposition of environmental stresses, may induce traits that protect plant components from repeated or rapidly induced conditions (Kozlowski and Pallardy 2002). Such studies are crucial for 'real world' perspectives on the response of plants to changes in their environment.

5.3.3 Magnitude of Osmotic Potential

Despite various mechanisms for the regulation of ψ_π, changes in tree species under stress are generally limited to between 0.2 and 0.8 MPa (Taiz and Zeiger 1998) with broad scale reviews into tree species exhibiting values consistently within this range (Table 5.2). ψ below -3.5 MPa are common for many tree species inhabiting low rainfall environments (Abrams 1990; Kozlowski and Pallardy 2002; Taiz and Zeiger 1998; Bell and Williams 1997; Kozlowski 1997; Merchant et al. 2009), therefore the magnitude of reductions in ψ_π are not sufficient to provide osmotic equilibrium with the prevailing conditions. While the potentially significant influence of sub-compartmentalisation is largely not known for many solutes, a logical hypothesis may be that changes in ψ_π do not play a major role in determining the growth and survival characteristics (hence distribution) of tree species, but rather help to 'expand the window' of positive turgor across diurnal changes in ψ.

5.3.4 Tissue Types

While the majority of studies investigating ψ_π focus on leaves, osmotic balance is critical in maintaining function of other plant tissues. Regulation of root tissue ψ_π is undoubtedly important for plants to acquire water from an increasingly drier and/or saltier soil profile. Many biochemical and molecular characterizations of this process are now available (Kozlowski and Pallardy 2002) including details of root and mychorrizal associations (Lehto and Zwiazek 2011). Phloem tissues are also thought to regulate ψ_π in response to external ψ with several studies

Table 5.2 Osmotic potential at full turgor among a range of tree species grown under 'stressed' and 'unstressed' conditions

Species	Location	Method	π_{FT} unstressed	π_{FT} stressed	Osmotic adjustment	Reference
Quercus alba	Controlled	Pressure volume	−1.47	−1.95	0.48	Parker and Pallardy (1987)
Quercus macrocarpa	Controlled	Pressure volume	−1.56	−2.1	0.54	Parker and Pallardy (1987)
Quercus stellata	Controlled	Pressure volume	−1.82	−2.08	0.26	Parker and Pallardy (1987)
Celtis occidentalis	Field	Pressure volume	−1.46	−2.67	1.21	Abrams (1986)
Quercus macrocarpa	Field	Pressure volume	−1.6	−2.24	0.64	Abrams (1986)
Quercus muehlenbergii	Field	Pressure volume	−1.69	−2.05	0.36	Abrams (1986)
Quercus velutina	Field	Pressure volume	−1.95	−2.26	0.31	Bahari et al. (1985)
Quercus alba	Field	Pressure volume	−1.89	−2.17	0.28	Bahari et al. (1985)
Quercus rubra	Field	Pressure volume	−1.86	−1.98	0.12	Bahari et al. (1985)
Cornus florida	Field	Pressure volume	−1.53	−1.74	0.21	Bahari et al. (1985)
Acer saccharum	Field	Pressure volume	−1.52	−1.55	0.03	Bahari et al. (1985)
Juniperus virginiana	Field	Pressure volume	−1.7	−1.8	0.1	Bahari et al. (1985)
Pinus banksiana	Controlled	Pressure volume	−1.084	−1.107	0.023	Fan et al. (1994)
Picea mariana	Controlled	Pressure volume	−1.266	−1.181	−0.085	Fan et al. (1994)
Populus tremuloides	Field	Pressure volume	−1.43	−1.97	0.54	Abrams (1988b)
Acer rubrum	Field	Pressure volume	−1.81	−1.85	0.04	Abrams (1988b)
Quercus ellipsoidalis	Field	Pressure volume	−0.83	−1.2	0.37	Abrams (1988b)
Erythropleum chlorostachys	Field	Pressure volume	−1.95	−2.23	0.28	Myers et al. (1997)
Xanthostemon paradoxus	Field	Pressure volume	−1.39	−1.5	0.11	Myers et al. (1997)
Planchonia careya	Field	Pressure volume	−1.62	−1.58	−0.04	Myers et al. (1997)
Terminalia ferdinandiana	Field	Pressure volume	−1.54	−1.48	−0.06	Myers et al. (1997)
Ilex opaca	Field	Pressure volume	−1.53	−1.98	0.45	Roberts and Knoerr (1977)
Acer rubrum	Field	Pressure volume	−1.59	−1.75	0.16	Roberts and Knoerr (1977)
Liquidamber styraciflua	Field	Pressure volume	−1.65	−1.88	0.23	Roberts and Knoerr (1977)
Liriodendron tulipifera	Field	Pressure volume	−1.49	−1.7	0.21	Roberts and Knoerr (1977)
Cornus florida	Field	Pressure volume	−1.48	−1.91	0.43	Roberts and Knoerr (1977)

characterizing changes in phloem sap ψ_π among tree species (Cernusak et al. 2003; Merchant et al. 2010a; Pate et al. 1998) including that of carbohydrate and raffinose family oligosaccharides (Merchant et al. 2010a,b).

5.3.5 Plasticity

The plasticity of traits that regulate ψ_π remains weakly characterized. This is surprising given the regulation of ψ_π, with a firm footing in biochemistry and molecular biology, represents a promising candidate for tree selection and environmental monitoring. Despite a plethora of screening studies seeking to develop such tools by characterizing both inter- and intra-specific variation in growth strategies under drought and salt stress (van der Moeszel et al. 1991; van der Moezel and Bell 1987; Sun and Dickinson 1993; Zohar and Schiller 1998), broad categorizations on chemical responses are rare. Such approaches are well served by an understanding of the identity and functional significance of solutes that accumulate in plant tissues across spatial and temporal scales. In addition to characterizing the general mechanisms responsible for regulation of ψ_π among the diversity of tree species, advances in metabolomic tools are now available to assist in identifying the chemical identity of these responses.

5.4 Candidate Solutes for the Regulation of Osmotic Potential in Tree Species

Relatively few solute classes have been shown to significantly influence ψ_π in tree species attributable mainly to the physiochemical properties required for solutes to perform this role. Stable osmotica must be compatible with the function of cellular components (Popp et al. 1997; Paul and Cockburn 1989; Palacio et al. 2007), and must be highly reduced to avoid damage to cellular structures. Such properties can be far reaching in metabolic networks, enabling trees to avoid cellular damage through the maintenance of metabolic function at constitutively low osmotic potentials. Osmotica should also be synthesized from readily accessible precursors and ideally, should be readily convertible to other more metabolically active and transportable compounds. A limited number of common cell constituents satisfy these criteria and are often found in tissues of higher plants.

5.4.1 Organic Acids

Organic acids that accumulate in plant tissues in response to abiotic stresses are usually either amino acids (e.g. proline) or amino acid derivatives (e.g. betaines). The magnitude of increases in organic acid concentration is often sufficient to suggest an osmolytic role (Prat and Fathi-Ettai 1990; Escobar-Gutierrez et al. 1998; Larher et al. 2009; Morabito et al. 1996) notwithstanding the potential and likely effects of sub-cellular compartmentalization. However, in cases where an osmolytic effect of organic acids is substantiated in the genus *Eucalyptus*, often reductions in growth are also observed (Morabito et al. 1994). Such responses, concurrent with the breakdown of cellular processes suggest that increases in concentration may be acute consequence of stress induction rather than an adaptive response. For example, increasing concentrations of proline and glycinebetaine in the initial phases of salt (NaCl) adaptation have been reported in *E. microcorys* tissue culture (Chen et al. 1998), with the subsequent reduction of these solutes during the remainder of the conditioning period. As is the case with many previous studies, increases in solute concentration reported on a fresh weight basis in both studies clouds our ability to distinguish increases in solute concentration attributable to inducible or constitutive phenomena.

Additional functions of organic acids in the amelioration of stress in plant tissues are subject to various hypotheses (e.g. Franco and Melo 2000). Hare et al. (1998) highlighted the common hypothesis that methylation (which is present in proline, MHP, DHP, and betaines) improves the protein stabilizing properties of osmolytes. Methylation incorporates a partial hydrophobicity as well as increasing the size of the hydration shell of small zwitterionic molecules (Hare et al. 1998) thus may stabilize tertiary protein structures through both direct and indirect mechanisms (such as 'preferential exclusion', Bohnert and Shen 1999). Similarly, Palfi et al. (1974) also suggest that the high level of hygroscopicity enhances any role of proline as an osmoprotectant.

5.4.2 Inorganic Ions

Inorganic ions often have significant influence over ψ_π in plant tissues although their role in regulation is less clear. A major limitation to the function of inorganic ions as regulators of ψ_π is their disruptive influence over cellular structures and the need to maintain trans-membrane electrochemical balance. It is widely argued that plants must separate toxic ions (such as Na^+ and Cl^-) from cellular processes to maintain cellular function as excessive ion accumulation disrupts membrane and protein integrity. Excessive concentrations of cations can also lead to nutrient deficiencies due to their inherent toxicities and competition for transport sites within the plasma membrane (Taiz and Zeiger 1998; Blumwald 2000) with a large and permanent efflux of potassium (Morabito et al. 1996) usually indicating damage to the limiting membranes (Chen et al. 1998).

The role of potassium in lowering leaf ψ_π in tree species is widely recognised on account of its ubiquitous presence in plant tissues (Taiz and Zeiger 1998). Potassium often constitutes over 50 % of ψ_π in plant tissues due to the combined effects of its high concentrations and low molecular weight (Eq. 5.1). While the contribution of potassium to the regulation of ψ_π is difficult to establish due to the effects of concentration and rapid exchange between plant compartments, rapid transmembrane movement of potassium has been implicated in osmo-sensing capabilities among many plants including that of trees (Liu et al. 2001). Further use of isotopic and other non-invasive techniques to detect the concentration and movement of potassium in plant tissues will undoubtedly prove informative.

5.4.3 Carbohydrates

Carbohydrates are frequently cited as being suitable osmotica and are major constituents of osmoregulation in expanded leaves of many species of higher plants (Morgan 1984). Many studies have sought to characterize non-structural carbohydrates (NSC) in both temperate (Hoch et al. 2003) and tropical (Wurth et al. 2005) trees in response to variation in environmental conditions. Of particular interest for osmotic regulation is the occurrence of non-reducing sugars, most commonly sucrose and raffinose family oligosaccharides (RFO) among a range of plant tissues.

The close association of carbohydrates and their participation in primary metabolism, low molecular weight, relative low toxicity, and potential for recycling back into plant metabolism renders them sensitive to environmental change leading to suggestions that synthesis and degradation may be genetically engineered to improve stress tolerance (Rajam et al. 1998). Fluctuations in pools of carbohydrates due to close association with primary metabolism reduce their effectiveness as stable osmotica, however, this close association simultaneously supports the notion that these compounds also act in the sequestration of damaging oxygen radicals (Orthen et al. 1994) and in other forms of diversion of excess photochemical energy (Hare et al. 1998). Many studies have shown accumulations of carbohydrates in response to environmental perturbations, although few give regard to the hypothesis by Turner and Jones (1980) that osmotic adjustment arises from an accumulation of solutes as a result of the continuation of photosynthesis after the cessation of leaf growth.

5.4.4   *Polyols*

Polyols are reduced forms of aldose and ketose sugars consisting of either the alditol (straight chain) or cyclitol (cyclic) forms. Polyols have been shown to accumulate in plant tissues in response to many environmental cues thus are regularly included in lists of stress metabolites in plant tissues. Chemo-taxonomical and physiological studies conducted by various authors illustrate that polyol distribution among trees follows strong taxonomical patterns (Bieleski and Briggs 2005; Merchant et al. 2007b; Pfundner 1993; Popp et al. 1997) suggesting large potential for use as selective markers for tree improvement programs.

The physiochemical properties of polyols provide putative evidence for their function in osmotic regulation. Polyols are non-charged, relatively inert solutes and are thought to assist in hydrating protein structures via a 'preferential exclusion' (Andersen et al. 2011) from membrane surfaces. Some cyclitols may also be mono- and di-methylated, a characteristic unique to plant tissues (Bieleski 1994; Nguyen and Lamant 1988; Popp et al. 1997) and it has been proposed as a mechanism for sequestration of excess photochemical energy arising from photorespiration (Hare et al. 1998) and improving potential as an osmoprotectant (Popp et al. 1997; Hare et al. 1998; Popp and Smirnoff 1995; Orthen et al. 2000; Orthen and Popp 2000). More generally, polyols can contribute up to 80 % of the carbon isolated from phloem sap (Moing et al. 1997), and hence significantly influence leaf carbon balance and primary metabolism. While much is known regarding the roles and distribution of polyols in plant tissues, few studies (Streeter et al. 2001) have incorporated this information to the development of plant breeding programs.

5.5   Conclusions

Changes in osmotic potential are an important chemical trait that functions in concert with physiological changes to regulate plant water relations. A limited number of chemical species are thought to function as cellular osmotica based on their occurrence in plant tissues, the chemical properties of the solute, and their likely interactions with cellular processes. Characteristics of the regulation of osmotic potential also differ between genera and such changes need to be considered in the context of plant adaptive and acclimative responses to changes in resource availability. Great potential remains for the use of osmotic adjustment as a selection criterion for plant improvement programs and as a tool for use in environmental monitoring and assessments of tree health.

References

Abrams MD, Knapp AK (1986) Seasonal water relations of 3 gallery forest hardwood species in northeast Kansas. For Sci 32:687–696

Abrams MD (1988a) Sources of variation in osmotic potentials with special reference to North-American tree species. For Sci 34:1030–1046

Abrams MD (1988b) Comparative water relations of 3 successional hardwood species in central Wisconsin. Tree Physiol 4:263–273

Abrams MD (1990) Adaptations and responses to drought in *Quercus* species of North-America. Tree Phys 7:227–238

Andersen HD, Wang CH, Arleth L, Peters GH, Westh P (2011) Reconciliation of opposing views on membrane-sugar interactions. Proc Natl Acad Sci 108:1874–1878

Bahari ZA, Pallardy SG, Parker WC (1985) Photosynthesis, water relations, and drought adaptation in 6 woody species of Oak-Hickory forests in central Missouri. For Sci 31:557–569

Bell DT, Williams JE (1997) Eucalypt ecophysiology. In: Williams J, Woirnarski J (eds) Eucalypt ecology. Cambridge University Press, Cambridge

Bieleski RL (1994) Pinitol is a major carbohydrate in leaves of some coastal plants indigenous to New-Zealand. New Zeal J Bot 32:73–78

Bieleski RL, Briggs BG (2005) Taxonomic patterns in the distribution of polyols within the *Proteaceae*. Aust J Bot 53(2):05–217

Blumwald E (2000) Sodium transport and salt tolerance in plants. Curr Opin Cell Biol 12:431–434

Bohnert HJ, Shen B (1999) Transformation and compatible solutes. Sci Hortic 78:237–260

Cameron IL, Kanal KM, Keener CR, Fullerton GD (1997) A mechanistic view of the non-ideal osmotic and motional behavior of intracellular water. Cell Biol Int 21:99–113

Cernusak LA, Arthur DJ, Pate JS, Farquhar GD (2003) Water relations link carbon and oxygen isotope discrimination to phloem sap sugar concentration in *Eucalyptus globulus*. Plant Physiol 13:1544–1554

Chen DM, Keiper FJ, De Filippis LF (1998) Physiological changes accompanying the induction of salt tolerance in *Eucalyptus microcorys* shoots in tissue culture. J Plant Physiol 152:555–563

Clayton-Greene KA (1983) The tissue water relationships of *Callitris columellaris*, *Eucalyptus melliodora* and *Eucalyptus microcarpa* investigated using the pressure-volume technique. Oecologia 57:368–373

Correia MJ, Torres F, Pereira JS (1989) Water and nutrient supply regimes and the water relations of juvenile leaves of *Eucalyptus globulus*. Tree Physiol 5:459–471

Escobar-Gutierrez AJ, Zipperlin B, Carbonne F, Moing A, Gaudillere JP (1998) Photosynthesis, carbon partitioning and metabolite content during drought stress in peach seedlings. Aust J Plant Physiol 25:197–205

Fan SH, Blake TJ, Blumwald E (1994) The relative contribution of elastic and osmotic adjustments to turgor maintenance of woody species. Physiol Plant 90:408–413

Franco OL, Melo FR (2000) Osmoprotectants – a plant strategy in response to osmotic stress. Russ J Plant Physiol 47:137–144

Gebre GM, Tschaplinski TJ, Shirshac TL (1998) Water relations of several hardwood species in response to throughfall manipulation in an upland oak forest during a wet year. Tree Phys 18:299–305

Grieve CM, Shannon MC (1999) Ion accumulation and distribution in shoot components of salt-stressed *Eucalyptus* clones. J Am Soc Hortic Sci 124:559–563

Hare PD, Cress WA, Van Staden J (1998) Dissecting the roles of osmolyte accumulation during stress. Plant Cell Environ 21:535–553

Hoch G, Richter A, Körner C (2003) Non-structural carbon compounds in temperate forest trees. Plant Cell Environ 26:1067–1081

Jones HG (1992) Plants and microclimate: a quantitative approach to environmental plant physiology, 2nd edn. Cambridge University Press, Cambridge

Kozlowski TT (1997) Responses of woody plants to flooding and salinity. Tree Physiol Mono 1: 1–17

Kozlowski TT, Pallardy SG (2002) Acclimation and adaptive responses of woody plants to environmental stresses. Bot Rev 68:270–334

Ladiges PY (1975) Some aspects of tissue water relations in three populations of *Eucalyptus viminalis* Labill. New Phytol 75:53–62

Larher FR, Lugan R, Gagneul D, Guyot S, Monnier C, Lespinasse Y, Bouchereau A (2009) A reassessment of the prevalent organic solutes constitutively accumulated and potentially involved in osmotic adjustment in pear leaves. Environ Exp Bot 66:230–241

Lehto T, Zwiazek JJ (2011) Ectomycorrhizas and water relations of trees: a review. Mycorrhiza 21:71–90

Lemcoff JH, Guarnaschelli AB, Garau AM, Basciauli ME, Ghersa CM (1994) Osmotic adjustment and its use as a selection criterion in *Eucalyptus* seedlings. Can J Forest Res 24:2404–2408

Lenz TI, Wright IJ, Westoby M (2006) Interrelations among pressure-volume curve traits across species and water availability gradients. Physiol Plant 127:423–433

Li CY (1998) Some aspects of leaf water relations in four provenances of *Eucalyptus microtheca* seedlings. For Ecol Manag 111:303–308

Liu WH, Fairbairn DJ, Reid RJ, Schachtman DP (2001) Characterization of two HKT1 homologues from *Eucalyptus camaldulensis* that display intrinsic osmosensing capability. Plant Physiol 127:283–294

Martinez EM, van der Moezel PG, Pearcepinto GVN, Bell DT (1991) Screening for salt and waterlogging tolerance in *Eucalyptus* and *Melaleuca* species. For Ecol Manage 40:27–37

Martinez EM, Cancela JJ, Cuesta TS, Neira XX (2011) Review. Use of psychrometers in field measurements of plant material: accuracy and handling difficulties. Span J Agric Res 9: 313–328

Merchant A, Arndt SK, Callister AN, Adams MA (2007a) Contrasting physiological responses of six *Eucalyptus* species to water deficit. Ann Bot 100:1507–1515

Merchant A, Ladiges PY, Adams MA (2007b) Quercitol links the physiology, taxonomy and evolution of 279 Eucalypt species. Glob Ecol Biogeog 16:810–819

Merchant A, Arndt SK, Callister A, Adams MA (2009) Quercitol plays a key role in stress tolerance of *Eucalyptus leptophylla* (F. Muell) in naturally occurring saline conditions. Environ Exp Bot 65:296–303

Merchant A, Tausz M, Keitel C, Adams MA (2010a) Relations of sugar composition and $\delta^{13}C$ in phloem sap to growth and physiological performance of *Eucalyptus globulus* (Labill). Plant Cell Environ 33:1361–1368

Merchant A, Peuke AD, Keitel C, Macfarlane C, Warren CR, Adams MA (2010b) Phloem sap and leaf $\delta^{13}C$, carbohydrates and amino acid concentrations in *Eucalyptus globulus* change systematically according to flooding and water deficit treatment. J Exp Bot 61:1785–1793

Moing A, Carbonne F, Zipperlin B, Svanella L, Gaudillere JP (1997) Phloem loading in peach: symplastic or apoplastic? Physiol Plant 101:489–496

Morabito D, Mills D, Prat D, Dizengremel P (1994) Response of clones of *Eucalyptus microtheca* to NaCl *in-vitro*. Tree Physiol 14:201–210

Morabito D, Jolivet Y, Prat D, Dizengremel P (1996) Differences in the physiological responses of two clones of *Eucalyptus microtheca* selected for their salt tolerance. Plant Sci 114:129–139

Morgan JM (1984) Osmoregulation and water stress in higher plants. Annu Rev Plant Physiol 35: 299–319

Myers BA, Duff GA, Eamus D, Fordyce IR, O'grady A, Williams RJ (1997) Seasonal variation in water relations of trees of differing leaf phenology in a wet-dry tropical savanna near Darwin, northern Australia. Aust J Bot 45:225–240

Myers BA, Neales TF (1986) Osmotic adjustment, induced by drought, in seedlings of three *Eucalyptus* species. Aust J Plant Physiol 13:597–603

Nguyen A, Lamant A (1988) Pinitol and *myo*-inositol accumulation in water stressed seedlings of maritime pine. Phytochemistry 27:3423–3427

Niknam SR, McComb J (2000) Salt tolerance screening of selected Australian woody species – a review. For Ecol Manag 139:1–19

Orthen B, Popp M (2000) Cyclitols as cryoprotectants for spinach and chickpea thylakoids. Environ Exp Bot 44:125–132

Orthen B, Popp M, Smirnoff N (1994) Hydroxyl radical scavenging properties of cyclitols. P Roy Soc Edinb B 102:269–272

Orthen B, Popp M, Barz W (2000) Cyclitol accumulation in suspended cells and intact plants of *Cicer arietinum* L. J Plant Physiol 156:40–45

Palacio S, Maestro M, Montserrat-Marti G (2007) Seasonal dynamics of non-structural carbohydrates in two species of Mediterranean sub-shrubs with different leaf phenology. Environ Exp Bot 59:34–42

Palfi G, Koves E, Bito M, Sebestyen R (1974) Role of amino-acids during water-stress in species accumulating proline. Phyton Int J Exp Bot 32:121–127

Parker WC, Pallardy SG (1987) The influence of resaturation method and tissue-type on pressure-volume analysis of Quercus-Alba L seedlings. J Exp Bot 38:535–549

Pate J, Shedley E, Arthur D, Adams M (1998) Spatial and temporal variations in phloem sap composition of plantation-grown *Eucalyptus globulus*. Oecologia 117:312–322

Paul MJ, Cockburn W (1989) Pinitol, a compatible solute in *Mesembryanthemum crystallinum* L? J Exp Bot 40:1093–1098

Pfundner G (1993) Vergleichende Untersuchungen zum Inhaltsstoffmuster neuweltlicher Trocken- und Salzpflanzen (Comparison of Metabolite patterns of plants from arid or saline habitats of the new world). Dissertation, University of Vienna, Vienna

Pita P, Pardos JA (2001) Growth, leaf morphology, water use and tissue water relations of *Eucalyptus globulus* clones in response to water deficit. Tree Physiol 21:599–607

Popp M, Smirnoff N (1995) Polyol accumulation and metabolism during water deficit. In: Smirnoff N (ed) Environmental and metabolism flexibility and acclimation – environmental plant biology. Bios Scientific, Oxford

Popp M, Lied W, Bierbaum U, Gross M, Grosse-Schulte T, Hams S, Oldenettel J, Schuler S, Wiese J (1997) Cyclitols – stable osmotica in trees. In: Rennenberg H, Eschrich W, Ziegler H (eds) Trees – contributions to modern tree physiology. Backhuys Publisher, The Hague, pp 257–270

Prat D, Fathi-Ettai RA (1990) Variation in organic and mineral components in young *Eucalyptus* seedlings under saline stress. Physiol Plant 79:479–486

Rajam MV, Dagar S, Waie B, Yadav JS, Kumav PA, Shoeb F, Kumna R (1998) Genetic engineering of polyamine and carbohydrate metabolism for osmotic stress tolerance in higher plants. J Biosci 23:473–482

Roberts SW, Knoerr KR (1977) Components of water potential estimated from xylem pressure measurements in 5 tree species. Oecologia 28:191–202

Schulte PJ, Hinckley TM (1985) A comparison of pressure volume curve data analysis techniques. J Exp Bot 36:1590–1602

Smith AM, Stitt M (2007) Coordination of carbon supply and plant growth. Plant Cell Environ 30:1126–1149

Stoneman GL, Turner NC, Dell B (1994) Leaf growth, photosynthesis and tissue water relations of greenhouse-grown *Eucalyptus-marginata* seedlings in response to water deficits. Tree Physiol 14:633–646

Streeter JG, Lohnes DG, Fioritto RJ (2001) Patterns of pinitol accumulation in soybean plants and relationships to drought tolerance. Plant Cell Environ 24:429–438

Sun D, Dickinson G (1993) Responses to salt stress of 16 *Eucalyptus* species, *Grevillea robusta*, *Lophostemon confertus* and *Pinus caribaea* var *hondurensis*. For Ecol Manag 60:1–14

Taiz L, Zeiger E (1998) Plant physiology. Sinauer Associates, Sunderland

Tardieu F, Simonneau T (1998) Variability among species of stomatal control under fluctuating soil water status and evaporative demand: modelling isohydric and anisohydric behaviours. J Exp Bot 49:419–432

Tomos D (2000) The plant cell pressure probe. Biotechnol Lett 22:437–442

Turner NC (1981) Techniques and experimental approaches for the measurement of plant water status. Plant Soil 58:339–366

Turner NC (1988) Measurement of plant water status by the pressure chamber technique. Irrig Sci 9:289–308

Turner NC, Jones MM (1980) Turgor maintenance by osmotic adjustment. A review and evaluation. In: Turner NC, Kramer PJ (eds) Adaptation of plants to water and high temperature stress. Wiley-InterScience, New York

van der Moezel PG, Bell DT (1987) Comparative seedling salt tolerance of several *Eucalyptus* and *Melaleuca* species from western Australia. Aust For Res 17:151–158

van der Moezel PG, Pearcepinto GVN, Bell DT (1991) Screening for salt and waterlogging tolerance in *Eucalyptus* and *Melaleuca* species. For Ecol Manage 40:27–37

White DA, Turner NC, Galbraith JH (2000) Leaf water relations and stomatal behavior of four allopatric *Eucalyptus* species planted in Mediterranean southwestern Australia. Tree Physiol 20:1157–1165

Wurth MKR, Pelaez-Riedl S, Wright SJ, Körner C (2005) Non-structural carbohydrate pools in a tropical forest. Oecologia 143:11–24

Zohar Y, Schiller G (1998) Growth and water use by selected seed sources of *Eucalyptus* under high water table and saline conditions. Agric Ecosyst Environ 69:265–277

Chapter 6
Ecophysiology of Long-Distance Water Transport in Trees

Hanno Richter and Silvia Kikuta

Abstract We give a short overview over some basics of tree water relations and the likely impact of future climate changes on the functionality of the soil-plant-atmosphere continuum. We start with a short account of the biophysics of water transport and explain some methods for determining relevant parameters in the field and in the laboratory. Important results are described next: the variable values of total water potential in crowns, the use of pressure-volume curves established in the lab for a detailed analysis of leaf water relations in the field, parameters indicating water stress, and the stability of water columns in the xylem. A look at the water relations of small plants shows that the numerical values of key parameters in seedlings or herbs are not much different from those in tall trees; this enigma may have to do with evolutionary selection for a cautious use of soil water reserves. There are still a number of weak spots in our understanding of water transport in trees. We have only a very general knowledge of root distribution over the soil layers and of fine root turnover. Also, cavitation in the different size classes of roots has not been investigated sufficiently. The most important unknown however is probably the distribution of resistances over the whole length of the xylem, which depends on xylem structure at different points in the plant. Finally we speculate about possible effects of future climate changes. There are a number of relevant climate factors which can influence plant water relations in both positive and negative ways; they are described in some detail. Changes of these factors will not be uniform, not even on small scales, let alone over continents or the whole globe. It depends on their combinations whether tree water stress will increase or decrease at a given site. It seems however safe to conclude that the existence of trees on our planet is far from being jeopardized: trees have existed uninterruptedly since the late Devonian, including long periods with climates far more extreme than the ones predicted for the next 100 years. This positive outlook does however not

H. Richter (✉) • S. Kikuta
Institute of Botany, University of Natural Resources and Life Sciences (BOKU), Vienna, Austria
e-mail: hanno.richter@boku.ac.at

M. Tausz and N. Grulke (eds.), *Trees in a Changing Environment*, Plant Ecophysiology 9, DOI 10.1007/978-94-017-9100-7_6, © Springer Science+Business Media Dordrecht 2014

pertain to the fate of single tree species, which may well succumb to increased competition by better adapted ones.

6.1 Introduction

Water transport in plants has fascinated researchers from the earliest times of physiological investigations up to the twenty-first century. It is still an active field, where novel experimental approaches have contributed to a deeper understanding of a complex issue even in the past few years.

Water relations date back to the beginning of plant life on land in the mid-Ordovician. Finding a supply of water in the soil and bringing it to the leaves, where green cells photosynthesize, was crucial for exploiting the resources of minerals and light on land, by far richer ones than those in the sea. A mechanism for transporting huge amounts of water from the roots through the stem into the leaves was not perfected until the late Devonian, when the first tall trees appeared. It is this transport which will be the subject of this chapter.

In green land plants, a unique trait is the direct coupling between photosynthetic production on the one hand and water loss to the dry atmosphere on the other. During the lifetime of a plant, the transpirational water loss may add up to 100 times the fresh weight of its body. Water for transpiration is thus the most important water requirement of a green land plant. All chemical reactions of water molecules in the cell, including those in photosynthesis, are negligible in comparison.

Of the many possible topics, we focus mainly on the following: biophysics of plant water relations, methods for studying physiological parameters relevant for transport, experiments on transport and their results, and the likely future stress on the transport system due to climate change. We refrain from detailed discussions and extensive citations. Pertinent monographs (Holbrook and Zwieniecki 2005; Meinzer et al. 2001; Tyree and Zimmermann 2002) are recommended for more in depth information.

6.2 Biophysics of Transport and the Cohesion – Tension Theory

The foundations of water transport are part of the theory of plant water relations, which describes the energy status of water at any point in the plant body and the forces acting on this water. In this framework, how can we describe the water status of a plant by measurable parameters? We have a set of two equations at our disposal which are well defined thermodynamically (Nobel 2005).

The first equation describes the demands of the soil-plant-atmosphere continuum (SPAC) in a steady-state situation:

$$(-)\Psi_t = (-)\Psi_S + (-)\Psi_G + (-)\Psi_F \qquad (6.1)$$

Where Ψ_t is total water potential, Ψ_S static substrate potential, Ψ_G gravitational potential, and Ψ_F frictional potential. The dimension of water potential is energy per volume; therefore, we should expect the proper SI units, $J\ m^{-3}$. Since "energy per unit volume" equals exactly "force per unit area" (that is, pressure), we can and do use pressure units for quantifying Ψ_t and its components. The numerical values are therefore given in the SI pressure units, megapascals (MPa).

The water potential of pure, distilled water under atmospheric pressure has been arbitrarily set at 0 MPa. The energy content of water in the SPAC is usually less. All the values in Eq. 6.1 thus become negative or 0 at the most; therefore the minus signs are indicated in brackets. Also, it should be kept in mind that they are of different variabilities with time.

Static substrate potential (Ψ_S) becomes less negative whenever rainfall or irrigation refill the soil, and more negative when evaporation and plant transpiration extract water. In the field, these are mostly slow to moderately rapid processes; in a small pot on a balcony, with a big plant in it, they can become very rapid. The second term, *gravitational potential* (Ψ_G), depends on the vertical distance between the absorbing roots and the measurement point. Work has to be done to hold water against gravity, but this leads to a loss of water potential of only 0.103 MPa per 10 m. Gravitational potential can be easily calculated from a height measurement, and it is completely invariant with time and climate. The last term, *frictional potential* (Ψ_F), is the most variable of all the parameters in Eq. 6.1. Moreover, this potential loss is directly caused by long-distance transport. Wherever water is moved by transpiration, passing through soil capillaries, narrow xylem conduits and the pit membranes connecting them, and finally into the pores of cell walls lining a substomatal cavity, there will be friction and a requirement for energy to overcome it. Frictional potential may be simply described as:

$$(-)\Psi_F = -\sum f_i.r_i \qquad (6.2)$$

This means that Ψ_F is the sum of products of the water fluxes (f_i) in different sections of the SPAC and the resistances (r_i) against this water transport. In other words, the higher the amount of water flowing per unit time towards transpiring leaves, and the higher the resistances, the more negative Ψ_t will become at any point along the pathway. A numerical solution of this equation is difficult even for a steady-state situation, but the principle should be kept in mind.

The second equation describes the responses of compartments in the plant body to the demands of the continuum:

$$(-)\Psi_t = (-)\Psi_o + (+/-)\Psi_p \qquad (6.3)$$

Again there is total water potential (Ψ_t) on the left side. The right side shows only two parameters for adjusting Ψ_t to the value preset by the demands. The first is osmotic potential (Ψ_o). This term becomes more negative with higher concentrations of solutes in the cell and plant. A wide range of different values co-exist in a plant, from the very concentrated solutions in vacuoles and protoplasm of living cells to the xylem sap, which has a composition close to distilled water and, therefore, an osmotic potential close to zero.

To these osmotic potentials we add the potential changes caused by hydrostatic pressure. They appear as pressure potential (Ψ_p). Pressure may be either an over-pressure or a tension, and is thus the only parameter which can both increase and reduce total water potential, depending on the compartment under study. Concentrations of osmotically active substances on the one hand and pressures on the other change at very different rates: concentration changes need time, pressures can be adjusted almost immediately.

One of the pressure phenomena is well known and undisputed: the turgor pressure in living cells. Here, the very negative osmotic potential of a concentrated cell sap is made less negative by a rapidly variable positive pressure adjusting Ψ_t to the fluctuating numerical values imposed by Eq. 6.1. We could easily imagine an opposite system where an invariant overpressure would be maintained by variable concentrations in the cell sap, leading to fluctuating osmotic potentials. No plant ever utilized such a mechanism, and the reason is obvious: changes in osmotic potential require time and metabolic energy for the translocation of substances by active transport, whereas pressure changes need no active transport of big molecules or charged ions, just rapid diffusion of water molecules across cell membranes perforated by aquaporins (Kaldenhoff et al. 2008). The energy required is supplied by the sun via transpiration and does not depend on metabolism.

For the xylem, theory forecasts negative pressures of variable magnitude, since xylem water is poor in solutes and its osmotic potential therefore negligible in most cases. Water in narrow capillaries is remarkably stable and may be subjected to high tensions without snapping. The cohesion of water molecules is very strong, and so is its adhesion to cell walls and cell wall capillaries. This is a consequence of hydrogen bond formation between neighboring molecules. Water can thus be sucked up from the soil reservoir to replace transpirational losses by the leaves. This process resembles the action of a technical suction pump. There is however a fundamental difference: the tube of a suction pump has a large diameter, whereas the water column in a tree stem of the same diameter is divided into thin capillaries, the tracheids and vessels with a diameter of between 10 and 500 μm. And in mesophyll cell walls there is another set of capillaries, extremely fine ones of the order of 10 nm. Where these walls line intercellular spaces or substomatal cavities, surface tension keeps air from entering the capillaries even when cell wall water is at very negative pressures.

Water evaporates from the walls of mesophyll cells, and the surface tensions in the narrow wall capillaries suffice to suck the chain of water molecules, as a replacement, against all resistances up into the crowns of the tallest trees. Under these conditions water in the xylem is in a metastable state of tension, but the lack of surfaces in contact with free air suppresses the formation of vapor bubbles, and therefore boiling. The water column can thus be lifted far higher than in the wide tube of a suction pump.

The solid theoretical foundation of the Cohesion-Tension Theory (CTT) based on these facts and elaborated since the 1890s adds to its credibility, as do a number of independent experimental findings (Holbrook et al. 1995). Nevertheless, the theory is not accepted by everybody; in particular, the postulated negative pressures have repeatedly met with disbelief. The history of the CTT and the discussions concerning its validity may be read in more detail in two web essays (Cruiziat and Richter 2006; Richter and Cruiziat 2006) connected to recent editions of the textbook Plant Physiology by Taiz and Zeiger. In addition, Tyree (1997) is recommended reading.

6.3   Methods and Instrumentation

Let us first turn to the parameters forming the two-equation system with Ψ_t at its center (Richter 1997). Which methods determine the parameters in these equations, what are their pros and cons?

Total water potential (Ψ_t): There are two instruments with radically differing measurement philosophies (Pallardy et al. 1991; Turner 1981): (i) The pressure chamber has a number of advantages. It is robust and equally well suited for field work and the lab, it does not depend on sophisticated electronics, and one can buy inexpensive versions or custom-made for special applications. The disadvantages are the weight of the equipment, the (small) danger inherent in working with compressed air, the destructive nature of the measurements, and the fact that there are apparently no pathways to automation. (ii) Thermocouple psychrometers are easily automated, often non-destructive, and ideally suited for the lab, however are very sensitive to temperature fluctuations. Thus, only the temperature-compensated Dixon-Tyree stem hygrometer is a field instrument, although a rather voluminous one suited for tree stems or thick branches only (Dixon and Tyree 1984; Vogt 2001).

The parameters on the right side of Eq. 6.1 start with *static substrate potential* (Ψ_S). Direct measurements utilize tensiometers or soil thermocouple psychrometers. Equation 6.1 opens up an indirect approach: where friction is negligible, Ψ_t (minus the easily calculated value for Ψ_G) will equal Ψ_S of the root zone. Thus, pressure chamber measurements at the end of the night (predawn), on resaturated plants with closed stomata and in energy equilibrium with the soil, should give values for Ψ_S in the root zone (Martínez-Vilalta et al. 2002). However, it is often a difficult question whether soil and plant are indeed in equilibrium. Another indirect approach is the calculation of Ψ_S from measurements of soil water content by time-domain reflectometry (TDR; Evett 2008) and the relationship between water content and water potential established with a pressure-plate apparatus (Cresswell et al. 2008).

As mentioned, *gravitational potential* (Ψ_G) needs only a simple height measurement for calculating its correct value. This does not contribute much to the total except in the crowns of very tall trees. Ψ_G is however completely invariant with time and meteorological conditions; leaves 40 m above ground will thus never experience total water potentials higher (less negative) than -0.41 MPa.

Frictional potential (Ψ_F) depends on both the water fluxes in the SPAC and the resistances against this water transport (cf. Eq. 6.2). In most cases the resistances are more or less constant over short timespans, while fluxes are always likely to change rapidly. The easiest way to find the momentary value for Ψ_F is to solve Eq. 6.1, where slowly changing Ψ_S and invariant Ψ_G are not hard to determine and Ψ_t can be measured.

There are experimental methods to determine the components of Ψ_F in Eq. 6.2. The fluxes in stems and branches of trees may be measured by variants of Bruno Huber's approach (Huber 1932). They all use heat as a tracer and quantify the fluxes over the whole depth of the sapwood or even in different layers from the outermost to the innermost conducting zones (often non-destructively: Čermák et al. 2004). Detailed flux data for the circumference of tree stems and different sapwood depths show their dependence on climate factors such as irradiation and temperature as well as on the complex hydraulic architecture of trees. For resistances, on the other hand, water flows are measured under a preset pressure head. This is a cumbersome and destructive approach which however provides valuable insights into the distribution of low-resistance sections and high-resistance "bottlenecks" from the roots to the transpiring leaves. The high-pressure flow meter (HPFM) gives resistance values for relatively small root systems, leaves, and shoots (Nardini et al. 2001), but not for entire tall trees.

For the two parameters on the right side of Eq. 6.3, we can make use of quite an impressive array of methods. However, the choice requires careful attention to select the appropriate ones for a given purpose. *Osmotic potential* (Ψ_o) is, after gravitational potential (Ψ_G), the parameter of the two-equation system known for the longest time. The first method, used from the beginning of the twentieth century onward, was cryoscopy of press saps, mostly from leaves or succulent stems. It is rather time-consuming and fraught with a number of error sources. These can be partly avoided by vapor-pressure osmometry of killed (preferably frozen and re-thawed) tissues. The common problem of these two approaches is the mixing of vacuolar sap, which is the target, with xylem and cell wall water poor in solutes; the values found are therefore not as negative as those inside the living cells (Kikuta and Richter 1992). *Pressure potential* (Ψ_p) requires at first a decision on the tissue compartment which we intend to study. There are good indirect methods for the pressures above atmospheric in the interior of living cells. They can be calculated from total water potential and osmotic potential, preferably from pressure-volume (PV) data. A direct approach is the cell pressure probe, where a small capillary is inserted into the vacuole and overpressures are registered with a pressure sensor (Steudle 2000). For the negative pressures inside the xylem, there is no direct method serving over the whole range of values encountered. Attempts to apply the cell pressure probe to the lumen of conducting elements have failed (Wei et al. 2001). Water inside the probe cavitates under modest tension; the negative pressures usually reached in trees in the field are therefore not measurable. The fact that the composition of xylem water is very close to distilled water gives us the possibility to use Ψ_t as a proxy for xylem pressure potential (Eq. 6.3). Under special circumstances, for instance on saline soils, the value thus found must be corrected by the osmotic potential of the xylem sap. Information on the whole range of possible combinations for Ψ_t, Ψ_o and Ψ_p in the living cells is available from the construction of pressure-volume (PV) curves in the lab (Sect. 6.4.2).

6.4 Some Significant Experimental Results on Tree Water Transport

6.4.1 Total Water Potentials and Water Potential Gradients in Trees

It is predicted by theory (Sect. 6.2) that there is no unique Ψ_t value for a whole tree or even for a crown. The reason is the branched pathway of water. Two leaves inserted somewhere on a tree are the endpoints of pathways which up to a certain point in the stem are in common but then depart from one another. Starting at the point of departure, resistances and especially fluxes may become very different. This is evident from diurnal measurements in crowns, highly branched systems, where there may be potential differences of up to 1 MPa between leaves at the same height on opposite sides, not stable ones, but fluctuating with the changing position of the sun.

One particular observation caused some confusion when the first results of pressure chamber measurements from tree crowns were published (Scholander et al. 1965). At that time the hydrostatic gradient of 0.103 MPa per 10 m was considered most important. To the consternation of researchers, leaves at a vertical distance of 10 m showed sometimes higher, sometimes lower differences than this, and there were even cases where leaves on upper branches of a crown had less negative potentials than those on the lower ones. This apparent enigma is solved by the insight that friction is numerically far more important than gravity for producing negative total water potentials, and that leaves at the end of different branches have partly independent pathways with different hydraulic resistances.

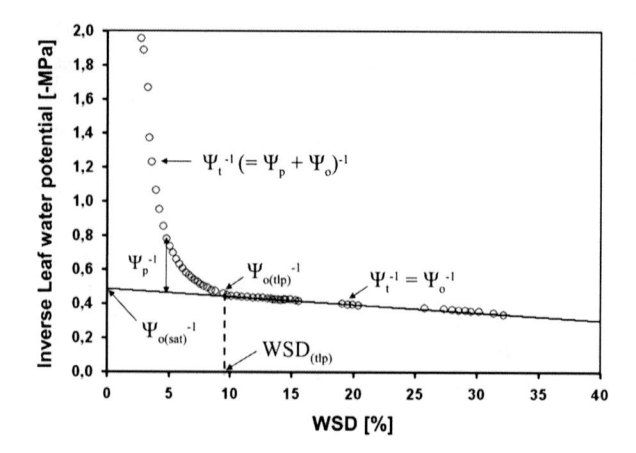

Fig. 6.1 Pressure-volume (PV) curve of a mature, fully saturated leaf of *Euonymus latifolius* showing the relationship between the inverse of leaf water potentials (Ψ_t^{-1}) and the water saturation deficit (WSD). Leaf water potential was measured with an L-51 Leaf Hygrometer (Wescor Inc.) connected to a CR7 Data Logger (Campbell Scientific Ltd.). In the curvilinear part of the graph Ψ_t^{-1} is composed of the inverse of pressure potential (Ψ_p^{-1}) and osmotic potential (Ψ_o^{-1}). In the linear part Ψ_t^{-1} equals Ψ_o^{-1}, since turgor has been lost. $\Psi_{o(tlp)}^{-1}$ is the inverse of osmotic potential at the turgor loss point, $WSD_{(tlp)}$ the water saturation deficit at the turgor loss point. $\Psi_{o(sat)}^{-1}$ is the inverse of osmotic potential at full saturation estimated by extrapolation of the straight osmotic line

6.4.2   Pressure-Volume (PV) Curves

The venerable determination of osmotic potentials on pressed sap or killed tissues includes a dilution error because of the pure water in xylem conduits (Sect. 6.3). This effect is avoided by the pressure-volume (PV) curve technique, a lab method, which has however its greatest benefit in combination with field measurements of Ψ_t. The PV curve (Fig. 6.1) shows all the combined values for Ψ_t and Ψ_o from full saturation to far below the turgor loss point (TLP). Thus, total water potentials measured in the field can be related to the osmotic counterbalance inside the living cells, which gives unique insights into the variability of turgor pressures.

PV curves describe the relationship between total water potential (Ψ_t) and water saturation deficit (WSD; Eq. 6.4) and are typically measured on single leaves (Brodribb and Holbrook 2006) or small twigs (Lo Gullo et al. 2000), which are first resaturated with distilled water. Weighing for saturated weight and determining Ψ_t, which is very close to 0, come next. Then the plant organ is allowed to dehydrate on the bench for a short period. Afterwards, the next pair of values for fresh weight and Ψ_t is measured, and so on. The series is broken off when Ψ_t becomes very low. At last, the organ is dried in an oven. From the data for saturated weight (SW), fresh weight (FW), and dry weight (DW), the water saturation deficit (WSD) can be calculated for the measurement points:

$$WSD = (SW - FW)/(SW - DW) \qquad (6.4)$$

If we plot the inverse of the corresponding water potentials against WSD, we get a graph which separates into two parts (Fig. 6.1). A steep curved portion is followed by a flat straight line. An analysis shows that the linear portion starts at the turgor loss point (TLP), where the pressure potential inside the living cells becomes zero. From then on, total water potential (Ψ_t) and osmotic potential (Ψ_o) are identical, their slow decline is due to the gradual concentration of sap inside the shrinking cells. The straight line may be extrapolated back to full saturation and the difference between the Ψ_t curve and the straight Ψ_o line in the region above the turgor loss point gives the values for turgor potential (Ψ_p).

We can now calculate Ψ_o and Ψ_p for all combinations of WSD and Ψ_t. When there are values for Ψ_t available, as from diurnal courses of total water potential measured in the field, the other three parameters can be calculated or read from a graph (Hinckley et al. 1983). We thus obtain diurnal courses for Ψ_o and WSD and, most importantly, for Ψ_p.

6.4.3 Parameters Indicating Stress

Historically, the differences in osmotic potential between plants on different sites were already recognized in the first half of the twentieth century, and at that time more negative values for Ψ_o were considered as indicative of increased stress. Since diverse strategies for coping with water stress have been found even for desert plants, which do not always utilize negative osmotic potentials, this reasoning is no longer considered valid.

Low water potentials in trees arise from different combinations of the three factors on the right side of Eq. 6.1. Total water potential of one individual leaf in a crown will become more negative whenever the soil gets drier or more water has to be transported to maintain transpiration. Therefore, it may be tempting to use values of Ψ_t as indicators for water stress situations. This reasoning is justified when we compare Ψ_t in leaves of similar age exposed to identical microclimatic conditions during one diurnal course. However, errors can arise with leaves at different developmental stages or leaves from different species. The best indicator for water stress is a lowering of water potentials to the vicinity of the turgor loss point (TLP). This parameter is indeed nearly identical in leaves of the same age and position in a tree crown, but will usually occur at different water potentials in different species, and in younger and older leaves of the same species, e.g. in 1- and 2-year old leaves or needles of evergreens. The cell saps of young leaves contain less solutes, turgor pressures at full saturation are therefore lower and turgor loss points are found at a less negative total water potential. Young leaves close stomata at a less negative Ψ_t, thus avoiding a drop below their TLP. This does certainly not indicate that their water status is better than in adult leaves with lower potentials or that they are less stressed (Karlic and Richter 1983). Why is turgor so important? First, it is the driving force for cellular expansion. This becomes clear from the Lockhart equation:

$$dV/dt = m(\Psi_p - Y) \tag{6.5}$$

Where dV/dt is the expansion rate of a growing cell, m is the extensibility of its cell wall, Ψ_p is turgor pressure, and Y a threshold value of turgor which has to be reached before extension starts. A lower turgor means a reduced expansion rate and, since the extensibility of the cell wall ceases after some time, a smaller final volume of the cell.

Second, leaf cells at the TLP begin to wilt. The protoplasm of non-turgescent, wilting cells is mechanically stressed between the cell wall and the shrinking vacuole; this stress is avoided, if possible. Stomatal closure is induced to prevent a further drop in water potential. However, it also reduces photosynthetic CO_2 uptake. The leaf has, so to speak, the choice between thirst and hunger.

6.4.4 Cavitation, Embolization, and Refilling

Although gas-free water in glass capillaries can sustain tensions of several dozen MPa, water in the xylem is far less stable. The reason lies in the construction of xylem elements. Their walls are perforated by pits, holes closed by a very thin membrane with rather wide pores. It is through these pit-membrane pores that water passes from one vessel or tracheid to the next in its course along the transport pathway. Pit-membrane pores are weak spots for the stability of the water column. Where they contact air-filled structures, air will be sucked into the water-filled conducting element when the pressure difference exceeds a given threshold value (Sperry and Hacke 2004). This causes cavitation, the sudden snapping of the water column (Tyree and Zimmermann 2002). Cavitating elements emit audible and ultrasonic signals, which provide a means for detailed studies. Sensitive microphones pick up the signals (Ritman and Milburn 1991). The number of cavitating elements may be related to water potentials at the point of measurement, and the wave form characteristics emitted give information on the size of the cavitating structures (Johnson et al. 2009; Rosner et al. 2009).

A near vacuum filled with water vapor is created in a cavitated element and water conduction interrupted, but the cell walls of the cavitated vessel or tracheid keep this bubble from spreading over the whole xylem. Later, air seeps into the vacuum, and emboli are formed, that is, gas-filled xylem elements (Hölttä et al. 2007). The vulnerability of an individual conducting element to cavitation depends on its surroundings (since an element cannot cavitate when it contacts only water-filled counterparts), its anatomy and its mechanical properties (Delzon et al. 2010). The vulnerability of the entire xylem column of intact stems, branches

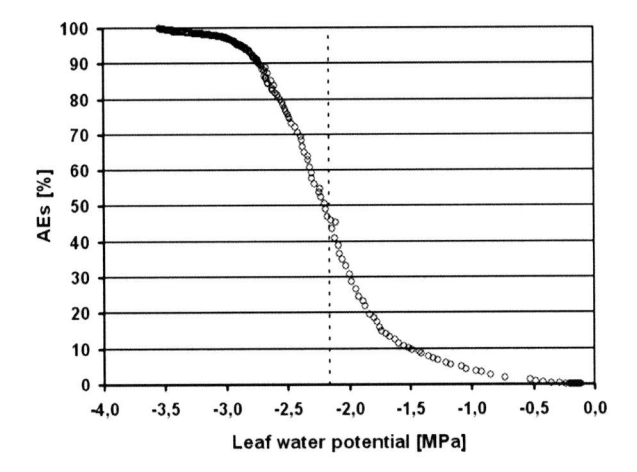

Fig. 6.2 Vulnerability curve of a mature, attached leaf of *Euonymus latifolius* showing the relationship between leaf water potentials and the percentage of cumulated numbers of ultrasonic acoustic signals (AEs) emitted during bench-drying from full saturation. Leaf water potentials were continuously measured with an L-51 Leaf Hygrometer connected to a CR7 Data Logger, ultrasonic acoustic emissions were detected with a I15I transducer connected to a 4615 Drought Stress Monitor (Physical Acoustics Corp.). The *dashed line* shows the osmotic potential at the turgor loss point ($\Psi_{o(tlp)}$) from a pressure-volume curve measured on the other leaf of the pair

and twigs is further limited by cortex and bark, which isolate the interior from the outside air to species-specific and developmentally variable degrees.

There are numerous data on vulnerability curves in the literature, that are plots of the loss of hydraulic conductivity (in percent of the maximum) or of cumulated numbers of ultrasonic signals against the increasingly negative water potentials from full saturation to severe stress (Fig. 6.2). Various methods are employed for inducing cavitation (Choat et al. 2010; Cochard et al. 2010), such as bench-drying, high-speed centrifugation, and air overpressure applied to the outside of twigs with pressure collars. The results are highly dependent on the methodology used, and numerical comparisons should probably be avoided where methods differ.

The question is whether non-embolized structures remain numerous enough to sustain water conduction, or whether refilling processes may restore the conductivity of embolized elements after some time. The answer seems to be yes in both cases. The wood of trees is constructed in a very redundant way, which means that the functional impairment of a good part of the conducting elements can be tolerated quite easily. However, refilling under tension will also occur, for instance in laurel (Hacke and Sperry 2003; Salleo et al. 2006). This process depends on metabolism. Although there are numerous attempts to suggest a plausible mechanism, up to now no final consensus has been reached (Clearwater and Goldstein 2005; Zwieniecki and Holbrook 2009). The difficulty to explain is that water must be transferred into a vacuum from the xylem water, which is at negative pressures, against a potential gradient.

6.5   Tall Trees and Small Herbs: A Useful Comparison

Resistances and fluxes result in gradually more negative values for frictional potential in the direction of the water flow. When we compare tall trees with small herbs, a riddle arises. The absolute values for Ψ_t in the leaves at the end of these pathways are quite similar, although the length of the pathway from the roots upwards may differ by factors of several hundreds. Why aren't herbs investing an additional, very small amount of material and energy into the construction of parallel xylem elements, thus building a more efficient transport pathway with lower resistances and potential losses? The reason may be an inherent selection advantage for higher resistances in herbs, but what does it consist of?

We think this advantage is in water economy. An over-efficient xylem with very low resistance would let a tiny herb transport huge amounts of water over the short distances in its body with hardly any loss of energy. Normally, a Ψ_t close to the TLP triggers stomatal closure. In the absence of this signal, transpirational water loss would become extreme during periods of high temperatures and low air humidity, e.g. at noon on a hot summer day. The amount of water molecules spent for the uptake of one CO_2 molecule would increase dramatically and water use efficiency (WUE), which is poor anyway, would become extremely low. The inherent resistances of the transport pathway are high enough in trees, but they have to also be maintained in seedlings and herbs, which have evolved from woody plants. Without this precaution, soil water would be rapidly depleted, with Ψ_S falling to very low values. This would result in stomatal closure and prolonged periods of low photosynthesis, thus reducing competitiveness of the herbs.

6.6   Weak Spots in Our Understanding of Water Transport in Trees

Long-distance water transport makes use of the dead conducting elements of the xylem. The amount and energy content of the water transported may be described by a formula analogous to Ohm's law in electricity (cf. Eq. 6.1). So here, too, there are potential differences, fluxes, and resistances. This could perhaps be misleading us to regarding water transport as a simple, purely physical problem. Water transport described in electrical terms, together with the extant morphological and anatomical information characterizing the pathway, could permit us to calculate parameters from simple data and to devise precise models. However, this reasoning is incorrect. Our knowledge has vast gaps, which are mostly due to our incomplete understanding of the resistances in the pathway. Our gaps in understanding these resistances follow.

Fine roots: These short-lived organs are the points of entry for water molecules into the plant body. The total resistance of the fine root system of a plant will obviously depend on two parameters: the transverse resistance of a single root from the surface to the xylem and the number of parallel roots taking up water. While there are good methods for assessing single root hydraulics (e.g. Knipfer et al. 2007), fine root dynamics are still underinvestigated. This is mainly a problem of methodology (Hendricks et al. 2006). Even apparently simple procedures, such as counting and mass determination, are difficult, tedious and imprecise. Different methods give vastly differing results. Root growth and dieback are therefore not fully understood. What is known suggests highly variable numbers of functioning fine roots over the year.

Embolization: The pathway for long-distance transport is reached in the root where the water enters the xylem, a continuous system of dead vessels and tracheids leading from the root through the stem to the leaves. Here, the water column becomes metastable under the prevailing tensions. Cavitation as well as embolization will have an impact on the resistance of the pathway. Numerous data on cavitation thresholds are available for stems and leaves of woody plants, they are scarce for herbs, and absent for minor roots.

Bottlenecks: By this term we understand short anatomical regions where the xylem elements are rearranged. Here, their average diameter, length, and wall thickness will often differ from the values in the main part of the pathway. In general, they are sites of higher resistances. Examples are the branching points of roots, the transition zone between root system and stem, and the nodes. Therefore, we can for instance observe a pronounced potential drop between the xylem of a twiglet and the leaves inserted. Information on bottlenecks is still incomplete because the HPFM has not yet been applied to all the relevant spots in trees.

We notice a lack of simple and efficient methods to determine all these resistances with the same ease and accuracy and at the same level of resolution as the fluxes. Thus it is still impossible to calculate the value for frictional potential in, say, a leaf or a needle from the available data for fluxes and resistances along the pathway to this individual organ.

Leaves: Transport of liquid water becomes even more complicated in a leaf (Sack and Holbrook 2006). A network of veins with differing diameters distributes water in the lamina. The minor veins are in close contact with the living mesophyll cells. The distribution of resistances in leaves is rather unclear yet and will certainly depend on species and developmental stage. However, living protoplasm surely plays an important role, since a strong reaction to changing light and temperatures has been demonstrated (Cochard et al. 2007). Involvement of aquaporins is shown by an increase in transport resistances after application of mercury salts, reversible by mercaptoethanol. We do not know exactly where the water is passing from the finest veins into the mesophyll cells and through them into the cell walls lining the intercellular air spaces and substomatal cavities.

From these cell walls, water molecules finally escape into the gaseous phase, pass through open stomata along the vapor pressure gradient and diffuse into the very dry free atmosphere.

6.7   Likely Effects of Climate Change on Tree Water Transport

Water transport is strongly influenced by the microclimate impinging on a tree. Since microclimate depends on macroclimate, global changes must have an impact on water transport. It is however still very difficult to predict the direction of future changes for the most relevant parameters, i.e., rainfall, air humidity, summer and winter temperatures, irradiance, and wind. This is true at the macroscopic scale of continents or landscapes, and even more so on the microscopic scale of sites or forest stands. We will therefore shortly discuss, for each of the climate factors mentioned, the effects of changes in both directions on water transport.

Rainfall refills the soil and specifically the wide pores which have been emptied first. They have less transport resistance than narrow pores, so friction will be less in saturated soils. Increased rainfall will keep stomata open longer for CO_2 uptake. Prolonged flooding will stress most plants and lead to root dieback. Reduced rainfall will lower static substrate potential (Ψ_S). Transpiration decreases because of high friction in the smaller soil pores and the stomatal closing reaction at the turgor loss point.

Transpiration is driven by vapor pressure deficit (VPD). With an increase in temperature, air can hold a higher weight of water until a relative humidity of 100 % is reached. The increase is not linear, but higher at elevated temperatures, so the differences between the vapor pressures at 100 % relative humidity and, say, 50 % will be by far higher at 25 °C than at 15 °C. An increase in temperature or a decrease in relative humidity will therefore increase transpiration, water flows, and tensions in the xylem. Decreasing temperature or higher relative humidity will reduce these parameters. For the globe as a whole, increased average temperatures are forecast for this century. It is not clear how this scenario will hold at continental scales Western Europe. For instance, some models predict lower temperatures for Western Europe because of an altered course of the Gulf Stream (Rahmstorf 1997).

Winter temperatures have probably exercised the strongest selection pressure on plants moving from the tropics north or south, or from sea level into high mountains. Among taxa evolved in the tropics there are comparatively few which tolerate the formation of ice within the plant. Recent studies (Mayr et al. 2006) have shown that water transport is a key factor in explaining how an alpine timberline is established. Frozen xylem cannot transport water, so the duration of this transport interruption in relation to water loss, which continues even at sub-zero temperatures, will be crucial for survival. Since xylem water in stems and branches may stay frozen when evergreen leaves or needles have already thawed and started to transpire, sudden warming by air or high springtime irradiation, followed by refreezing during the night, will stress trees even more than a continuous cold period in winter.

Changing patterns of wind directions could steer air masses at unusual temperature or humidity towards a site, with consequences for VPD. Changing wind speeds will have a minor impact limited to an increase or decrease in calms unable to remove the water-vapor boundary layer from the leaf surface.

Irradiation depends first on solar radiation, which is not influenced by terrestrial climate. The second factor however, cloud cover, might be altered. Clouds reduce leaf temperature, transpiration, and transport requirements, while a cloudless sky increases transpiration.

Will climate change have a major impact on the capacity for water transport in trees? The answer is probably no, barring extreme scenarios. Xylem transport is rather adaptable and resilient. Strain on the transport system of trees is nothing new: our planet has gone through periods of extreme climate change, far more pronounced ones than now predicted for the next century, and fossil evidence shows that trees never disappeared. This is not to deny, of course, that altered transport requirements will influence the competitive success of species. Changes in the tree floras of landscapes or sites are quite likely, but as a result of interacting physiological processes, not of the breakdown of a single system (McDowell et al. 2008).

In conclusion, theoretical considerations and experimental results show us that plants have developed an efficient way of securing their water supply by means of an elaborate mechanism for long-distance transport, a fact which becomes especially conspicuous in tall trees.

References

Brodribb TJ, Holbrook NM (2006) Declining hydraulic efficiency as transpiring leaves desiccate: two types of response. Plant Cell Environ 29:2205–2215

Čermák J, Kučera J, Nadezhdina N (2004) Sap flow measurements with some thermodynamic methods, flow integration within trees and scaling up from sample trees to entire forest stands. Trees 18:529–546

Choat B, Drayton WM, Brodersen C, Matthews MA, Shackel KA, Wada H, McElrone AJ (2010) Measurement of vulnerability to water stress-induced cavitation in grapevine: a comparison of four techniques applied to a long-vesseled species. Plant Cell Environ 33:1502–1512

Clearwater MJ, Goldstein G (2005) Embolism repair and long distance water transport. In: Holbrook NM, Zwieniecki MA (eds) Vascular transport in plants, Physiological ecology series. Elsevier, Amsterdam

Cochard H, Venisse J-S, Barigah TS, Brunel N, Herbette S, Guilliot A, Tyree MT, Sakr S (2007) Putative role of aquaporins in variable hydraulic conductance of leaves in response to light. Plant Physiol 143:122–133

Cochard H, Herbette S, Barigah T, Badel E, Ennajeh M, Vilagrosa A (2010) Does sample length influence the shape of xylem embolism vulnerability curves? A test with the Cavitron spinning technique. Plant Cell Environ 33:1543–1552

Cresswell HP, Green TW, McKenzie NJ (2008) The adequacy of pressure plate apparatus for determining soil water retention. Soil Sci Soc Am J 72:41–49

Cruiziat P, Richter H (2006) The cohesion-tension theory at work. In: Taiz L, Zeiger E (eds) Plant physiology, 4th edn. Sinauer Associates, Sunderland

Delzon S, Douthe C, Sala A, Cochard H (2010) Mechanism of water-stress induced cavitation in conifers: bordered pit structure and function support the hypothesis of seal capillary-seeding. Plant Cell Environ 33:2101–2111

Dixon MA, Tyree MT (1984) A new stem hygrometer, corrected for temperature gradients and calibrated against the pressure bomb. Plant Cell Environ 7:693–697

Evett SR (2008) Soil water measurement: time domain reflectometry. In: Trimble SW, Stewart BA, Howell TA (eds) Encyclopedia of water science, 2nd edn. CRC Press, Boca Raton

Hacke UG, Sperry JS (2003) Limits to xylem refilling under negative pressure in *Laurus nobilis* and *Acer negundo*. Plant Cell Environ 26:303–311

Hendricks JJ, Hendrick RL, Wilson CA, Mitchell RJ, Pecot SD, Guo D (2006) Assessing the patterns and controls of fine root dynamics: an empirical test and methodological review. J Ecol 94:40–57

Hinckley TM, Duhme F, Hinckley AR, Richter H (1983) Drought relations of shrub species: assessment of the mechanisms of drought resistance. Oecologia 59:344–350

Holbrook NM, Zwieniecki MA (2005) Vascular transport in plants, Physiological ecology series. Elsevier, Amsterdam

Holbrook NM, Burns MJ, Field CB (1995) Negative xylem pressures in plants: a test of the balancing pressure technique. Science 270:1193–1194

Hölttä T, Vesala T, Nikinmaa E (2007) A model of bubble growth leading to xylem conduit embolism. J Theor Biol 249:111–123

Huber B (1932) Beobachtung und Messung pflanzlicher Saftströme. Ber Deut Bot Ges 50:89–109

Johnson DM, Meinzer FC, Woodruff DR, McCulloh KA (2009) Leaf xylem embolism, detected acoustically and by cryo-SEM, corresponds to decreases in leaf hydraulic conductance in four evergreen species. Plant Cell Environ 32:828–836

Kaldenhoff R, Ribas-Carbo M, Flexas Sans J, Lovisolo C, Heckwolf M, Uehlein N (2008) Aquaporins and plant water balance. Plant Cell Environ 31:658–666

Karlic H, Richter H (1983) Developmental effects on leaf water relations of two evergreen shrubs (*Prunus laurocerasus* L. and *Ilex aquifolium* L.). Flora 173:143–150

Kikuta SB, Richter H (1992) Leaf discs or press saps? A comparison of techniques for the determination of osmotic potentials in freeze-thawed leaf material. J Exp Bot 43:1039–1044

Knipfer T, Das D, Steudle E (2007) During measurements of root hydraulics with pressure probes, the contribution of unstirred layers is minimized in the pressure relaxation mode: comparison with pressure clamp and high-pressure flowmeter. Plant Cell Environ 30:845–860

Lo Gullo MA, Trifilò P, Raimondo F (2000) Hydraulic architecture and water relations of *Spartium junceum* branches affected by a mycoplasm disease. Plant Cell Environ 23:1079–1088

Martínez-Vilalta J, Prat E, Oliveras I, Piñol J (2002) Xylem hydraulic properties of roots and stems of nine Mediterranean woody species. Oecologia 133:19–29

Mayr S, Hacke U, Schmid P, Schwienbacher F, Gruber A (2006) Frost drought in conifers at the alpine timberline: xylem dysfunction and adaptations. Ecology 87:3175–3185

McDowell N, Pockman WT, Allen CD, Breshears DD, Cobb N, Kolb T, Plaut J, Sperry J, West A, Williams DG, Yepez EA (2008) Mechanisms of plant survival and mortality during drought: why do some plants survive while others succumb to drought? New Phytol 178:719–739

Meinzer FC, Clearwater MJ, Goldstein G (2001) Water transport in trees: current perspectives, new insights and some controversies. Environ Exp Bot 45:239–262

Nardini A, Tyree MT, Salleo S (2001) Xylem cavitation in the leaf of *Prunus laurocerasus* and its impact on leaf hydraulics. Plant Physiol 125:1700–1709

Nobel PS (2005) Physicochemical and environmental plant physiology, 3rd edn. Elsevier, Amsterdam

Pallardy SG, Pereira JS, Parker WC (1991) Measuring the state of water in tree systems. In: Lassoie JP, Hinckley TM (eds) Techniques and approaches in forest tree ecophysiology. CRC Press, Boca Raton

Rahmstorf S (1997) Risk of sea change in the Atlantic. Nature 388:825–826

Richter H (1997) Water relations of plants in the field: some comments on the measurement of selected parameters. J Exp Bot 48:1–7

Richter H, Cruiziat P (2006) A brief history of the study of water movement in the xylem. In: Taiz L, Zeiger E (eds) Plant physiology, 4th edn. Sinauer Associates, Sunderland

Ritman KT, Milburn JA (1991) Monitoring of ultrasonic and audible emissions from plants with or without vessels. J Exp Bot 42:123–130

Rosner S, Karlsson B, Konnerth J, Hansmann C (2009) Shrinkage processes in standard-size Norway spruce wood specimens with different vulnerability to cavitation. Tree Physiol 29:1419–1431

Sack L, Holbrook NM (2006) Leaf hydraulics. Annu Rev Plant Biol 57:361–381

Salleo S, Trifilò P, Lo Gullo MA (2006) Phloem as a possible major determinant of rapid cavitation reversal in stems of *Laurus nobilis* (laurel). Funct Plant Biol 33:1063–1074

Scholander PF, Hammel HT, Bradstreet ED, Hemmingsen EA (1965) Sap pressure in vascular plants. Science 148:339–346

Sperry JS, Hacke UG (2004) Analysis of circular bordered pit function I. Angiosperm vessels with homogenous pit membranes. Am J Bot 91:369–385

Steudle E (2000) Water uptake by plant roots: an integration of views. Plant Soil 226:45–56

Turner NC (1981) Techniques and experimental approaches for the measurement of plant water status. Plant Soil 58:339–366

Tyree MT (1997) The cohesion–tension theory of sap ascent: current controversies. J Exp Bot 48:1753–1765

Tyree MT, Zimmermann MH (2002) Xylem structure and the ascent of sap, 2nd edn, Springer series in wood science. Springer, Berlin

Vogt UK (2001) Hydraulic vulnerability, vessel refilling, and seasonal courses of stem water potential of *Sorbus aucuparia* L. and *Sambucus nigra* L. J Exp Bot 52:1527–1536

Wei C, Steudle E, Tyree MT, Lintilhac PM (2001) The essentials of direct xylem pressure measurement. Plant Cell Environ 24:549–555

Zwieniecki MA, Holbrook NM (2009) Confronting Maxwell's demon: biophysics of xylem embolism repair. Opinion. Trends Plant Sci 14:530–534

Chapter 7
Forest Trees Under Air Pollution as a Factor of Climate Change

Rainer Matyssek, Alessandra R. Kozovits, Jörg-Peter Schnitzler, Hans Pretzsch, Jochen Dieler, and Gerhard Wieser

Abstract Air pollution and climate change are inherently linked to each other. After introducing into the presently prevalent air pollutants and their relevance for forest tree and ecosystem performance, the account focuses on nitrogen deposition and tropospheric ozone (O_3), the latter being regarded as potentially most detrimental to vegetation, and hence, as negating carbon sink strength and storage. Mechanisms of O_3 action in trees and stands are highlighted, stressing interactions with other abiotic and biotic factors, including volatile organic compounds, as a fundamental pre-requisite for understanding O_3 effects. O_3 is emphasized as a globally effective agent of climate change, regarding relevance for forest productivity, in particular, at hot spots of air pollution in the southern hemisphere, prognosticated for the upcoming decades. Adaptation capacities of forest trees are discussed in view of the rapidity in the progression of environmental change.

R. Matyssek (✉)
Ecophysiology of Plants, Technische Universität München, von Carlowitz-Platz 2, 85354 Freising, Germany
e-mail: matyssek@wzw.tum.de

A.R. Kozovits
Department of Biodiversity, Evolution and Environment, Federal University of Ouro Preto, Campus Morro do Cruzeiro, Ouro Preto, MG 35400-000, Brazil

J.-P. Schnitzler
Department of Environmental Engineering, Institute of Biochemical Plant Pathology, Helmholtz Zentrum München, Ingolstädter Landstr. 1, 85764 Neuherberg, Germany

H. Pretzsch • J. Dieler
Forest Growth and Yield Science, Technische Universität München, Hans-Carl-von-Carlowitz-Platz 2, 85354 Freising, Germany

G. Wieser
Department of Natural Hazards and Alpine Timberline, BFW, Innsbruck, Austria

M. Tausz and N. Grulke (eds.), *Trees in a Changing Environment*,
Plant Ecophysiology 9, DOI 10.1007/978-94-017-9100-7_7,

7.1 Air Pollution: A Component of Climate Change

Air pollution and climate change are inherently linked to each other, as both have the same origin: human activity. The link arises from the combustion of fossil energy resources and land-use changes such as clear-cutting and burning of ecosystems, which release, as major constituents, carbon dioxide and monoxide (CO_2 and CO, respectively), nitrogen oxides (NO_x) and, with currently decreasing relevance, sulphur dioxide (SO_2). Such trace gas emissions add to those from intense agricultural practices, in particular, ammonia (NH_3; IPCC 2007). The resulting high nitrogen load on ecosystems can lead to enhanced release of nitrous oxide (N_2O). In addition, biogenic volatile organic compounds (BVOCs) increase the mixture of reactants in the atmosphere (Guenther et al. 1995; Laothawornkitkul et al. 2009), predominantly originating from natural sources.

Although BVOCs are emitted by all biota, they are largely produced and emitted by forests and shrublands accounting for about 90 % of the natural emissions. BVOCs are important in the global C cycle. With an atmospheric carbon flux of $1,200$ Tg C yr^{-1}, they represent approximately 1 % of the total C exchange between biota and the atmosphere (Lal 1999). The contribution of VOCs from anthropogenic sources to the total VOCs emission accounts for approximately 150 Tg C yr^{-1}, indicating that on global scale biogenic emissions dominate. Anthropogenic VOC emissions relate to fuel production and handling as well as to fuel combustion in car traffic. Another component released from industries is halogenated hydrocarbons. As such trace gases accumulate in the atmosphere, their chemical interactions are increased under high irradiance. In particular, short-wave radiation as energy source, production of NO_x as catalysers, and VOCs as substrate and precursors result in the photo-chemical formation of ground-level tropospheric ozone (O_3), a secondary air pollutant (Stockwell et al. 1997). NO_x and VOCs are crucial in driving O_3 regimes, as both have the capacity of lowering or enhancing O_3 levels, depending on the circumstantial physico-chemical conditions of the atmosphere and on ecosystem characteristics.

At this stage, the major determinants of climate warming are represented, i.e., the greenhouse gases CO_2, O_3, N_2O, and halogenated hydrocarbons. As warming favours the evaporation of water, the accumulation of water vapour as another important greenhouse gas is promoted in the atmosphere. In view of the current atmospheric conditions, water vapour accounts for about 66 % of global warming, CO_2 and O_3 by about 15 % and 10 %, respectively, and both N_2O and CH_4 by about 3 % each. In addition, halogenated hydrocarbons are known to be responsible for the destruction of the O_3 layer in the stratosphere ("ozone hole" formation), opening the protective shield for terrestrial life against the deleterious UV-B radiation in sunlight. Although the physico-chemical mechanisms determining the regimes of atmospheric trace gases are still poorly understood, their relevance both as air pollutants and agents of climate change is evident. Some exceptions are CO_2 and nitrogen compounds, as they act primarily as plant fertilizers. However, if available in excess and in imbalance with other nutrients, N compounds are to be regarded as pollutants, as they can perturb ecosystem processes, accumulate and eventually become harmful. Either way, trace gases act on and are affected by ecosystems, both directly and through effects on climate.

For these reasons, forests are currently facing significant pressures, while being ecologically crucial for carbon (C) storage in a world of increasing atmospheric CO_2 concentration with consequences for climate warming. Forests with coverage of about 30 % of the terrestrial surface area comprise about 85 % of the C stored in the phytomass (Saugier et al. 2001). Not to be overlooked is the C storage in soils, which can range between 20 % of the entire ecosystem C pool in tropical rain forests to 70 % in boreal coniferous forests. In contrast to forests, croplands have smaller belowground C pools and are generally C sources because of agricultural management (Luyssaert et al. 2008; Schulze et al. 2009). Forests and other vegetation, nevertheless, do not have the C sequestration capacity to counter-balance anthropogenic CO_2 emissions (burning of fossil fuels and ecosystems). They are, however, important for mitigating the atmospheric CO_2 enrichment, including the sustainable use of wood production in lieu of other energy sources. At present, C fixation by terrestrial ecosystems globally equals the C release through forest destruction (Körner 2003).

It is crucial to prevent the transformation of forests from a C sink into a C source, as could occur with degradation due to air pollution and climate change. Determinants of this quest are (i) air quality as defined by atmospheric CO_2 concentration, nitrogen (N) deposition (Nadelhoffer et al. 1999), and O_3 exposure (Sitch et al. 2007) in relation to precursor availability (VOCs, NOx), (ii) climate change influencing site conditions through altered regimes of insolation, air temperature and precipitation (Myneni et al. 1997; Wamelink et al. 2009), and (iii) soil conditions through effects of air pollution and climate change (Poorter and Navas 2003; Wamelink et al. 2009; De Vries and Posch 2011).

Given the focus on air pollution in this chapter, atmospheric CO_2 in elevated concentration, although being of anthropogenic origin, is least representative of a gaseous pollutant from the viewpoint of trees. The extent of growth stimulation by CO_2 is mediated through the availabilities of other resources, such as irradiance, nutrient elements and water (apart from plant-internal factors), and if such effects become limited, then by interaction with other pollutants like ground-level O_3 (see below) rather than by CO_2 per se. Elevated CO_2, if in imbalance with plant nutrition, can nevertheless affect food chains in ecosystems and host-parasite relationships. However, such effects lie within the range of ecological responsiveness, and it appears to be a matter of definition, what kind of quality or extent such effects should attain to view them as adverse. Hence, CO_2 effects on trees per se, which have been reviewed extensively elsewhere (Ceulemans et al. 1999; Saxe et al. 1998; Karnosky et al. 2003; Körner 2003), will not be a subject of this chapter, however, elevated CO_2 will be highlighted, if interacting with actual pollutants in the impact on or the response of trees.

More pronounced is the ambiguity of the N compounds amongst the trace gases in terms of being fertilizer or pollutant. An outline will be presented on the contrasting ways of action in trees and forest ecosystems. Although a severe pollutant in the past, SO_2 has been lost in importance since the 1980s, which had been released from the burning of coal (in particular, brown coal) in parts of the industrialized world (e.g. in Central Europe). The reasons for this change are the shift from coal to mineral oil and petroleum gas for energy production, use of heating oil of low sulphur (S) content, and desulphurization of industrial waste gases. In some areas, S deposition has become that low so that S fertilization is required again in agricultural crop production.

Conversely, with concentrations well above pre-industrial levels, O_3 is regarded as the potentially most detrimental air pollutant to vegetation (Matyssek and Sandermann 2003; Fowler et al. 2008; Wittig et al. 2007, 2009). Empirical evidence suggests chronically enhanced O_3 to negate the C sink strength of forest ecosystems to an extent similar to the (limited) stimulation by the increasing atmospheric CO_2 concentration (Karnosky et al. 2003, 2007; Grams et al. 1999, 2007; Wittig et al. 2009). Findings are scarce and circumstantial, being defined by genotype, phenology and ontogeny as well as ecophysiological traits (e.g. tree and foliage type, trace gas interactions of BVOCs efflux), and factors of climate change like nitrogen (N) deposition and water availability (Matyssek et al. 2010a, b). Poorly validated modelling predicts chronically enhanced O_3 stress to negate the C sink strength of forest ecosystems by the end of this century at the global scale, as O_3 regimes are expected to stay high or further increase (Sitch et al. 2007; Dentener et al. 2006; Vingarzan 2004). As a consequence, the radiative forcing of the atmosphere by reduced C fixation is expected to be significant. Relationships of such a kind again demonstrate air pollution as an intrinsic component of climate change (Fowler et al. 2008). The global dimension is supported, in addition, by the observation that plumes of enhanced O_3 levels are transported across and between continents (Newell and Evans 2000; Derwent et al. 2004). Such a global view on tropospheric O_3 has attained awareness only since recently (Keating et al. 2004; Matyssek et al. 2010a).

Given the paramount role O_3 currently plays in air pollution at global scale, the major focus of this chapter will be on tropospheric O_3 and its impact on forest trees and ecosystems. As O_3 today is to be understood as part of climate change scenarios, emphasis will be directed to interactions with other factors of relevance in a changing environment. This will constitute most of the presentation after introducing into principles of O_3 action in trees. Global dimensions and relevance of O_3 regimes, including the southern hemisphere, will be examined before concluding on capacities in forest trees of coping with and adapting to a changing environment under air pollution impact.

7.2 Principles of Pollutant Impact on Trees

7.2.1 Nitrogen

Natural forest sites are typically limited in N availability, and accordingly, managed forests have been planted on sites too poor in nutrition to allow for profitable agricultural crop production. Exceptions are N-demanding, short-rotation tree plantations for biogenic energy production, but in general, high N input as introduced from anthropogenic sources initially stimulates forest productivity as a fertilizer (Aber et al. 1998). This is relevant for industrialized countries, where such sources are the combustion of fossil energy resources (release of NO_x and other N oxides) and emissions from intense agricultural practices (ammonia) with high N demands and turnover. Such agricultural practices are based on the Haber process, which converts atmospheric N_2 into ammonium in chemical fertilizer production. As N emissions are not transported in the atmosphere across long distances, high N deposition transport is typically associated within the region of high N release. Since pre-industrial times, N flux to continents has doubled on average (Galloway et al. 2004). As a result, increasing N availability to plants in the atmosphere and soils drives N uptake and the stimulation of photosynthesis and net primary production, provided other nutrient elements, water, insolation and temperature are not limiting. Gaseous N compounds (except for N_2) taken up through stomata may cover about 20 % of the metabolic N demand of trees (Fig. 7.1a).

The stimulation of net primary production (NPP) is based on the high biosynthetic demands of the CO_2-fixing enzyme Rubisco and chlorophyll for nitrogen. However, luxurious N supply, which may arise from human activity, leads to an over-saturation of forest trees and ecosystems (hyper-eutrophication), and to the acidification of soils (Aber et al. 1998), which is regarded to be one driver of forest decline.

In more detail, high N deposition, typically characterized by high ammonium/nitrate ratios, leads to ammonium accumulation in soils, expelling other nutritional cations from soil particles (as ion exchangers) into the soil solution. Such cations are prone to leaching, impoverishing the site in nutrients, while soil micro-organisms make use of the excess ammonium for their population growth. This leads to high proliferation of nitrate as an outcome of nitrification and favours the decomposition of organic soil matter. As a result, respiratory C loss from soils can be enhanced, so that forest ecosystems may turn into C sources, promoting rather than mitigating climate warming. The mobile nitrate ion is readily leached out of the ecosystem, contaminating groundwater and springs to the extent that preclude their use as drinking water (Schulze et al. 2005). In such springs, nitrate levels have increased by about four times since the 1930s in Central Europe. High nitrate levels in soils also stimulate denitrification, a process that releases N_2O as a driver of global warming (Aber et al. 1998). In addition, isotopic signatures of the leached nitrate demonstrate that a high proportion has not entered metabolic pathways of the vegetation, corroborating the N over-saturation of forest ecosystems (Durka

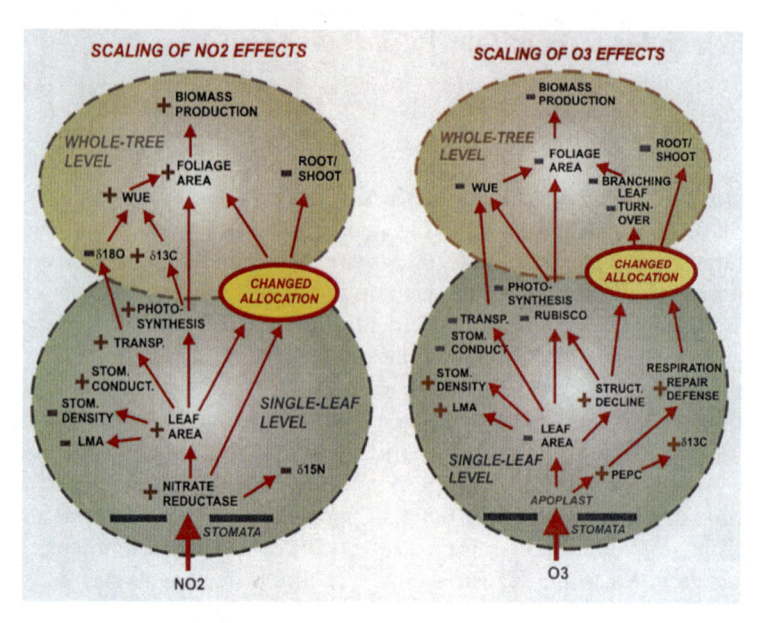

Fig. 7.1 Scaling principles of NO_2 and O_3 effects in juvenile trees under chamber conditions, exemplifying (**a**) *Populus x euramericana*/NO_2 (Siegwolf et al. 2001), and (**b**) *Betula pendula*/O_3 (After Kolb and Matyssek 2001 and Matyssek and Sandermann 2003; "+" stimulation, "−" limitation, *arrows* vectors of action. (**a**) *Populus* grown as cuttings under NO_2 exposure oriented at high urban situations (100 nl NO_2 l^{-1}) or in the absence of NO_2 as control (duration of experiment 15 weeks): Upon NO_2 uptake through stomata, stimulated nitrate reductase activity (NaR) reflects metabolic NO_2 incorporation, confirmed by respective change in the $^{15}N/^{14}N$ signature of leaves. Subsequent effects are increase in leaf size, decrease in stomatal density and leaf mass per area (LMA = inverse of SLA) as well as increase of whole-plant foliage area at unchanged leaf formation rate. Leaf area-related photosynthesis, stomatal conductance, transpiration and water-use efficiency (WUE) are increased (the latter parameters being corroborated by decrease in $^{18}O/^{16}O$ and increase in $^{13}C/^{12}C$ ratios) along with whole-plant biomass production, whereas C allocation changes at the expense of root growth. Overall, NO_2 uptake caused plant performance to resemble that by N fertilization from the soil (Mooney and Winner 1991, Stitt and Schulze 1994), irrespective of the N supply by the soil. *Betula* grown as cuttings, summarizing several experiments under up to 100 nl O_3 nl^{-1} or under charcoal-filtered air as control (duration of experiment growing season). After entering the leaf through stomata, O_3 induces repair and defense mechanisms (including raised respiratory activity) with effects on resource allocation; stimulation of the enzyme PEP carboxylase (PEPC) is involved, and tissue ^{13}C content is increased by this enzyme rather than by effects of leaf gas exchange. Leaf size can be reduced and LMA and stomatal density increased, but transpiration is often decreased by partial stomatal closure. Tissue collapse also reduces photosynthesis in parallel to inhibition through O_3-induced biochemical signals, leading to low WUE and perturbed translocation processes in resource allocation. Also, disruption of phloem structure and function can occur. Although R/S stays low, whole-plant biomass production declines because of reduced leaf size, photosynthesis, branching, and leaf turnover, but increased premature leaf loss and respiration. Also, WUE of biomass production decreases. Note, however, that the selected processes and their linkage strongly depend on nutrition, water availability and insolation as well as on the genotype and co-occurring biotic stress (e.g. by competitors). See Matyssek and Sandermann (2003) for graphical extension by the cell level, and Sandermann and Matyssek (2004) for further details on mechanistic scaling)

et al. 1994). The enhanced nitrate availability along with the loss of alkaline cations (Ca, Mg), and hence, of buffering capacity, causes soil acidification, which in turn facilitates acidic cations such as aluminum (Al) to enter the soil solution. Also, Al is toxic in freshwater and has been discussed as a disturbant of ecosystem processes (Marscher 1991).

At this level of ecosystem N saturation, tree mortality may increase, and the once stimulated NPP is now suppressed, as the proportions among nutrient elements gradually shifts to an imbalance dominated by nitrogen (Aber et al. 1998). However, a balanced nutrient supply is crucial for tree growth. Imbalance is indicated by lowered Ca/Al and Mg/Al ratios (reinforced by a high affinity of roots to passive Al uptake) as indicators of forest decline. A shortage in Ca can restrict radial stem growth (Lautner et al. 2007), limiting water transport capacity. In this way, the development of tree foliage may be reduced, which can lead to progressive crown transparency. In addition, leaves may become yellowish as a consequence of Mg limitation (Schulze et al. 1989). Mg is crucial for chlorophyll functionality so that shortage reduces photosynthesis and C gains and is accompanied by the loss of chlorophyll. In particular, advanced needle age classes of evergreen coniferous tree species can turn yellow, as the retranslocation of Mg from old needles to new represent a strong sink during needle growth. Further adverse effects of luxurious N availability may be an increasing attractiveness to herbivores and pathogens and an enhanced susceptibility to early and late frosts, as high specific leaf area is induced during differentiation at the expense of leaf robustness. High N supply in addition favours aboveground relative to belowground organs in whole-tree C allocation, so that trees may become susceptible to soil drought.

At the global scale, the annual consumption of N fertilizer has steadily increased since the 1950s throughout the year 1990 from 10 to about 80×10^6 t yr^{-1} (Millenium Ecosystem Assessment 2005). At that time, the economic collapse of the former Soviet Union and other eastern European countries led to reduced use into the early twenty-first century. In other industrialized countries, the consumption stagnated at somewhat more than 20×10^6 t yr^{-1} since the early 1980s. This contrasts, however, with the developing countries, where fertilizer use is still increasing, representing the highest proportion in the global fertilizer consumption. To date, the amounts of N_2 converted into ammonium by terrestrial ecosystems and by the Haber process have become similar (Galloway et al. 2004). Another doubling of N flux to the continents by the Haber process is expected by the year 2050.

7.2.2 Ozone

Ozone is occurring both in cities and rural landscapes of industrialized countries at levels that exceed natural, pre-industrial background concentrations (10–20 ppb) by a factor of 2–5. During recent decades, O_3 has increased by 1–2 % per year (Stockwell et al. 1997). For trees and forests, O_3 is regarded as an air pollutant of high risk (Fowler et al. 2008). Nevertheless, involvement in forest decline is still a matter of debate, as knowledge about O_3 effects, due to logistic reasons, is predominantly based on controlled experiments in chambers with juvenile individual trees. Such findings can hardly be validated for adult trees at forest sites with multiple concurrent stressors and contrasting ontogenetic specificities (Kolb and Matyssek 2001). The fundamental principles of O_3 action have been described here rather than their relevance for mature trees growing under field conditions (see overviews by Matyssek and Sandermann 2003; Matyssek et al. 2010a; Niinemets 2010 and Fig. 7.1b).

7.2.2.1 Juvenile Trees, Chamber Conditions

Cell Level

Ozone and its oxidative derivatives which originate from its rapid decomposition after entering the plant via stomata primarily act on the apoplast of the leaf mesophyll and the plasmalemma adjacent to the cell walls (Sandermann et al. 1997). The apoplast contains antioxidants as defence metabolites (e.g. ascorbate) which need to be regenerated by the cells, while signal chains are initiated by the plasmalemma (as long as it is resistant to injury), which can enhance the defence capacity or may induce membrane and chloroplast injury, and as a result, decrease photosynthesis. In addition, cellular processes (e.g., translocation of assimilates) may be disrupted and leaf differentiation may be shifted towards reduced specific leaf area (SLA) and leaf size (Pääkkönen et al. 1998a; Matyssek 2001). A broad spectrum of gene expression is involved that controls the synthesis of lignins and callose along with "programmed cell death" in leaves (i.e. induction of necrosis for confining oxidative injury; Kangasjärvi et al. 2005), as well as a broad range of metabolites for processes of defence and repair (e.g. synthesis of pathogen-related proteins, as pathogen attack also stimulates oxidative stress; Hahlbrock et al. 2003; Matyssek et al. 2005). The additional metabolic effort may be reflected by enhanced respiratory activity.

Leaf-Level

Although O_3 stress can cause stomatal closure, the photosynthetic water-use efficiency (WUE) often declines, as relative to the restricted transpiration reduction in photosynthesis may be higher in proportion (Matyssek et al. 1995; Fig. 7.1b). In addition to cell-internal processes (see above), one reason is that collapse of mesophyll cells collapsed causes membrane injury (Günthardt-Goerg et al. 1993; Matyssek 2001). Progressive tissue destruction eventually hinders the assimilate translocation in the leaves. Also, phloem loading can be impeded, as injury eventually leads to premature leaf loss.

Whole-Tree Level

Premature leaf loss, in addition to inhibited photosynthesis at the cell and leaf level (see above), reduces the foliage area and, as a consequence, the whole-tree C gain (Maurer and Matyssek 1997; Fig. 7.1b). The reduced C gain along with perturbed assimilate translocation and enhanced respiration lowers the growth rate and biomass production (Karnosky et al. 1996; at reduced WUE related to whole-tree water consumption, according to the lowered WUE in leaf gas exchange; see above) and alters the C allocation between the tree organs (Matyssek 2001). Often, root growth appears to be more limited than shoot growth (Andersen 2003), although leaf sizes and branching may be reduced. The lowered C gain can be at the expense of competitiveness relative to neighbouring plants, as space occupation and exploitation for resources is mitigated upon reduced above and belowground growth (Kozovits et al. 2005).

Tree responses are strongly influenced, however, by multiple stresses, in particular, competing plants, parasites, mycorrhization, as well as plant-related factors like genotype and age (Coleman et al. 1995; Maurer and Matyssek 1997; Matyssek et al. 2005). Crucial to these responses are O_3 dose uptake, the time span of uptake (Reich 1987), and the dynamics of O_3 levels along with the sensitivity of tree per unit dose (i.e. the 'effective O_3 dose', as determined by the capacities of repair and defence, the seasonal course and further environmental stress; Musselman et al. 2006; Matyssek et al. 2008). Given the significance of the effective O_3 dose, it is not justified to merely conclude O_3 sensitivity from foliage type (evergreeness vs. deciduousness with low vs. high O_3 flux, respectively) (Calatayud et al. 2010). The interaction of internal and external variations determines O_3 uptake, C gain and reserve storage as pre-requisites for repair and defence capacities, and eventually, O_3 sensitivity (Wieser and Matyssek 2007). O_3 is no longer believed to be a 'tree killer' at the short-term scale. However, long-term O_3 impact may weaken the trees' tolerance of further stresses and induce respective pre-dispositions, which can eventually become lethal. Such kinds of mechanisms are discussed for regions of chronically high O_3 regimes (such as California/USA; Miller and McBride 1999). Hence, considering principles of O_3 action in trees must not neglect the complexity of O_3 impact and responses, within the context of abiotic and biotic stressors and other factors. The topics will be pursued in Sect. 7.3.

7.2.2.2 Adult Trees and Stand Conditions

Most data have been obtained from juvenile trees in closed chambers (see above), and therefore are biased by unnatural growth conditions and altered plant sensitivity to O_3 (Kolb and Matyssek 2001; Stockwell et al. 1997). Their value for extrapolation to forests, therefore, is limited. In the following, key results from a unique 8-year free-air O_3-fumigation experiment on mature beech and spruce trees (*Fagus sylvatica* and *Picea abies*, respectively), both being ecologically and economically important climax tree species in Central Europe, will be discussed (Fig. 7.2; see

Fig. 7.2 Response of whole-stem productivity as related to the stem allometry of adult *Fagus sylvatica* and *Picea abies* trees at Kranzberg Forest, exposed to the local ambient O_3 regime ($1 \times O_3$, control) or the experimental twice-ambient O_3 regime ($2 \times O_3$). Contour functions represent means of $n = 5$ individuals in each species ($ba_{1.3}$ = basal area at 1.3 m height; after Pretzsch et al. 2010)

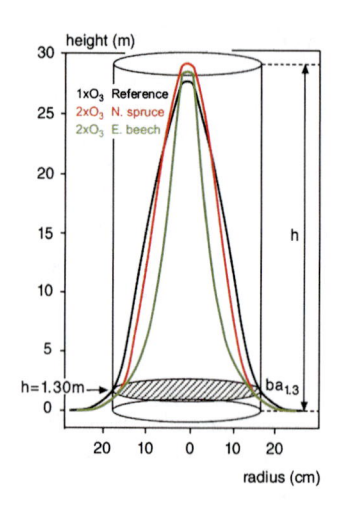

overviews in Matyssek et al. 2010a, b). A novel canopy O_3 exposure methodology was employed that allowed whole-plant assessment of in situ growing forest trees under twice-ambient O_3 levels ($2 \times O_3$; Nunn et al. 2002; Werner and Fabian 2002; Karnosky et al. 2007) without enclosure in chambers (therefore, no micro-climatic bias on O_3 sensitivity). The treatment $2 \times O_3$ was restricted to < 150 nl l^{-1} to prevent acute injury. The $2 \times O_3$ regime was based on monitoring the ambient O_3 regime at the site ($1 \times O_3$, control) continuously throughout the eight growing seasons of 2000 through 2007.

Volume increments of whole stems revealed a deleterious effect of $2 \times O_3$ in beech throughout the study period (Pretzsch et al. 2010; Fig. 7.2), based on increment analysis along the entire stem length by making use of diameter-height allometric relationships, which determined the individual's stem shape and volume. At the stand level, the decline in stem wood productivity attributable to O_3 was 10 m^3 ha^{-1} yr^{-1}, on average over the 8-year study period, representing a 44 % loss relative to the $1 \times O_3$ beech trees. Such an effect was absent in spruce, where O_3-changed diameter-height allometry and compensated for the reduction in radial increment at breast height which was similar as for beech, resulting in an increase in total stand productivity by 0.5 m^3 ha^{-1} yr^{-1} under $2 \times O_3$ (Pretzsch et al. 2010). The O_3 limitation in beech did not become apparent, however, when restricting the analysis to radial breast height increments, according to conventional forestry practices (Wipfler et al. 2005; Pretzsch et al. 2010).

In contrast to prevalent knowledge, elevated O_3 did not affect mesophyll photosynthesis of beech directly, but reduced net CO_2 assimilation by stomatal closure in both sun and shade foliage (Kitao et al. 2009; Fig. 7.3) to some moderate extent, so that the WUE of leaves was increased. In the absence of drought and at probable enhancement of ethylene formation (Nunn et al. 2005b), O_3-induced stomatal closure was associated with the induction of the abscisic acid (ABA) biosynthesis gene *NCED1* (9-cis-epoxycarotenoid dioxygenase 1), whose transcripts were more abundant in $2 \times O_3$ (Jehnes et al. 2007, cf. Wilkinson and Davies 2010). No consistent differences in expression were detected for other genes, but at periods of exceptionally high O_3 levels, there were indications of oxidative stress and of challenge to plant defences (cf. Dizengremel et al. 2009). Lowered starch and sucrose levels indicated increased glycolysis (Blumenröther et al. 2007; Einig et al. 1997) and were consistent with enhanced leaf dark respiration (Kitao et al. 2009), fuelling defence and repair under $2 \times O_3$. Nevertheless, minor macroscopic leaf injury and shortening of the growing season (by up to 10 days; Nunn et al. 2002, 2005a) was not prevented. Also in spruce, photosynthetic limitation occurred under $2 \times O_3$ at reduced stomatal conductance, although such effects were mainly restricted to shade crowns (Nunn et al. 2005b). Although O_3 does not physically penetrate soil, under $2 \times O_3$ soil respiration was increased (Nikolova et al. 2010) around respective beech and spruce trees. Hence in beech, the C storage capacity was reduced in the entire tree-soil system. The respiratory effect was accompanied by increased diversity of mycorrhizal associations and enhanced annual production of fine roots under $2 \times O_3$ (Nikolova et al. 2010; Grebenc and Kraigher 2007; Haberer et al. 2007). Stimulated root growth under O_3 stress is remarkable, in contrast to findings from chamber studies (Matyssek and Sandermann 2003). In adult beech, O_3-induced disturbance of the phytohormonal relationships of cytokinins as mediators in shoot-root communication appear to be responsible (Winwood et al. 2007), because O_3-mediated cytokinin destruction in leaves led to declining re-translocation belowground via the phloem – a change known to relieve inhibition of fine-root production (Riefler et al. 2006).

The measured decline in beech stem growth validates predictions from modelling that elevated O_3 will result in a substantial reduction of C sink strength in trees (Sitch et al. 2007), also suggesting that the most commonly used stem growth assessment method (i.e. being restricted to breast height) is insufficient for assessing the impact of environmental stress. The reported empirical study for the first time demonstrated whole-stem growth of adult trees of a climax species to be restricted under natural stand conditions and enhanced O_3 impact. In addition, adult and juvenile trees appear to be similarly sensitive to O_3, although physiological mechanisms may differ (Niinemets 2010). The study also underscores the need to understand the O_3 risks for individual species when modelling tree and stand performance in climate change. Along with reduced stem growth of adult beech under $2 \times O_3$, the spectrum of further O_3 responses reflected O_3 stress to be metabolically experienced by the trees. In the long term, weakening of trees in terms of their tolerance of further environmental stress, therefore, cannot be excluded.

7.3   Modifying Factors of O_3 Impact

Air pollution is one component of climate change. Among other anthropenic pollutants, tropospheric O_3 is the ecologically most significant compound, given its toxic potential for plants and widely spread occurrence at enhanced concentrations (see Sect. 7.1). O_3 impact, therefore, must be understood in concert with other factors of relevance in a changing environment (addressed in this section), because multiple interactions determine the plants' sensitivity to stress (Mooney and Winner 1991; Skärby et al. 1998; Matyssek and Innes 1999). Because of this, principles of O_3 action in trees, as highlighted in Sect. 7.2, may be moderated or even masked (Matyssek and Sandermann 2003). Still, our judgement largely relies on findings from young trees and chamber experiments, although evidence has increased recently on tree performance at advanced ontogenetic stages and under ecologically relevant field conditions (Kubiske et al. 2007; Matyssek et al. 2010a, b). Section 7.3.1 will highlight interactions of O_3 with temperature, drought, and irradiance, followed by Sect. 7.3.2 on the relevance of nutrition. Sections 7.3.3 and 7.3.4 will address such interactions with VOCs and CH_4, and between O_3 and CO_2 respectively. Biotic influences on the O_3 response of trees will be highlighted in Sect. 7.3.5, comprising the significance of tree genotype and effects by competition, host-pathogen/herbivore relationships and mycorrhizospheric interactions.

7.3.1   Ozone Interactions with Drought, Temperature and Irradiance

Temperature increase and, as a consequence, altered spatio-temporal patterns of precipitation are facets of climate change (Scarascia-Mugnozza et al. 2001; Gessler et al. 2007), which are basically driven by the anthropogenic release of CO_2 and other 'climate-effective' gases including ground-level O_3 at enhanced concentrations (Fabian 2002; IPCC 2007). Altered precipitation and, hence, cloudiness affects light availability, which in turn is crucial both for the O_3 formation (Stockwell et al. 1997) and the sensitivity of plants to O_3 (Matyssek et al. 2008). As in concert with light, water availability and temperature determine O_3 sensitivity. In the following, these three abiotic factors will be jointly viewed in relation to O_3 impact.

7.3.1.1   Temperature

Associating climate change with temperature typically implies expected increase of the latter (IPCC 2007). However, risks associated with enhanced temperature and O_3 regimes are rather uncertain (Matyssek and Sandermann 2003). Although ecologically relevant, this factorial combination has few investigations, given the

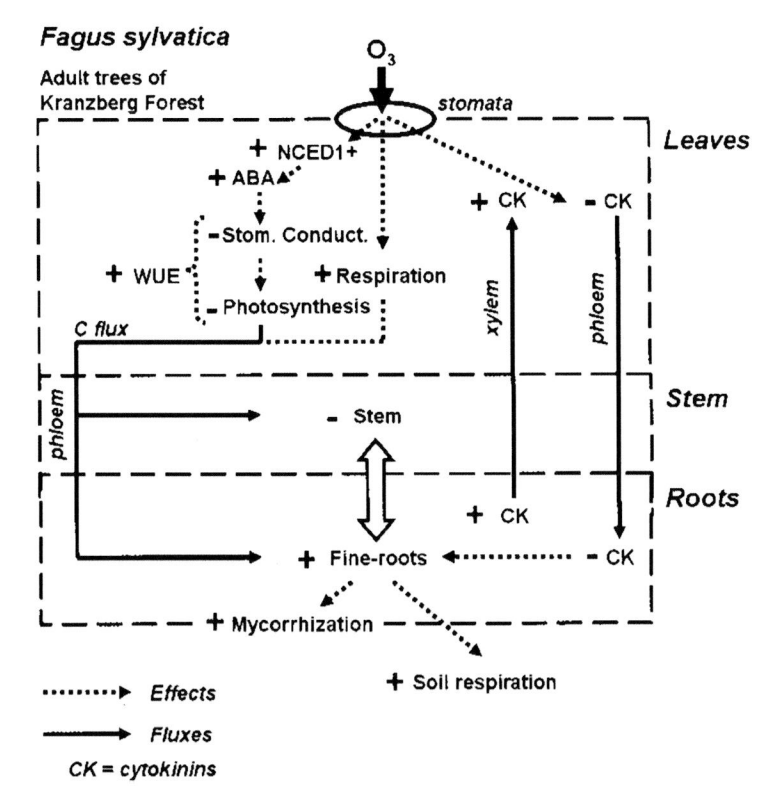

Fig. 7.3 Pathways of O_3 action in adult individuals of *Fagus sylvatica* at Kranzberg Forest (near Freising, Germany; Matyssek et al. 2010b). CK = active cytokinin; upon destruction in leaves under $2 \times O_3$, export of mobile isopentenyladenine-type CK to the roots is decreased, diminishing the suppression of fine-root growth (Winwood et al. 2007, Riefler et al. 2007). The sink induced in this way for carbon belowground apparently competes with stem growth

susceptibility of trees to spring and autumn frosts under prolonged growing seasons and, hence, O_3 impact, or pests profiting from mild winters (Karnosky et al. 2001). O_3 stress in combination with low temperature has been accounted for (Skärby et al. 1998). Tree injury during winter has been debated as a consequence of O_3 effects during the preceding summer that eventually weaken frost hardening (Chappelka et al. 1990). Such an O_3 action might be mediated through the proportion of unsaturated fatty acids arising from lipid synthesis, given their importance in preventing frost injury of membranes (Heath 1980; Wolfenden and Mansfield 1991), and through the availability of glycoproteins and carbohydrates as cryoprotectants. Such demands relate to the trees' carbon balance may have been curtailed *via* the prior summer's O_3 impact (Wieser and Matyssek 2007). Starch must sufficiently be converted into raffinose and sucrose for freezing-point depression of cell solutions (Sheppard 1994), and in fact, Alscher et al. (1989) observed delay in raffinose formation under O_3 exposure. Reduced accumulation of starch and glucose under O_3 stress, however, does not necessarily lower sucrose availability (Barnes et al. 1995), and some cryoprotectants may even be increased (e.g. pinitol in *Pinus sylvestris*, Landolt et al. 1989). In addition, desiccation and photoinhibition during winter can relate to preceding O_3 impact in summer (Mikkelsen and Ro-Poulsen 1995). Although low-temperature regimes appear to be atypical of climate change, they are of relevance in combination with enhanced O_3 levels at high altitude (Matyssek et al. 1997a) and high geographic latitude (Karlsson et al. 2009). Metabolic limitation of detoxification and repair by low temperature can reduce the O_3 tolerance of trees, although temperature-limited stomatal opening may restrict O_3 uptake (Klingberg et al. 2010). However, measurements of frost hardiness during winter dormancy at the inner-continental alpine timberline suggested that the maximum frost resistance of *Picea abies* was related to minimum air temperature throughout the winter rather than to mean and peak ambient O_3 concentration of the preceding growing season (Wieser 1999).

7.3.1.2 Drought

The extraordinarily warm and dry year of 2003 (Ciais et al. 2005) provided a rare episode in Central Europe of prolonged water shortage during the growing season in combination with distinctly enhanced O_3 exposure (Matyssek et al. 2010b). The Kranzberg free-air O_3 fumigation experiment on adult forest trees (see Sect. 7.2) made use of this circumstance (Löw et al. 2006, 2007; Nikolova et al. 2009, 2010; Matyssek et al. 2010b). Progressive soil drought was shown to decouple O_3 uptake from O_3 exposure in beech (Matyssek et al. 2006, cf. Panek et al. 2002). The O_3 uptake was reduced because early summer drought-driven stomatal closure pre-empted O_3-driven closure. As a result, seasonal O_3 uptake in 2003 was slightly less than the average for the humid years, even though O_3 exposure was greater by 41 %. Apparently, the assumption upon which current protection policy in Europe is based, that O_3 exposure correlates reliably with O_3 uptake, was not substantiated (cf. Wieser and Tausz 2007). Drought overruled the O_3 impact on photosynthesis, causing distinct reduction of CO_2 uptake at decreasing sensitivity to O_3 (Löw et al. 2007). Drought also nullified the O_3-driven stimulation of belowground activity (see Sect. 7.2), although this result is within the context of only one study of O_3/drought interactions in belowground processes (Andersen 2003). In 2003, the prolonged water limitation rather than O_3 stress limited both the radial and whole-stem volume increment of the beech trees (Pretzsch and Dieler 2010).

The findings from the Kranzberg experiment corroborate previous evidence that both stresses (i.e. by drought and O_3) have the capacity of lowering stomatal apertures. Nevertheless, drought-induced stomatal closure can limit O_3 uptake even under high O_3 levels, whereas the O_3 influx may be high under the low O_3 levels of humid weather or moist soil conditions (Fig. 7.4; Wieser and Havranek 1993, 1995; Pääkkönen et al. 1998b; Retzlaff et al. 2000; Paoletti 2006). The exclusion of O_3 under drought, however, is 'paid' for by restricted CO_2 uptake (Panek and Goldstein 2001). Conversely, O_3 can prime trees to drought through inducing sluggishness in stomatal regulation and, by this, inefficient control of transpiration (Keller and Häsler 1987; Barnes et al. 1990a, b; Pearson and Mansfield 1993; Karlsson et al. 1995) or even enhancement of stomatal conductance during drought (Oksanen 2003). Altered 'mechanics' in the stomatal apparatus appear to be responsible due to reduced cell wall lignification and membrane impairment under O_3 stress (Heath and Taylor 1997; Maier-Maercker 1998). Given resource limitation caused by O_3 and drought, both stressors reduce lignification during organ differentiation (Kivimäenpää et al. 2003). In *Betula*, drought and O_3 stress were additive in reducing foliage area, but increasing epidermal cell wall width and tannin-like depositions in the vacuoles are indications of O_3 defence (Pääkkönen et al. 1998a). The extent to which drought-induced abscisic acid (ABA) affects stomatal sensitivity under O_3 stress is still uncertain, but appears to result from interaction with ethylene formation (Wilkinson and Davies 2010), although ABA synthesis may be stimulated by O_3 (see Sect. 7.2.2). Injury in cell membranes may modify, through effects on ion pumps, the stomatal sensitivity to ABA under drought (Torsethaugen et al. 1999). 'After-effects' of drought on stomatal conductance and O_3 uptake can also occur (Karlsson et al. 2000).

Although both O_3 and drought favour ROS production, interactive effects on detoxification apparently depend on the experimental conditions and analysed parameters (Pääkkönen et al. 1998a). Relative to drought impact alone, only concurrent exposure to elevated O_3 increased ascorbate levels (Kronfuß et al. 1998). Similarly, needle age is crucial, as shown for *Pinus halepensis* (Gerant et al. 1996; Alonso et al. 2001), in which O_3 tended to lower the resistance to drought. Given that both stressors induce ROS and the need for detoxification (both inciting defence-related proteins; Pääkkönen et al. 1998b), one stressor may prime for improved defence against the other, depending on the sequence of their occurrence. The drought tolerance of evergreen Mediterranean plants in this way may inherently foster their O_3 tolerance (Nali et al. 2004). Decrease in RubisCO activity under O_3 and drought stress was less pronounced than under O_3 stress alone (Pelloux et al. 2001; Fontaine et al. 2003). O_3 exposure lowered antioxidant capacity, reduced chlorotic mottling, but induced earlier needle loss in *Pinus jeffreyi* growing in mesic relative to xeric sites. In xeric sites, mottling was related to high drought-induced ROS production perhaps strengthening defence, which was not necessarily driven by stomatal O_3 uptake (Grulke et al. 2003).

Both O_3 and drought, alone each or in combination, have the potential for restricting the carbon sink strength of trees under climate change scenarios (Sitch et al. 2007). However, O_3/drought interactions are largely variable, depending circumstantially on plant-intrinsic and external factors (including the intensities of the two stressors relative to each other).

7.3.1.3 Irradiance

Provided that the steady increase in atmospheric CO_2 concentration leads to climate warming (IPCC 2007), in regions of induced progressive drought, coupled with

Fig. 7.4 Half hour means
of ambient air O_3
concentration (*top*),
stomatal conductance for O_3
(*middle*) and O_3 flux
(*bottom*) into short-shoot
needles of *Larix decidua* at
the alpine timberline during
daylight hours (PPFD
>600 µmol m^{-2} s^{-1}, *open
symbols*) and during dawn
and at night (PPFD
<100 µmol m^{-2} s^{-1}, *solid
symbols*) in relation to leaf
to air water vapor deficit
(Δw; after Wieser and
Havranek 1996)

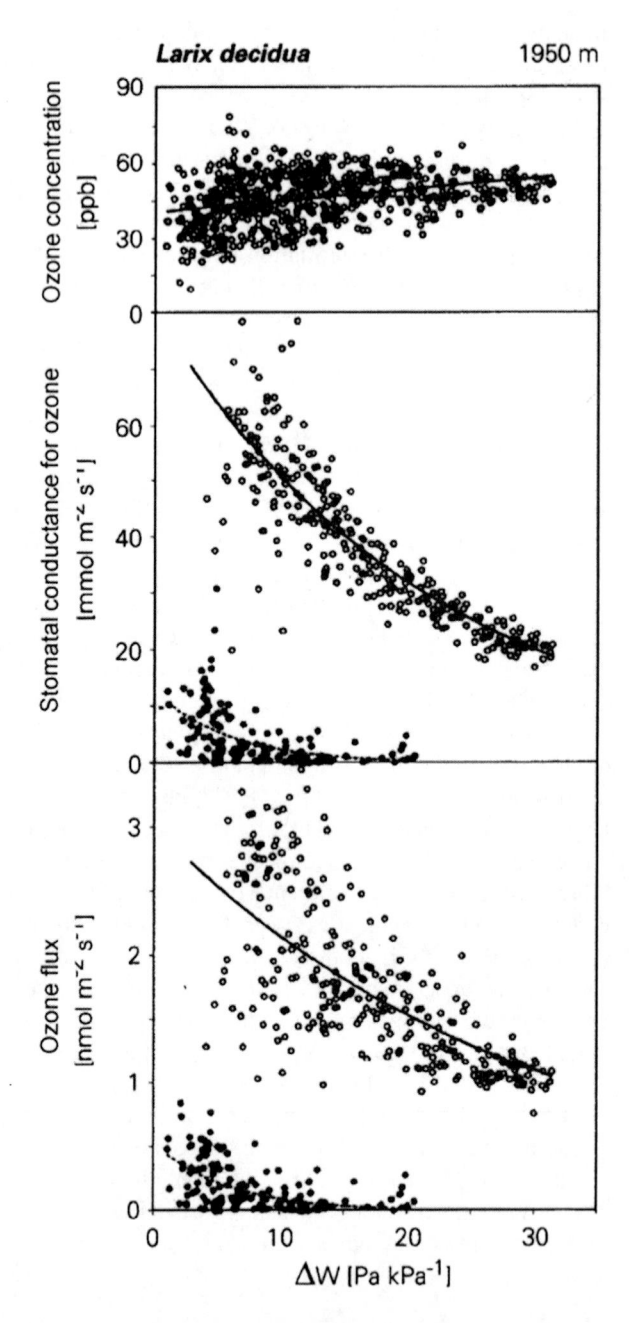

reduced cloudiness, insolation will increase. High irradiance promotes O_3 formation (Stockwell et al. 1997), and in addition, the injurious potential of O_3 to plants is
increased under high light. The reason is that oxidant-driven chloroplast

degradation upon O_3 uptake is exacerbated by the high photosynthetic energy capture which eventually turns, via formation of singlet oxygen, into photoinhibitory bleaching (Heath and Taylor 1997). Oxidant production during photosynthesis requires detoxification, which under undisturbed conditions is accommodated by the photosynthetic energy supply (Schupp and Rennenberg 1988). Findings suggest that the light-demanding character of pioneer tree species predisposes them to high O_3 sensitivity, although comparisons across species and genotypes are inconsistent (Kolb and Matyssek 2001; Matyssek et al. 2010a). However, if light becomes limiting as in the shade or at high geographic latitude, detoxification may become insufficient to cope with the additional oxidative stress upon O_3 uptake (Foyer et al. 1991). Evidence indicates that low-light conditions increase sensitivity to O_3 (Volin et al. 1993; Tjoelker et al. 1995; Kolb et al. 1997). In *Betula* (Matyssek et al. 1995), the same O_3 exposure at night (including dawn and dusk) led to more distinct decline in biomass production than during the daylight hours, as open stomata at night allowed O_3 influx. High O_3 sensitivity of shade leaves (Tjoelker et al. 1995) may relate to their low palisade *versus* spongy parenchyma ratio (and therefore reduced antioxidant capacity, cf. Matyssek et al. 2010a) and low structural mesophyll compactness. The latter alleviates O_3 diffusion within the intercellular space and increases the exposure of palisade cells relative to sun leaves (Bennett et al. 1992). Shaded leaves may in addition have a higher ratio of O_3 versus CO_2 uptake (Fredericksen et al. 1996a, b). As the latter is a proxy of defence capacity, ratios relating agent uptake to assimilatory capacities may provide estimates of O_3 tolerance (Wieser et al. 2002).

7.3.2 Nutrition

Forest ecosystems tend to be oversaturated with N in regions of the industrialized world, where N depositions are enhanced (Durka et al. 1994). High N depositions are drivers of global change scenarios (IPCC 2007), so that other air pollution effects such as enhanced tropospheric O_3 regimes need to be viewed in relation to N availability and nutrition in general. The water relations of trees are influenced by nutrition, and in particular low N availability, which tends to ensure high sensitivity in stomatal control (Schulze 1994). Under high N availability and climate change scenarios that are characterized by high O_3 regimes and water limitation to an extent that still permits stomatal opening (favored by the high N supply), there is also high O_3 influx into leaves (see Sect. 7.3.1). If N is the only nutrient to vary, high supply tends to decrease O_3 injury and premature leaf loss (Pääkkönen and Holopainen 1995; Greitner and Winner 1989), and high N deposition is believed to ameliorate O_3 effects (Takemoto et al. 2001). Conversely, it was shown for *Fagus creneta* leaves that at high N availability O_3 stress reduced the N allocation to the soluble protein fraction in conjunction with promoting protein degradation, so that in total a higher O_3-induced protein loss was observed than under low N availability (Yamaguchi et al. 2010).

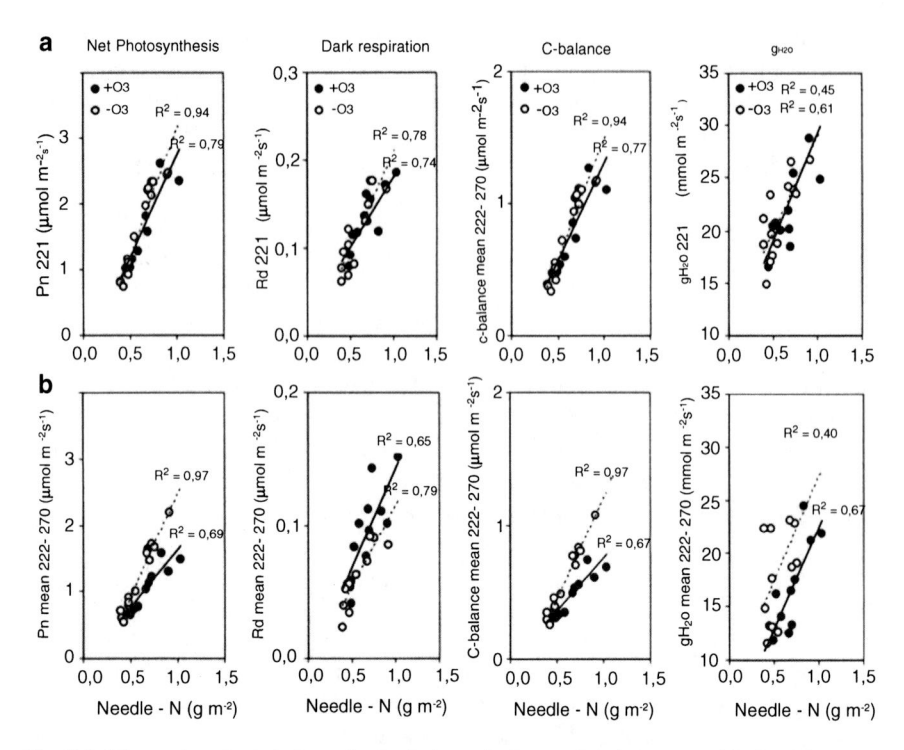

Fig. 7.5 Mean rates of net photosynthesis, dark respiration, carbon balance, and leaf conductance (g_{H2O}) related to needle nitrogen levels of *Picea abies* seedlings expose to O_3 free air ($-O_3$) and twofold ambient O_3 concentration ($+O_3$) from August 10, throughout September 27, 1999, related to needle nitrogen content of the needles. (**a**) On day 221 i.e. prior to O_3 exposure, and (**b**) mean rates of the same seedlings estimated for the whole fumigation period (mean 222–270). Each data point in B represents the mean of 5,300 measurements taken every 12 min). Note: linear regressions were similar in $-O_3$ and $+O_3$ plants prior to O_3 exposure (day 221), but differed throughout the entire experiment (*solid line* = $+O_3$, *dotted line* = $-O_3$ plants) (After Havranek, Wieser and Tausz, unpublished data)

Viewing nutrient availability in general, chronic O_3 exposure lowered the stomatal conductance in spruce (*Picea abies*; Fig. 7.5) and in *Betula pendula* only at low nutrition (Maurer et al. 1997) by restricting O_3 influx. However, low nutrition did not necessarily prevent O_3 injury (Bielenberg et al. 2001), although the antioxidative defense capacity against O_3 was enhanced and premature leaf loss delayed (Fredericksen et al. 1995; Maurer et al. 1997; Polle et al. 2000; Matyssek 2001). The maintenance of O_3-injured foliage can substantially diminish photosynthesis, increase respiratory costs, and lower the whole-tree carbon balance (Wieser and Matyssek 2007; Fig. 7.5). As a consequence, WUE of biomass production can be lowered (Maurer and Matyssek 1997), and root growth limited rather than stimulated at low nutrition (cf. Mooney and Winner 1991).

Conversely, high nutrient supply does not necessarily enhance O_3 tolerance, as also shown for *Betula* (Maurer and Matyssek 1997; Matyssek et al. 1997b), and may foster limitation of root growth and leaf loss (Grulke and Balduman 1999). Surprisingly, periodic soil drought with water potentials down to -0.5 MPa did not significantly affect the biomass production and partitioning of spruce, independent of fertilization (Fig. 7.6). O_3 effects on nutrition have predominantly been found to be inconsistent and may be confounded by processes of senescence and re-translocation or depend on the specific nutrient under consideration (Matyssek et al. 1995; Polle et al. 2000). The role of nutrition in driving leaf turnover, i.e., replacement of O_3-injured leaves, may be determined by the different seasonal dynamics of determinate *versus* indeterminate shoot growth (Tjoelker and Luxmoore 1991; Laurence et al. 1994; Matyssek 2001; Kolb and Matyssek 2001). An interacting third component under climate change is drought, as nutrient uptake may be affected both through O_3-caused limitation of belowground C allocation and soil drought (see below and Kreuzwieser and Gessler 2010). Multiple factorial interactions contribute to inconsistencies in trees' response to O_3.

7.3.3 Ozone Interactions with VOCs

As mentioned earlier, BVOCs are reactive molecules and play an important role in air quality and climate dynamics in the lower troposphere of anthropogenically polluted areas (Fig. 7.7). The group of BVOCs embraces a large variety of chemicals, including alkanes, alkenes, alcohols, ketones, aldehydes, ethers, esters and carboxylic acid (Kesselmeier and Staudt 1999; Laothawornkitkul et al. 2009). The dominate and most important BVOCs for O_3 formation are isoprene (2-methyl-1,3-butadiene, a C5 compound) and monoterpenes (e.g. α-pinene, β-pinene, camphor, linalool, C10 compounds). The reactive terpenes react with hydroxyl radicals, NO_3 radicals, and O_3 in a similar way to most VOCs of anthropogenic origin (Atkinson and Arey 1998).

On a global scale, BVOCs quantitatively dominate anthropogenic VOC emissions. In industrial areas (e.g. most of Europe, North America, or Eastern China), however, VOC emissions from fossil fuel combustion, direct release from industrial processing of chemicals, and waste are the main sources of non-methane hydrocarbons (NMHCs) emitted into the atmosphere (Fuentes et al. 2000). The world's estimated annual emission of isoprene and monoterpenes from vegetation is about 200–500 Mt of C, an amount equivalent to annual global CH_4 emissions (Guenther et al. 1995; Keppler et al. 2006, 2009).

Volatile terpenes can react directly with O_3, and become degraded mainly to methyl vinyl ketone, methacrolein and formaldehyde. In the presence of a sufficient concentration of NO_x, BVOCs directly contribute to the formation of tropospheric O_3 (Thompson 1992). The dynamics of O_3, hydroxyl radical (OH·), carbon monoxide (CO), methane (CH_4) formation/breakdown and secondary aerosol formation cannot be explained without taking into account the plant-produced reactive compounds (Biesenthal et al. 1997; Tunved et al. 2006; Kiendler-Scharr et al. 2009; Fig. 7.7).

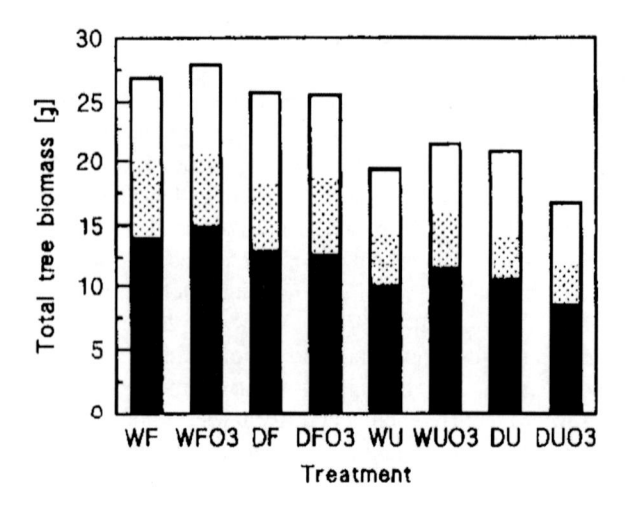

Fig. 7.6 Biomass partitioning between roots (*solid bars*), shoots (*hatched bars*), and needles (*open bars*) of fertilized (*F*) or unfertilized (*U*) *Picea abies* seedlings, grown either under non-limiting water supply (*W*) or being submitted to three moderated drought cycles with soil water potentials down to −0.5 MPa (D). Each factorial combination was exposed, in addition, to either O_3 free air or 2.5 × ambient air O_3 (O3) concentration for 48 days (After Havranek, Wieser and Tausz, unpublished data)

Beside their relevance for tropospheric O_3 formation, isoprene as well as mono- and sequiterpenes (highly reactive semi-volatile C15 terpenes) influence global warming. Due to their unsaturated double bonds, they are highly reactive and thus significantly reduce the OH· concentration in the lower troposphere, with consequences on the decomposition rate of methane (Bell et al. 2003). Model calculations predict that BVOCs, mainly isoprene and monoterpene, can increase the surface concentration of O_3 by up to 60 % and increase the atmospheric lifetime of methane by about 14 % (Poisson et al. 2000; Fig. 7.7).

To quantify the importance of BVOCs for tropospheric O_3 formation under future climate, the regulation and driving forces of BVOC emission must be considered. BVOCs emissions from plants are controlled by multiple abiotic and biotic factors that might be differentially affected under future conditions (Peñuelas and Staudt 2010). The main environmental drivers of BVOC emissions are temperature, light (Kesselmeier and Staudt 1999), drought (Brilli et al. 2007), ambient O_3 (Loreto and Velikova 2001), as well as atmospheric CO_2 concentration (Rosenstiel et al. 2003; Fig. 7.7). Plant BVOC emissions are also largely induceable by both abiotic and biotic stresses (for review see Loreto and Schnitzler 2010; Holopainen and Gershenzon 2010; Dicke and Baldwin 2010). Low atmospheric CO_2 concentrations, as present in historical times, are associated with high isoprene and monoterpene emission rates, whereas the high atmospheric CO_2 concentrations in the future will suppress volatile terpene emissions (Peñuelas and Staudt 2010). Drought reduces isoprene and monoterpene emissions only when the stress is severe and almost completely inhibits photosynthesis (Loreto and Schnitzler

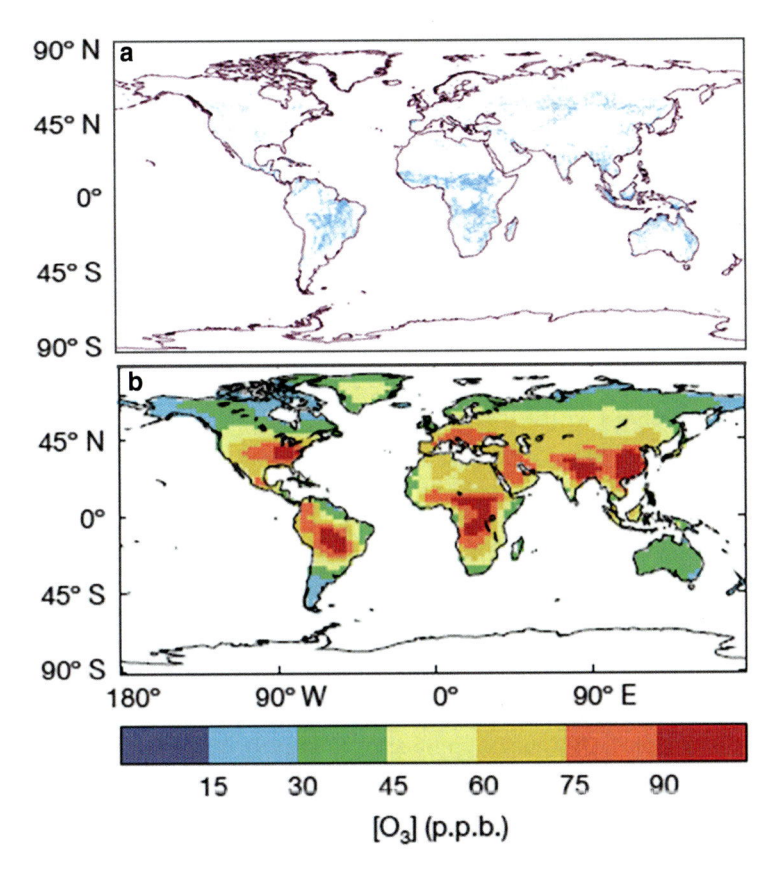

Fig. 7.7 (**a**) Burnt area distribution given by the ATSR satellite sensor (Modified from Monks et al. 2009); (**b**) modelled mean surface ozone concentrations in p.p.b. (=nl l^{-1}) averaged over June, July and August (JJA) for the year 2100 under the SRES A2 emissions scenario (Modified from Sitch et al. 2007)

2010). Temperature strongly influences the activity of the enzymes, which catalyze the synthesis of many BVOCs (Monson et al. 1992), and increasing light directly influences the isoprene and monoterpene emission by enhancing photosynthesis (Sharkey et al. 2001).

Besides all abiotic drivers, plants produce a diversity of BVOCs in response to biotic stress, as for example, by herbivore feeding for direct defence or to attract herbivore enemies (Dicke et al. 2009; Unsicker et al. 2009). At the moment, one can only speculate whether these spatially and temporally highly transient bursts of BVOCs are important for O_3 formation. Conversely, evidence emerges that O_3 influences the lifetime of stress-induced BVOCs. By this, O_3 changes the spatial distribution of these signal molecules as well as their specific emission patterns with consequences for plant-insect interactions (Holopainen and Gershenzon 2010).

The increasing scarcity of fossil fuels, the combustion of fossil fuels causing increases in greenhouse gas (GHG) emissions, and growing interest in renewable energy sources has placed an enormous pressure on land use change in the tropics (Wiedinmyer et al. 2006; also see Sect. 7.4). In all scenarios for the future, energy fuels from biomass, so-called biofuels, and/or other energy sources from biomass (i.e. biogas, thermal recycling), play an important role (Demirbas 2008). To what extent biomass can contribute to a significant regional and global reduction of GHG emissions depends critically on several factors such as land availability and land use conflicts, sustainable crop production management, cost and actual carbon neutrality, and environmental safety.

Recent comprehensive analyses of biosphere-atmosphere exchange of BVOCs and atmospheric composition, together with atmospheric chemistry modeling (Hewitt et al. 2009) demonstrated that conversion of tropical rainforest to oil palm plantations in Malaysia resulted in much greater BVOC emissions that led to O_3 formation. In combination with increased NOx emissions, severe ground-level O_3 pollution (>100 ppbv) can be reached. This study clearly shows the need to quantify current and future effects of land use change on air quality when assessing the 'environmental safety' of palm oil and other biofuel crops and woody biomass species. Keeping in mind that all fast growing tree species (*Populus*, *Salix*, *Eucalyptus*) are among the strongest emitters of reactive isoprene and monoterpenes (Kesselmeier and Staudt 1999), the large scale extention (millions of hectares) of woody biomass plantations expected in near future will put a strong pressure on air pollution strategies. Of course, air quality is only one single consideration in assessing the consequences of biofuel production. Effects such as GHG emissions and climate change, deforestation, biomass burring, biodiversity and water availability must also be considered (Fig. 7.7).

7.3.4 O_3 Interactions with CO_2

The environment is currently influenced by the gradually increasing atmospheric CO_2 concentration, which sets the context for examining the action of enhanced tropospheric O_3 regimes. Contrasting with enhanced O_3 levels, elevated CO_2 may stimulate tree growth (Ceulemans et al. 1999), unless growth conditions, i.e. the supply of other resources, constrain plant development (Norby et al. 1999; Scarascia-Mugnozza et al. 2001). The question is whether high CO_2 availability may ameliorate adverse O_3 effects in trees (Karnosky et al. 2001), or conversely, if enhanced O_3 levels may neutralize enhanced carbon gains under elevated CO_2. Ameliorating CO_2 effects were observed regarding macroscopic foliar injury (Karnosky et al. 1999), photosynthesis and/or growth under O_3 stress (Manes et al. 1998; Volin et al. 1998), although dependence on nutrition, water supply or seasonal influences was apparent (see above, and Lippert et al. 1996; Grams et al. 1999; Lütz et al. 2000). High CO_2 may favor detoxification and repair (cf. Grams and Matyssek 1999) or lower stomatal conductance and O_3 influx

(Klingberg et al. 2010), perhaps at the expense of responsiveness to moisture deficits (Broadmeadow et al. 1999). Bias may result from experimental restrictions on root growth (e.g. tree growth in pots), inciting down-regulation of photosynthesis and associated stomatal closure (Saxe et al. 1998).

Regarding leaf gas exchange, the effects of elevated O_3 and/or CO_2 regimes apparently are highly variable, even amongst individuals of same species (see Sect. 7.3.5; Matyssek et al. 2010a), as reported from the Aspen FACE ('Free Air CO_2 Enrichment') experiment (Karnosky et al. 2005). Early-successional species (e.g. trembling aspen, paper birch) appeared, in general, to be more responsive than late-successional ones (e.g. sugar maple). As exposed within Aspen FACE, O_3 effects limited photosynthesis and biomass production, and elevated CO_2 incited stimulation. Co-exposure of both gases at high levels demonstrated O_3 to offset positive effects of CO_2 (Percy et al. 2002; King et al. 2005). In *Betula*, negative response to O_3 may only become effective when concurrently exposed to elevated CO_2, although the overall response still reflected stimulation (Kubiske et al. 2006). However, such a response contrasted with that of aspen, where actions of elevated O_3 and CO_2 resulted in neutralization. Tree responses to elevated O_3 and/or CO_2 can strongly be modified by competition with neighbouring plants, as will be detailed in Sect. 7.3.5. Such observations also hold for juvenile trees of European beech and *Picea abies*, where irrespective of competition, spruce tended to profit from elevated CO_2, whereas beech was more sensitive to O_3 (Kozovits et al. 2005a, b). Whole-tree biomass partitioning in response to the gas regimes followed allometric rules relating to changes in growth rates rather than reflected metabolic re-adjustments in allocation.

O_3 tolerance appears to be promoted by the trees' capacity for creating C sinks that may facilitate compensation growth of leaves or lammas shoots in replacement of injured structures, if favorable conditions in nutrient and CO_2 supply are also available (Kolb and Matyssek 2001). Given the multiple external and internal influences on the O_3/CO_2 interaction in trees, the ameliorating CO_2 effects are rather unlikely to occur throughout a tree's entire lifespan (Karnosky et al. 2001), as elucidated in detail by Körner (2006).

7.3.5 O_3 Interactions with Biotic Factors

Experimental research on O_3 stress has typically been designed as if trees merely existed in abiotic environments (i.e. without competitors, consumers or mycorrhizae, Matyssek and Sandermann 2003). In addition, focus has been on species rather than genotype sensitivity, often ignoring further effects by seasonality or the trees' history prior to enhanced O_3 impact (Karnosky et al. 2005; Niinemets 2010), and, most importantly, the plants' capacity for deploying pronounced plasticity in response. Consequently, variability in tree response to O_3 stress has been observed to be high, at times leading to contradictory findings. In some of the studies reported above, however, variability was apparently related to biotic influences. One typical example is the genus *Populus*, within which a broad range of O_3 sensitivity at the genetic level exists, with some genotypes even being stimulated by elevated O_3, regardless of the CO_2 regime (Karnosky et al. 2007). Also Coleman et al. (1995) reported on high variability in photosynthetic O_3 response amongst genotypes of trembling aspen. During the Aspen FACE experiment, some genotypes were limited in growth under enhanced O_3, whereas others were insensitive or showed amelioration of adverse O_3 impact under elevated CO_2 (Kubiske et al. 2007). Hence, ranking of O_3 sensitivity should not be done by species but by genotypes, as variability might be higher within than across species (Vanderheyden et al. 2001).

Another biotic factor driving variability in O_3/CO_2 response is competition. In juvenile beech, growth in mixture with juvenile spruce exacerbated the O_3 limitation on biomass production (Kozovits et al. 2005a, b). In parallel, spruce profited from the weakness of beech under high O_3, as the extent of suppression of beech appears to increase towards low soil pH (Körner 2006; Spinnler et al. 2002). A number of findings underscore competition to govern tree sensitivity to O_3 or CO_2 regimes (Körner 2006; Poorter and Navas 2003; Barbo et al. 1998, 2002; McDonald et al. 2002). Beech failed, contrasting with spruce, by losing its ability to efficiently occupy aboveground space through biomass investment (Kozovits et al. 2005a, b; Reiter et al. 2005; Grams and Andersen 2007). Belowground, beech became less effective in mixture in competing for N and water with again spruce becoming the profiteer (Grams and Matyssek 2010). Suppression of trembling aspen by competing paper birch was enhanced under elevated O_3 or CO_2, however, such an effect was absent with sugar maple as a competitor (Kubiske et al. 2007). Some aspen genotypes grown under competition may change from weak to strong competitors, if grown under enhanced O_3 exposure, or with other O_3-sensitive clones (Karnosky et al. 2003).

Both O_3 and CO_2 have the capacity of influencing the plants' C/N ratio (Maurer and Matyssek 1997; Hättenschwieler et al. 1996; Lindroth 2010). As a result, both foodchains and, within a tree, the balance between growth and competitiveness versus stress defense may become affected, for example against pathogens and herbivores (Herms and Mattson 1992; Matyssek et al. 2005). However, pathways of effects of host-parasite relationships on trees under air pollution and climate change, perhaps even as drivers of the trees' sensitivity to the abiotic agents, have remained an issue of debate (Manning and von Tiedemann 1995). The superiority of spruce in N acquisition addressed above when growing together with beech under O_3 stress was even favoured in the presence of the root rot pathogen *Phytophthora citricola*, which infected both tree species to different extents, but increased the daily N uptake per unit of fine-root biomass in spruce (Luedemann et al. 2005, 2009). The N demand of spruce was likely enhanced by stress defence (Matyssek et al. 2005; Grams and Matyssek 2010), although in contrast with beech, hardening by O_3 against the pathogen did not occur (given that both the abiotic and biotic agent induce oxidative stress to plants; Matyssek and Sandermann 2003). Hardening by O_3 was seen, however, in adult beech of Kranzberg Forest against the leaf endophyte *Apiognomonia errabunda*, which can cause beech blight disease (Bahnweg et al. 2005; Olbrich et al. 2010): During summer, the degree of leaf infection remained low under enhanced O_3. At Aspen-FACE, changes in chemical and structural properties of leaf cuticles under elevated O_3 (Percy et al. 2002; Mankovska et al. 2005) tended to influence fungal infection (Karnosky et al. 2002) and favoured insect attack, as defence metabolites were found to decline (Percy et al. 2002; Kopper and Lindroth 2002). However, the extent of such effects depends on the insect species, so that ecological contexts cannot yet be generalized (Lindroth 2010). Remarkably, O_3 can restrict air-borne plant-to-plant communication and interfere with multi-trophic relationships by oxidizing plant-emitted BVOCs (Blande et al. 2010; Holopainen and Gershenzon 2010), the enhancement of which is a signal of parasitic attack beyond mediating the extent of stand-level social stress (Pinto et al. 2010).

Regarding belowground biotic interactions, most O_3 exposure studies have neglected mycorrhization. As mycorrhizae represent strong C sinks in whole-tree allocation, incorrect conclusions about root/shoot ratios under O_3 stress may be a consequence (Andersen and Rygiewicz 1991). Reduced C flux to the root may limit mycorrhizal fungi under aboveground O_3 impact (Andersen et al. 1991; Kytöviita et al. 1999). It is conceivable that fungal sink strength may overrule the C demand of the foliage for defence, so that mycorrhization may turn into a 'burden' in trees under O_3 stress. Remarkably, root respiration and the fungal and bacterial biomass in the rhizosphere were enhanced in seedlings of *Pinus ponderosa* under O_3 stress (Andersen and Scagel 1997), which appears to be consistent with the findings about soil respiration and mycorrhization underneath the O_3-stressed adult beech and spruce trees at Kranzberg Forest, as a result of indirect effects through whole-tree allocation and/or hormonal relationships (see Sect. 7.2). Effects on mycorrhization under O_3 stress (Ericsson et al. 1996) can be relevant for the trees' metabolic defense capacity (Gehring et al. 1997; Langebartels et al. 1997; Bonello et al. 1993).

It is evident that responses to air pollution in concert with climate change can only be understood when viewing trees and ecosystem under the influence of the diverse kinds of biotic interactions (apart from the relevance of abiotic agents; cf. Lindroth 2010) – which is the natural stage of plant life, ecology and evolution. Complex examples are the involvement of mycorrhizae and soil microorganisms in tree nutrition under O_3 stress and variable water availability, including flooding under some climate change scenarios (Bardossy and Caspary 1990; Gessler et al. 2007; Kreuzwieser and Gessler 2010). What about the trees' resistance under chronic environmental stress? It is conceivable that resistance and fitness are weakened, if chronic O_3 impact 'over-strains' the defense capacity. Successional changes and effects on species composition may be a consequence (Miller and McBride 1999). In such a case, O_3 may prime trees to other stressors, both biotic and abiotic ones, which will act at the expense of sensitive genotypes. In conifers, priming is mediated through reduced resin pressure in the stem under O_3-limited photosynthesis, by this alleviating bark beetle attack (Pronos et al. 1999). Resulting changes in the gene pool can eventually affect the food webs and fire ecology of forest ecosystems (Miller and McBride 1999; Grulke et al. 2009).

7.4 Global Dimensions

Many regions of the globe are currently exposed to levels of surface O_3 and other air pollutants that are high enough to promote plant damage in natural ecosystems, cultivated forests and agricuktural fields, also affecting human health (Fowler et al. 2008; Dentener et al. 2006). Reductions in crop yield, pasture and in forest productivity have important economic consequences, being responsible for losses of several billion dollars per annum in the USA, EU and East Asia (e.g. Wang et al. 2009; Reilly et al. 2007). Such detrimental effects of O_3 have caused countries to establish increasingly stringent air quality policies. However, due to the high mobility of air pollutants in the atmosphere, plumes of anthropogenic O_3 and its precursors can be transported into the free troposphere and across country boundaries to expand over large geographical scales. Regions located very distant from the pollutant source are now suffering from enhanced O_3 background levels or episodes (e.g. Ireland within the lee of North America, Derwent et al. 2004). In parallel, new 'hot spots' of air pollution appear to arise or exacerbate, which demand enhanced awareness about their global significance (Sitch et al. 2007). It is evident that pollution in the twenty-first century is not restricted by national boundaries and can only be tackled through transboundary emission regulation strategies with a global perspective (Akimoto 2003).

7.4.1 Atmospheric Long-Range Transport

For being transported over long distances and becoming an international concern, pollutants need to have lifetimes of more than 1 week in the atmosphere (Akimoto 2003). The lifetime of O_3 in the troposphere varies with altitude and season, and is determined by sinks for deposition. The efficiency of O_3 production per available NO_x molecule increases with elevation. The O_3 lifetime in the continental boundary layer is only a few days in summer, but several weeks in the free atmosphere (Fiore et al. 2002), and up to 2 months in winter, which is sufficiently long to allow intercontinental and hemispherical transport (Akimoto 2003). Given convection sufficient for driving vertical transport, O_3 and precursors may rapidly reach the upper troposphere within less than 1 h, where the dry and cold environment slows chemical destruction and extends pollutant lifetime. Prolonged lifetime and the high wind speed in the upper troposphere carry air pollutants over long distances, being conducive to episodic O_3 pollution events and increasing background concentrations far downwind from the initial O_3 source (Monk et al. 2009; Derwent et al. 2004).

Employing a global chemical transport model, Wang et al. (2009) estimated the effects of anthropogenic emissions (including sources such as fuel combustion, industrial exhausts and fertilizer use, but excluding contributions of open fires) from Canada and Mexico on surface US ozone concentrations. As a result, O_3 background levels at northeast US sites were shown to vary between 23.1 and 28.9 nl l^{-1} in summer, and were increased by 4.6–10.8 nl l^{-1} due to Canadian influence. At southwest sites, levels were higher (30.6–45.0 nl l^{-1}) and enhanced by 3.8–10.3 nl l^{-1} due to Mexican emissions. Modelling showed strong correlation between high O_3 episodes in the USA and the elevation of the pollution source in Canada and Mexico. In parallel, North America releases 34 % of the global anthropogenic NO_x emissions, of which 80 % originate from the USA. The weather system crossing the North Atlantic Ocean transports North American pollutants with levels high enough to enhance surface O_3 over Europe, on average, by about 8.2 nl l^{-1} (Derwent et al. 2004), reaching up to 12.2 nl l^{-1} in Ireland, due to its westerly location. Asian and European O_3 production contribute about 5 and 20 nl l^{-1}, respectively, to surface O_3 mixing ratios over Europe. In turn, air masses from Europe that arrive at eastern Siberia contain O_3 levels that are higher by 2–3 nl l^{-1} than those transferred from other regions (Akimoto 2003) and also affect the Mediterranean Basin including North Africa (Stohl et al. 2002). Even though the contribution of long-range transport is small in some regions, they may raise background O_3 above air quality standards.

7.4.2 Biomass Burning, Climate Change and O_3 Pollution

Biomass burning (Fig. 7.7) is considered as the major source of aerosol and trace gas formation, including O_3 precursors, contributing globally about 50 % of CO and 15 % of surface NO_x emissions per year to the atmosphere (Bowman et al. 2009; IPCC 2007). As pollutants from biomass burning may enter the atmosphere at high elevations and then are rapidly transported in the free troposphere, effects on air chemistry may become substantiated in regions very distant from sources (Mieville et al. 2010). In the Southern Hemisphere, polluted air masses originating from biomass burning in Brazil can cross the tropical Atlantic and Africa, eventually reaching the Indian Ocean (Singh et al. 1996) and Australia (Gloudemans et al. 2006). Emissions of CO, CH_4 and BVOCs in combination with NO_x are conducive to the photochemical formation of tropospheric ozone (Figs. 7.7 and 7.8). In fact, in many regions of the world, high O_3 levels have been related to vegetation fires (Langmann et al. 2009). In Europe, high O_3 episodes were associated with fire events in Alaska (Real et al. 2007), and burning of boreal forests in Alaska and Canada enhanced O_3 levels in Texas (e.g. in July 2004, Morris et al. 2006). CO and O_3 levels were enhanced along the US west coast upon Siberian forest fires during summer 2003 (Bertschi and Jaffe 2005).

Emissions from biomass burning typically are seasonal, being high from December to February in the northern tropics, as in west and central Africa. In

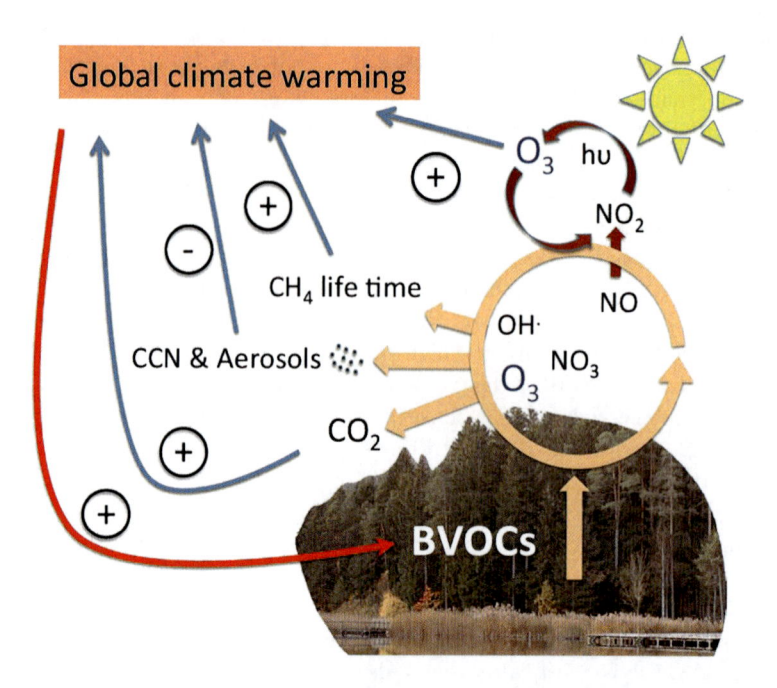

Fig. 7.8 Schematic coupling of BVOC emissions to atmospheric processes like ozone formation and climate change: increase in temperature will enhance BVOC emissions (+). Increased emission of BVOCS will enhance aerosol and CCN (cloud condensation nuclei) formation. Enhanced aerosol and CCN concentrations will decrease temperature (–) as a result of increased reflection of sunlight from low clouds. In presence of NO_X BVOCs degradation will enhance O_3 formation with indirect positive feedback on climate warming (+). Other positive feedbacks are ethane lengthening lifetime (+), CO_2 production (+) and release of latent heat of water condensation (Figure modified after Penuelas and Staudt 2010)

the southern tropics, as in South Africa, South America and Australia, fires are concentrated in August and September (Mieville et al. 2010). In many regions of the northern hemisphere, fire suppression policies, effective fire fighting and the end of intense deforestation had led to rather stable emission levels since the 1920s. From the 1980s onwards, however, emissions have increased rapidly, especially in South America and Indonesia (Fig. 7.7), as a result of deforestation and expansion of agricultural areas (Mieville et al. 2010).

In addition to precautionary prescribed fires, climate change also appears to increase wildfire frequency in some regions (Westerling et al. 2006) and thus, creates a feedback on climate, as biomass burning contributes relevantly to the global budgets of greenhouse gases and aerosol particles. Currently, direct radiative forcing in the atmosphere related globally to aerosols from vegetation fires is estimated as about 0.03 Wm^2 (IPCC 2007). Losses of vegetation cover due to fires and the associated smoke impact may also alter climate directly and indirectly by changes in the local absorption of solar radiation, soil water holding capacity, surface evaporation, transpiration, photosynthesis and BVOC emissions. As a result, the formation of convective clouds and precipitation is suppressed, altering the hydrological cycle. Moreover, aerosol particles from vegetation fires also influence droplet sizes, inhibiting the onset of precipitation up to heights of 7.5 km. Such delay in precipitation initiation by polluted air was reported by Rosenfeld (1999) and Andreae et al. (2004) in the Amazonian region. A persistent smoke layer over extensive tropical areas can be of relevance for the radiation balance and hydrologic cycling at the regional and global scale.

Fire frequency also increases, apart from direct human activity, in response to climate change, as in turn biomass burning promotes climate change and enhances O_3 production. Enhanced O_3 can reduce forest health and increases the susceptibility to wildfires, as shown for the San Bernardino Mountains in California (Grulke et al. 2009). Pathways are through changes in C allocation by high O_3 impact and N deposition, leading to enhanced leaf and branch turnover and decreased litter decomposability, which results in accumulating layers of combustibles. In addition, diminished C investments into the roots in parallel to impeded stomatal regulation under O_3 stress causes imbalances in water relations, which promotes susceptibility to drought and beetle attack. The declining trees eventually favor wildfires, with putative effects on regional and global C balances and biodiversity (Bowman et al. 2009).

7.4.3   Climate Change and O_3 Pollution During Upcoming Decades

Background O_3 levels have more than doubled in many parts of the northern hemisphere since the industrial revolution of the early nineteenth century as a result of fossil fuel and biomass burning, having reached, on average, above 40 nl l^{-1} today (Sitch et al. 2007). In most countries, legislation aims to reduce O_3 precursors by the year 2050 to levels of year 2000. Nevertheless, surface O_3 concentrations are expected to increase throughout the twenty-first century, mainly due to population growth, land use change, ongoing industrialization and economic growth in developing regions. However, modeling predictions depend on presumed emission scenarios (Sitch et al. 2007; Dentener et al. 2006). Rapidly developing countries in Asia and Latin America are expected to surpass average O_3 levels of 40 nl l^{-1}. The increasing air pollution in such countries has become a global concern, initiating legislation for emission control.

Using 26 global atmospheric chemistry models (GCMs) and three different maximum emissions scenarios, Dentener et al. (2006) estimated changes in the global atmosphere between 2000 and 2030. The IPCC SRES A2 is the most pessimistic scenario considering that pollutant emissions will continue to increase in future as it did the last decades. In contrast, the CLE (Current Legislation) takes into account the reduction in air pollution emissions at the regional and global scales due to efficient implementation of current air quality legislation around the world. The MFR (Maximum technologically Feasible Reduction) scenario is the most optimistic, considering a higher reduction in air pollution due to use of all current available and feasible technologies. In 2000, annual mean surface O_3 levels ranged between 40 and 50 nl l^{-1} over large parts of North America, Southern Europe and Asia, whereas in many regions of the Southern Hemisphere, levels were much lower, at around 15–25 nl l^{-1}. Under the high-emission SRES A2 scenario, however, some regions in Latin America, Africa and Asia should experience the highest increases in surface O_3 levels, namely by 5–15 nl l^{-1}, with the global mean increasing by 2030 by about 4.3 nl l^{-1} (Dentener et al. 2006). In comparison, considering the effectiveness of current legislation worldwide (CLE scenario), the global surface O_3 mean is calculated to still increase by 1.7 nl l^{-1}. In the northern and southern hemisphere corresponding increases would range around 2.3 and 0.6 nl l^{-1}, respectively. Huge O_3 enhancements are expected for South (by 7.2 nl l^{-1}) and Southeast Asia (by 3.8 nl l^{-1}). Increases should be at 1.3 nl l^{-1} in USA, 1.8 nl l^{-1} in Europe, 0.5 nl l^{-1} in South America and 1.4 nl l^{-1} in southern Africa. The more optimistic positive MFR scenario predicts reductions of mean O_3 levels by -1.4 nl l^{-1}.

A similar modeling approach was used by Sitch et al. (2007), which estimated possible impacts of future tropospheric O_3 levels on plant primary production, C storage and climate change considering scenarios of low and high plant sensitivity to O_3. Without efficient implementation of air pollutant control worldwide (under SRES A2), by 2100 summer mean monthly O_3 concentrations in the northern hemisphere might be above 40 nl l^{-1} almost everywhere, but exceed 70 nl l^{-1} in many regions between 30° of southern and 44° of northern latitude (e.g. western and central Eurasia, western and eastern North America, Brazil, central and south-western Africa, and East Asia; Fig. 7.8). Considering the negative effects of O_3 on plant physiology even under elevated CO_2 (see sections above), such a scenario is highly precarious, suggesting future C sequestration and storage to become strongly impaired. During the twenty-first century, the global GPP (Gross Primary Production) is expected to become reduced by 14–23 % due to O_3 impact, corresponding to a significant suppression of the global land-C sink strength (by 143–263 Pg C). Large reductions in GPP might occur in North America, Europe, China, India, and in tropical and subtropical ecosystems of South America and Africa. Due to the lack of databases on plant sensitivity to O_3 in tropical ecosystems, however, modeling suffers from significant uncertainties in extrapolating their responses from North American and European plant species.

7.4.4 Scenarios of the Southern Hemisphere

Negative O_3 impacts on natural and cultivated forest and crop systems have been widely studied in the northern hemisphere, especially in Europe and North America (Matyssek et al. 2010; Karnosky et al. 2007; Hayes et al. 2007). Very few respective studies, however, have been undertaken in the southern hemisphere (Pina and

Moraes 2010; Bulbovas et al. 2010; Furlan et al. 2008; Van Tienhoven and Scholes 2003; Emberson et al. 2001) despite increasing air pollution emissions and the potential vulnerability of regional ecosystem services, as food security, crop and forest productivity (van Dingenen et al. 2009), C sequestration and air quality, are altogether relevant for human health. In some countries, however, investigations of O_3 precursor emissions and related sources have been initiated recently (Sowden et al. 2007).

Published information about surface O_3 in Africa and Brazil and other south-American countries indicated that many urban areas experience surface levels greater than 40 nl l^{-1} (Josipovic et al. 2010; Zunckel et al. 2004; Muramoto et al. 2003; Sánchez-Ccoyllo et al. 2006; Olcese and Toselli 1998). In Namibia, Botswana and South Africa, mean surface O_3 levels exhibit strong diurnal and seasonal variations, relating to multiple sources of precursors. Maximum levels typically occur during spring (August to November) with means between 40 and 60 nl l^{-1}, but means may reach more than 90 nl l^{-1} as in October 2000, mainly due to intense biomass burning. Moreover, in some regions, the onset of summer rains with associated spring growth and BVOC and N emissions (Sowden et al. 2007) from vegetation and wetted soil, respectively, might be important causes of maximum O_3 levels. Conversely, in the highly industrialized South African Highveld region, emissions of O_3 precursors mostly result from human activity (e.g. mining, Zunckel et al. 2004). Many of the areas with mean O_3 levels surpassing 40 nl l^{-1} in winter are relevant for maize farming in South Africa and Zimbabwe, which is of great public concern. Van Tienhoven and Scholes (2003) summarize information on plant sensitivity in South Africa to air pollutants, mainly SO_2, fluoride and O_3. However, data have been missing to date about O_3 effects on native or cultivated woody plants.

In Brazil, studies are concentrated on metropolitan and industrialized regions, e.g., São Paulo and Rio Grande do Sul, where O_3 is considered as the most serious air pollutant (Teixeir et al. 2009; Muramoto et al. 2003; Sánchez-Ccoyllo et al. 2006). São Paulo metropolitan region has more than 17 million inhabitants and about 6 million registered vehicles, which are the main sources of air pollution, followed by industrial processes, waste burning and fuel storage. In 2001, the State of São Paulo O_3 standard (set at 82 nl l^{-1} for 1 h) established by the Brazilian National Ambient Air Quality Monitoring Program was exceeded 22 % of the time, and several times exceeded the attention level of 102 ppb, reaching maximum hourly O_3 levels of 180 nl l^{-1}. Exceedances last from August through March and occur during warm days with relative humidity of less than 60 %, being common conditions for most of the year.

Long-range O_3 transport has been measured within Brazilian territory throughout latitudes of 35–5 S (Boian and Kirchhoff 2005). In Paraná State of Southern Brazil, high O_3 events with maximum levels of 93–173 nl l^{-1} (average around 89 nl l^{-1}) were observed during the dry season. Such episodes were a consequence of transport from regions with intense biomass burning, as Para (Amazonian Rain Forest region), Mato Grosso, Mato Grosso do Sul and Tocantins (Savanna-Rain forest transition regions). Given the high levels of O_3 and precursors measured in the urban areas of Africa and Brazil, deforestation and seasonal biomass burning were identified as the most relevant pollution sources. Actually, 75 % of the total amounts of CO_2, CO and NO_x released in Brazil are thought to originate from the conversion of native forests into pasture and crop land.

In addition to the related loss of vegetation cover due to fires, it is seems that the remaining vegetation suffers from the associated high levels of O_3. Responses of tree species of the Atlantic Rain Forest to elevated O_3 levels have been based on macroscopic foliar injury, relative growth rate, carbon assimilation, stomatal conductance, superoxide dismutase activity, ascorbic acid as well as anthocyanins and tannins concentrations (e.g., Pina and Moraes 2010; Bulbovas et al. 2010; Rezende and Furlan 2009; Furlan et al. 2007, 2008; Moraes et al. 2006). The native tree species have proven as sensitive to current and predicted O_3 levels in southeastern Brazil, and as suitable bio-indicators of O_3 impact in the tropics.

7.5   Capacities for Adaptation in Forest Trees

Trees are a story of success in evolution. Since about 400 Ma, the geological era of Silurian, trees have existed on Earth, recorded already as fossils shortly after vascular plants had left the swamps, conquering continental land surface. The evolutionary secret of success is their capability of forming stems, overtopping other kinds of plants and extending into crowns that limit or even expel competitors by casting shade. Plants with such a capability early became superior, as competition for light is crucial for plant survival. Given the existence over many millions of years, the tree life form has withstood all environmental challenges and catastrophes that higher terrestrial life needed to overcome, although the appearance and anatomy of trees has gone through many modifications and ecological specializations, and through this, achieving adaptive diversification. One may conclude that trees are resilient in surviving environmental stress, arranging with it through adaptation (Niinemets 2010).

What is true for the entire life form also holds for the individual tree's capability of stress endurance. This is crucial for every plant, being bound to its site without an option of escape. However, stress endurance is of paramount importance for trees, given their prolonged lifetimes, often spanning centuries or even millennia. Without such a capability, evolution would never have succeeded with trees, as we know them today. Given such an evolutionary perspective, the chances of trees should be good in coping with – and adapting to – climate change and air pollution.

Although conclusions of such kind appear to be compelling at a first glance, one important point must be mentioned: It is the rapidity of the current progression in environmental change indicated by growth trends (Pretzsch 1996) as compared to the mostly prolonged environmental transitions, stretching over thousands to millions of years during Earth history (neither mass extinctions of species during Earth history necessarily were instantaneous). Given this current rapidity, a genetic fixation of new adaptations to changed site conditions cannot become effective, although the least adaptive individuals of present populations may be lost on a short-term scale (Miller and McBride 1999). At best, the capacity of trees in physiological acclimation may warrant survival – which, conversely, may be a strength of trees (Niinemets 2010).

Regarding O_3 impact as an oxidative stressor, trees may well be prepared. Ground-level O_3 in low concentration has always been an ecological factor in natural environments, causing oxidative stress, similar to that occurring through energetic overflow during photosynthesis, or under pathogen impact or drought. Plants have evolved defence against oxidative stress, with the kind of stressor rather being a secondary issue (Iriti and Faoro 2009) so that the crucial question focuses on whether or when the defence capacity may collapse.

As pointed out in this chapter, defence capacity as reflected by stress tolerance is a matter of the ecological scenario that the tree is part of, as well as physiological performance that the tree is able to sustain. Therefore, understanding the stress tolerance of trees never can be linked to mono-factorial causes and explanations, but results from multi-factorial analysis (Niinemets 2010). The factorial concert comprises tree-internal processes (genotype, metabolism, ontogenetic and phenological status) as well as abiotic (climate, soil) and biotic influences (competition, parasitism, mutualism) – in particular, the latter having been neglected in research. Also other pollutants are part of this concert: N deposition and nutrient and water availabilities, irradiance, and elevated CO_2. Therefore, the trees' stress response and buffering capacity as outcomes from multiple interactions, metabolic plasticity, and as determinants of stress tolerance and survival can be quite variable and hard to predict, in particular, for new scenarios of climate change. Stressors may be of variable relevance depending on tree ontogeny, while trees may become buffered against stress by accumulating reserve storage for defence as growing old and large (Niinemets 2010). In any case, altered tree performance does affect competitive interactions and food webs through tree-pathogen and -herbivore relationships with potential consequences for species successions and structure in plant communities, and hence, biodiversity. During such processes, trees may not be killed by one factor instantaneously, but as a gradual outcome from specific multiple stressors, as some factors may predispose to – or even harden against – the impact of others. Hence, the interactive scenario may be more crucial for tree survival than the impact of single factors per se, while acting ecophysiological mechanisms are similar to those during evolution, currently involving processes, however, of climate change, including air pollution.

As the predictability of tree performance is limited because of the complexity of the ecologically relevant mechanisms pointed out above, difficulties even increase, if episodic events in environmental impact occur (e.g. severe drought, late or early frost, pest calamities). Modelling will fail, given such unforeseen coincidental effects, in predicting the subsequent development and in giving a long-term perspective. In total, however, one may conclude, that trees are not principally endangered by climate change and air pollution. They have ample capacities of acclimation. Also a changing environment will provide sites suitable for tree existence, provided species migration *via* seed proliferation can keep pace with the environmental change, whereas previous sites may be given up. Pre-requisites for survival, acclimation and potential long-term adaptation need to be viewed in dependence on the specific case as being embedded into the overall factorial scenario of given ecosystems. It is likely, however, that the structure and function-ality of forest ecosystems, declining at some sites while establishing at others, will differ profoundly under climate change including air pollution from what we are presently acquainted at the regional scale, including "deterioration" under human view, based on our current and historical ecological experience.

References

Aber JD, McDowell W, Nadelhoffer K, Magill A, Berntsen G, Kamakea M, McNulty S, Currie W, Rustad L, Fernandez I (1998) Nitrogen saturation in temperate forest ecosystems: hypothesis revisited. BioScience 48:921–934

Akimoto H (2003) Global air quality and pollution. Science 302:1716

Alonso R, Elvira S, Castillo FJ, Gimeno BS (2001) Interactive effects of ozone and drought stress on pigments and activities of antioxidative enzymes in *Pinus halepensis*. Plant Cell Environ 24:905–916

Alscher RG, Amundson RG, Cumming JR, Fellows S, Fincher J, Rubin G, Van Leuken P, Weinstein LH (1989) Seasonal changes in pigments, carbohydrates and growth of red spruce as affected by ozone. New Phytol 113:211–223

Andersen CP (2003) Source-sink balance and carbon allocation below ground in plants exposed to ozone. New Phytol 157:213–228

Andersen CP, Rygiewicz PT (1991) Stress interactions and mycorrhizal plant response: under-standing carbon allocation priorities. Environ Pollut 73:217–244

Andersen CP, Scagel CF (1997) Nutrient availability alters belowground respiration of ozone-exposed ponderosa pine. Tree Physiol 17:377–387

Andersen CP, Hogsett WE, Wessling R, Plocher M (1991) Ozone decreases spring root growth and root carbohydrate content in ponderosa pine the year following exposure. Can J For Res 21:1288–1291

Andreae MO, Rosenfeld D, Artaxo P, Costa AA, Frank GP, Longo KM, Silva-Dias MAF (2004) Smoking rain clouds over the Amazon. Science 303:1337–1342

Atkinson R, Arey J (1998) Atmospheric chemistry of biogenic organic compounds. Account Chem Res 31:574–583

Bahnweg G, Heller W, Stich S, Knappe C, Betz G, Heerdt C, Kehr RD, Ernst D, Langebartels C, Nunn AJ, Rothenburger J, Schubert R, Wallis P, Muller-Starck G, Werner H, Matyssek R, Sandermann H (2005) Beech leaf colonization by the endophyte *Apiognomonia errabunda* dramatically depends on light exposure and climatic conditions. Plant Biol 7:659–669

Barbo DN, Chappelka AH, Somers GL, Miller-Goodman MS, Stolte K (1998) Diversity of an early successional plant community as influenced by ozone. New Phytol 138:653–662

Barbo DN, Chappelka AH, Somers GL, Miller-Goodman MS, Stolte K (2002) Ozone impacts on loblolly pine (Pinus taeda L.) grown in a competitive environment. Environ Pollut 116:27–36

Bardossy A, Caspary HJ (1990) Detection of climate change in Europe by analyzing European atmospheric circulation patterns from 1881 to 1989. Theor Appl Climatol 42:155–167

Barnes JD, Eamus D, Brown KA (1990a) The influence of ozone, acid mist and soil nutrient status on Norway spruce (*Picea abies* (L.) Karst): I. Plant-water relations. New Phytol 114:713–720

Barnes JD, Eamus D, Brown KA (1990b) The influence of ozone, acid mist and soil nutrient status on Norway spruce (*Picea abies* (L.) Karst): II. Photosynthesis, dark respiration and soluble carbohydrates of trees during late autumn. New Phytol 115:149–156

Barnes JD, Pfirrmann T, Steiner K, Lütz C, Busch U, Küchenhoff H, Payer H-D (1995) Effects of elevated CO_2, O_3, and K deficiency on Norway spruce (*Picea abies* [L.] Karst.). II. Seasonal changes in photosynthesis and non-structural carbohydrate content. Plant Cell Environ 18:1345–1357

Bell N, Heard DE, Pilling MJ, Tomlin AS (2003) Atmospheric lifetime as a probe of radical chemistry in the boundary layer. Atmos Environ 37:2193–2205

Bennett JP, Rassat P, Berrang P, Karnosky DF (1992) Relationships between leaf anatomy and ozone sensitivity of *Fraxinus pennsylvanica* Marsh. and *Prunus serotina* Ehrh. J Exp Bot 32:33–41

Bertschi IT, Jaffe DA (2005) Long-range transport of ozone, carbon monoxide, and aerosols in the NE Pacific troposphere during the summer of 2003: observations of smoke plumes from Asian boreal fires. J Geophys Res 110:D05303

Bielenberg DG, Lynch JP, Pell EJ (2001) A decline in nitrogen availability affects plant response to ozone. New Phytol 151:413–425

Biesenthal TA, Wu TAQ, Shepson PB, Wiebe HA, Anlauf KG, MacKay GI (1997) A study of relationships between isoprene, its oxidation products, and ozone, in the lower Fraser valley, BC. Atmos Environ 31:2049–2058

Blande JD, Holopainen JK, Li T (2010) Air pollution impedes plant-to-plant communication by volatiles. Ecol Lett 13:1172–1181

Blumenröther MC, Löw M, Matyssek R, Orwald W (2007) Flux-based response of sucrose and starch in leaves of adult beech trees (*Fagus sylvatica* L.) under chronic free-air O_3 fumigation. Plant Biol 9:207–214

Boian C, Kirchhoff VWJH (2005) Surface ozone enhancements in the south of Brazil owing to large-scale air mass transport. Atmos Environ 39:6140–6146

Bonello P, Heller W, Sandermann H (1993) Ozone effects on root-disease susceptibility and defence responses in mycorrhizal and non-mycorrhizal seedlings of Scots pine (*Pinus sylvestris* L.). New Phytol 124:653–663

Bowman DM, Balch JS, Artaxo JK, Bond P, Carlson WJ, Cochrane JM, D'Antonio MA, DeFries CM, Doyle RS, Harrison JC, Johnston SP, Keeley FH, Krawchuk IF, Kull MA, Marston CA, Moritz JB, Prentice MA, Roos CI, Scott AC, Swetnam TW, van der Werf GR, Pyne SJ (2009) Fire in the earth system. Science 324:481–484

Brilli F, Barta C, Fortunati A, Lerdau M, Loreto F, Centritto M (2007) Response of isoprene emission and carbon metabolism to drought in white poplar (*Populus alba*) saplings. New Phytol 175:244–254

Broadmeadow MSJ, Heath J, Randle TJ (1999) Environmental limitations to O_3 uptake – some key results from young trees growing at elevated CO_2 concentrations. Water Air Soil Pollut 116:99–310

Bulbovas P, Moraes RM, Rinaldi MCS, Cunha AL, Delitti WBC, Domingos M (2010) Leaf antioxidant fluctuations and growth responses in saplings of *Caesalpinia echinata* Lam. (Brazilwood) under an urban stressing environment. Ecotox Environ Safe 73:664–670

Calatayud V, Marco F, Cervero J, Sánchez-Peña G, Sanz MJ (2010) Contrasting ozone sensitivity in related evergreen and deciduous shrubs. Environ Pollut 158:3580–3587

Ceulemans R, Janssens IA, Jach ME (1999) Effects of CO_2 enrichment on trees and forests: lessons to be learned in view of future ecosystem studies. Ann Bot 84:577–590

Chappelka AH, Kush JS, Meldahl RS, Lockaby BG (1990) An ozone-low temperature interaction in loblolly pine (*Pinus taeda* L.). New Phytol 114:721–726

Ciais P, Reichstein M, Viovy N, Granier A, Ogée J, Allard V, Aubinet M, Buchmann N, Bernhofer C, Carrara A, Chevallier F, De Noblet N, Friend AD, Friedlingstein P, Gruenwald T, Heinesch B, Keronen P, Knohl A, Krinner G, Loustau D, Manca G, Matteucci G, Miglietta F, Ourcival JM, Papale D, Pilegaard K, Rambal S, Seufert G, Soussana JF, Sanz MJ, Schulze ED, Vesala T, Valentini R (2005) Europe-wide reduction in primary productivity caused by the heat and drought in 2003. Nature 437:529–534

Coleman MD, Dickson RE, Isebrands JG, Karnosky DF (1995) Carbon allocation and partitioning in aspen clones varying in sensitivity to tropospheric ozone. Tree Physiol 15:721–726

De Vries W, Posch M (2011) Modelling the impact of nitrogen deposition, climate change and nutrient limitations on tree carbon sequestration in Europe for the period 1900–2050. Environ Pollut 159:2289–2299

Demirbas A (2008) Biofuels sources, biofuel policy, biofuel economy and global biofuel projections. Energy Convers Manag 49:2106–2116

Dentener F, Stevenson D, Ellingsen K, van Noije T, Schultz M, Amann M, Atherton C, Bell N, Bergmann D, Bey I, Bouwman L, Butler T, Cofala J, Collins B, Drevet J, Doherty R, Eickhout B, Eskes H, Fiore A, Gauss M, Hauglustaine D, Horowitz L, Isaksen ISA, Josse B, Lawrence M, Krol M, Lamarque JF, Montanaro V, Muller JF, Peuch VH, Pitari G, Pyle J, Rast S, Rodriguez J, Sanderson M, Savage NH, Shindell D, Strahan S, Szopa S, Sudo K, Van Dingenen R, Wild O, Zeng G (2006) The global atmospheric environment for the next generation. Environ Sci Technol 40:3586–3594

Derwent RG, Stevenson DS, Collins WJ, Johnson CE (2004) Intercontinental transport and the origins of the ozone observed at surface sites in Europe. Atmos Environ 38:1891–1901

Dicke M, Baldwin IT (2010) The evolutionary context for herbivore-induced plant volatiles: beyond the cry for help. Trends Plant Sci 15:167–175

Dizengremel P, Le Thiec D, Hasenfratz-Sauder MP, Vaultier MN, Bagard M, Jolivet Y (2009) Metabolic-dependent changes in plant cell redox power after ozone exposure. Plant Biol 11:35–42

Durka W, Schulze ED, Gebauer G, Voerkelius S (1994) Effects of forest decline on uptake and leaching of deposited nitrate determined from ^{15}N and ^{18}O measurements. Nature 372:765–767

Einig W, Lauxmann U, Hauch B, Hampp R, Landolt W, Maurer S, Matyssek R (1997) Ozone-induced accumulation of carbohydrates changes enzyme activities of carbohydrate metabolism in birch leaves. New Phytol 137:673–680

Emberson LD, Ashmore MR, Murray F, Kuylenstierna JCI, Percy KE, Izuta T, Zheng Y, Shimizu H, Sheu BH, Liu CP, Agrawal M, Wahid A, Abdel-Latif NM, van Tienhoven M, de Bauer LI, Domingos M (2001) Impacts of air pollutants on vegetation in developing countries. Water Air Soil Pollut 130:107–118

Ericsson T, Rytter L, Vapaavuori E (1996) Physiology and allocation in trees. Biomass Bioenergy 11:115–127

Fabian P (2002) Leben im Treibhaus – Unser Klimasystem und was wir daraus machen. Springer, Berlin, p 258

Fiore AM, Jacob DJ, Bey I, Yantosca RM, Field BD, Fusco AC, Wilkinson JG (2002) Background ozone over the United States in summer: origin, trend, and contribution to pollution episodes. J Geophys Res 107:D4275

Fontaine V, Cabané M, Dizengremel P (2003) Regulation of phosphoenolpyruvate carboxylase in *Pinus halepensis* needles submitted to ozone and water stress. Physiol Plant 117:445–452

Fowler D, Amann M, Anderson R, Ashmore M, Cox P, Depledge M, Derwent D, Grennfelt P, Hewitt N, Hov O, Jenkin M, Kelly F, Liss P, Pilling M, Pyle J, Slingo J, Stefenson D (2008) Ground-level ozone in the 21st century: future trends, impacts and policy implications. The Royal Society Policy Document

Foyer CH, Lelandais M, Edwards EA, Mullineaux PM (1991) The role of ascorbate in plants, interactions with photosynthesis, and regulatory significance. In: Pell E, Steffen K (eds) Active oxygen/oxidative stress and plant metabolism. American Society of Plant Physiologists, Rockville

Fredericksen TS, Joyce BJ, Skelly JM, Steiner KC, Kolb TE, Kouterick KB, Savage JE, Snyder KR (1995) Physiology, morphology, and ozone uptake of leaves of black cherry seedlings, saplings, and canopy trees. Environ Pollut 89:273–283

Fredericksen TS, Skelly JM, Steiner KC, Kolb TE, Kouterick KB (1996a) Size-mediated foliar response to ozone in black cherry trees. Environ Pollut 91:53–63

Fredericksen TS, Skelly JM, Snyder KR, Steiner KC, Kolb TE (1996b) Predicting ozone uptake from meteorological and environmental variables. J Air Waste Manage Assoc 46:464–469

Frieldingstein P, Fung I, Holland E, John J, Brasseur G, Erickson D, Schimel D (1005) On the contribution of CO_2 fertilization to the missing biospheric sink. Glob Biogeochem Cycl 9:541–556

Fuentes JD, Lerdau M, Atkinson R, Baldocchi, Bottenheim JW, Ciccioli P, Lamp C, Geron C, Gu L, Gunether A, Sharkey TD, Stockwell W (2000) Biogenic hydrocarbons in the atmospheric boundary layer: a review. Bull Am Meteorol Soc 81:1537–1575

Furlan CM, Moraes RM, Bulbovas P, Domingos M, Salatino A, Sanz MJ (2007) Psidium guajava 'Paluma' (the guava plant) as a new bio-indicator of ozone in the tropics. Environ Pollut 147:691–695

Furlan CM, Moraes RM, Bulbovas P, Sanz MJ, Domingos M, Salatino A (2008) *Tibouchina pulchra* (Cham.) Cogn., a native Atlantic forest species, as a bio-indicator of ozone: visible injury. Environ Pollut 152:361–365

Galloway JN, Dentener FJ, Capone DG, Boyer EW, Howarth RW, Seitzinger SP, Asner GP, Cleveland CC, Green PA, Holland EA, Karl DM, Michaels AF, Porter JH, Townsend AR, Vörösmarty CJ (2004) Nitrogen cycles: past, present, and future. Biogeochemistry 70:153–226

Gehring CA, Cobb NS, Whitman TG (1997) Three way interactions among ectomycorrhizal mutualists, scale insects, and resistant and susceptible pinyon pines. Amer Nat 149:824–841

Gerant D, Podor M, Grieu P, Afif D, Cornu S, Morabito D, Banvoy J, Robin C, Dizengremel P (1996) Carbon metabolism, enzyme activities and carbon partitioning in *Pinus halepensis* Mill. to mild drought and ozone. J Plant Physiol 148:142

Gessler A, Keitel C, Kreuzwieser J, Matyssek R, Seiler W, Rennenberg H (2007) Potential risks for European beech (*Fagus sylvatica* L.) in a changing climate. Trees 21:1–11

Gloudemans AMS, Krol MC, Meirink JF, de Laat ATJ, van der Werf GR, Schrijver H, van den Broek MMP, Aben I (2006) Evidence for long-range transport of carbon monoxide in the Southern Hemisphere from SCIAMACHY observations. Geophys Res Lett 33:L16807

Grams TEE, Andersen CP (2007) Competition for resources in trees: physiological versus morphological plasticity In: Esser K, Lüttge U, Beyschlag W, Murata J (eds) Progress in botany. Springer-Verlag, Berlin

Grams TEE, Matyssek R (1999) Elevated CO_2 counteracts the limitation by chronic ozone exposure on photosynthesis in *Fagus sylvatica* L.: comparison between chlorophyll fluorescence and leaf gas exchange. Phyton Ann Rei Bot 39:31–40

Grams TEE, Matyssek R (2010) Stable isotope signatures reflect competitiveness between trees under changed CO_2/O_3 regimes. Environ Pollut 158:1036–1042

Grams TEE, Anegg S, Häberle K-H, Langebartels C, Matyssek R (1999) Interactions of chronic exposure to elevated CO_2 and O_3 levels in the photosynthetic light and dark reactions of European beech (*Fagus sylvatica*). New Phytol 144:95–107

Grams TEE, Kozovits AR, Häberle K-H, Matyssek R, Dawson TE (2007) Combining $\delta^{13}C$ and $\delta^{18}O$ analyses to unravel competition, CO_2 and O_3 effects on the physiological performance of different-aged trees. Plant Cell Environ 30:1023–1034

Grebenc T, Kraigher H (2007) Changes in the community of ectomycorrhizal fungi and increased fine root number under adult beech trees chronically fumigated with double ambient ozone concentration. Plant Biol 9:279–287

Greitner CS, Winner WE (1989) Nutrient effects on responses of willow and alder to ozone. In: Olson RK, Lefohn AS (eds) Transaction: effects of air pollution on Western Forests. Air and Waste Management Association, Anaheim

Grulke NE, Balduman L (1999) Deciduous conifers: high N deposition and O_3 exposure effects on growth and biomass allocation in Ponderosa pine. Water Air Soil Pollut 116:235–248

Grulke NE, Johnson R, Esperanza A, Jones D, Nguyen T, Posch S, Tausz M (2003) Canopy transpiration of Jeffrey pine in mesic and xeric microsites: O_3 uptake and injury response. Trees 17:292–298

Grulke NE, Minnich RA, Paine TD, Seybold SJ, Chavez DJ, Fenn ME, Riggan PJ, Dunn A (2009) Air pollution increases forest susceptibility to wildfires: a case study in the San Bernardino Mountains in Southern California. In: Bytnerowicz A, Arbaugh M, Riebau A, Andersen C (eds) Developments in environmental science, vol 8. Elsevier, Oxford/UK

Guenther AB, Hewitt CN, Erickson D, Fall R, Geron C, Graedel T, Harley P, Klinger L, Lerdau M, McKay WA, Pierce T, Scholes B, Steinbrecher R, Tallamraju R, Taylor J, Zimmerman P (1995) A global model of natural volatile organic compound emissions. J Geophys Res 100:8873–8892

Günthardt-Goerg MS, Matyssek R, Scheidegger C, Keller T (1993) Differentiation and structural decline in the leaves and bark of birch (*Betula pendula*) under low ozone concentrations. Trees 7:104–114

Haberer K, Grebenc T, Alexou M, Gessler A, Kraigher H, Rennenberg H (2007) Effects of long-term free-air ozone fumigation on $\delta^{15}N$ and total N in *Fagus sylvatica* and associated mycorrhizal fungi. Plant Biol 9:242–252

Hahlbrock K, Bednarek P, Ciolkowski I, Hamberger B, Heise A, Liedgens H, Logemann E, Nürnberger T, Schmelzer E, Somssich IE, Tan J (2003) Non-self recognition, transcriptional reprogramming, and secondary metabolite accumulation during plant/pathogen interactions. Proc Natl Acad Sci U S A 100:14569–14576

Hättenschwiler S, Schweingruber FH, Körner C (1996) Tree ring responses to elevated CO_2 and increased N deposition in *Picea abies*. Plant Cell Environ 19:1369–1378

Hayes F, Mills G, Harmens H, Norris D (2007) Evidence of widespread ozone damage to vegetation in Europe (1990–2006). Programme Coordinating Centre for the ICP Vegetation, Centre for Ecology and Hydrology, Bangor, http://www.icpvegetation.ceh.ac.uk

Heath RL (1980) Initial events in injury to plants by air pollutants. Annu Rev Plant Physiol 31:395–401

Heath RL, Taylor GE (1997) Physiological processes and plant responses to ozone exposure. In: Sandermann H, Wellburn AR, Heath RL (eds) Forest decline and ozone, a comparison of controlled chamber and field experiments, Ecological Studies 127. Springer-Verlag, Berlin

Herms DA, Mattson WJ (1992) The dilemma of plants: to grow or defend. Q Rev Biol 67:283–335

Hewitt CN, MacKenzie AR, Di Carlo P, Di Marco CF, Dorsey JR, Evans M, Fowler D, Gallagher MW, Hopkins JR, Jones CE, Langford B, Lee JD, Lewis AC, Lim SF, McQuaid J, Misztal P, Moller SJ, Monks PS, Nemitz E, Oram DE, Owen SM, Phillips GJ, Pugh TAM, Pyle JA, Reeves CE, Ryder J, Sion J, Skiba U, Stewart DJ (2009) Nitrogen management is essential to prevent tropical oil palm plantations from causing ground-level ozone pollution. Proc Natl Acad Sci U S A 106:18447–18451

Holopainen JK, Gershenzon J (2010) Multiple stress factors and the emission of plant VOCs. Trends Plant Sci 15:176–184

IPCC (2007) Climate change 2007: synthesis report. Summary for policymakers. WG1, AR4. http://ipcc-wg1.ucar.edu/wg1/wg1-report.html

Iriti M, Faoro F (2009) Chemical diversity and defence metabolism: how plants cope with pathogens and ozone pollution. Int J Mol Sci 10:3371–3399

Jehnes S, Betz G, Bahnweg G, Haberer K, Sandermann H, Rennenberg H (2007) Tree internal signalling and defence reactions under ozone exposure in sun and shade leaves of European Beech (*Fagus sylvatica* L.) trees. Plant Biol 9:253–264

Josipovic M, Annegarn HJ, Kneen MA, Piernaar JJ, Piketh SJ (2010) Concentrations, distributions and critical level exceedance assessment of SO_2, NO_2 and O_3 in South Africa. Environ Monit Assess 171:181–196

Kangasjärvi J, Jaspers P, Kollist H (2005) Signalling and cell death in ozone-exposed plants. Plant Cell Environ 28:1021–1036

Karlsson PE, Medin E-L, Wickström H, Selldén G, Wallin G, Ottoson S, Skärby L (1995) Ozone and drought stress: interactive effects on the growth and physiology of Norway spruce (*Picea abies* L. Karst.). Water Air Soil Pollut 85:1325–1330

Karlsson PE, Pleijel H, Karlsson GP, Medin EL, Skärby L (2000) Simulations of stomatal conductance and ozone uptake to Norway spruce saplings in open-top chambers. Environ Pollut 109:443–451

Karlsson PE, Pleijel H, Simpson D (2009) Ozone exposure and impacts on vegetation in the Nordic and Baltic countries. Ambio 38:402–405

Karnosky DF, Gagnon ZE, Dickson RE, Coleman MD, Lee EH, Isebrands JG (1996) Changes in growth, leaf abscission, and biomass associated with seasonal tropospheric ozone exposure of Populus tremuloides clones and seedlings. Can J For Res 26:23–37

Karnosky DF, Mankovska B, Percy K, Dickson DE, Podila GK, Sober A, Noormets G, Hendrey MD, Coleman M, Kubiske M, Pregitzer KS, Isebrands JG (1999) Effects of tropospheric O_3 on trembling aspen and interaction with CO_2: results from a O_3 gradient and a FACE experiment. Water Air Soil Pollut 116:311–322

Karnosky DE, Oksanen E, Dickson RE, Isebrands JG (2001) Impacts of interacting greenhouse gases on forest ecosystems. In: Karnosky DF, Scarascia-Mugnozza G, Ceulemans R, Innes JL (eds) The impacts of carbon dioxide and other greenhouse gases on forest ecosystems. CABI Press, Wallingford

Karnosky DF, Percy KE, Xiang BX, Callan B, Noormets A, Mankovska B, Hopkin A, Sober J, Jones W, Dickson RE, Isebrands JG (2002) Interacting elevated CO_2 and tropospheric O_3 predisposes aspen (*Populus tremuloides* Michx.) to infection by rust (*Melampsora medusae f. sp tremuloidae*). Glob Change Biol 8:329–338

Karnosky DF, Zak D, Pregnitzer K, Awmack C, Bockheim J, Dickson R, Hendrey G, Host G, King J, Kopper B, Kruger E, Kubiske M, Lindroth R, Mattson W, McDonald E, Noormets A, Oksanen E, Parsons W, Percy K, Podila G, Riemenschneider D, Sharma P, Thakur R, Sober A, Sober J, Jones W, Anttonen S, Vapaavuori E, Manskovska B, Heilman W, Isebrands J (2003) Tropospheric O_3 moderates responses of temperate hardwood forests to elevated CO_2: a synthesis of molecular to ecosystem results from the Aspen FACE project. Funct Ecol 17:289–304

Karnosky DF, Pregitzer KS, Zak DR, Kubiske ME, Hendrey GR, Weinstein D, Nosal M, Percy KE (2005) Scaling ozone responses of forest trees to the ecosystem level in a changing climate. Plant Cell Environ 28:965–981

Karnosky DF, Werner H, Holopainen T, Percy K, Oksanen T, Oksanen E, Heerdt C, Fabian P, Nagy J, Heilman W, Cox R, Nelson N, Matyssek R (2007) Free-Air exposure systems to scale up ozone research to mature trees. Plant Biol 9:181–190

Keating TW, West JJ, Farrell AE (2004) Prospects for international management of inter-continental air pollution transport. In: Stohl A (ed) The intercontinental transport of air pollution. Springer, Berlin

Keller T, Häsler R (1987) The influence of a fall fumigation with ozone on the stomatal behavior of spruce and fir. Oecologia 64:284–286

Keppler F, Hamilton JTG, Brar M, Röckmann T (2006) Methane emissions from terrestrial plants under aerobic conditions. Nature 439:187–191

Keppler F, Boros M, Frankenberg C, Lelieveld J, McLeod A, Pirttilä AM, Röckmann T, Schnitzler JP (2009) Methane formation in aerobic environments. Environ Chem 6:459–465

Kesselmeier J, Staudt M (1999) Biogenic volatile organic compounds (VOC): an overview on emission, physiology and ecology. J Atmos Chem 33:23–88

Kiendler-Scharr A, Wildt J, Dal Maso M, Hohaus T, Kleist E, Mentel TF, Tillmann R, Uerlings R, Schurr U, Wahner A (2009) New particle formation in forests inhibited by isoprene emissions. Nature 461:381–384

King JS, Kubiske ME, Pregitzer KS, Hendrey GR, McDonald EP, Giardina CP, Quinn VS, Karnosky DF (2005) Tropospheric ozone compromises net primary production in young stands of trembling aspen, paper birch, and sugar maple in response to elevated CO_2. New Phytol 168:623–636

Kitao M, Löw M, Heerd C, Grams TEE, Häberle K-H, Matyssek R (2009) Effects of chronic elevated ozone exposure on gas exchange responses of adult beech trees (*Fagus sylvatica*) as related to the within-canopy light gradient. Environ Pollut 157:537–544

Kivimäenpää M, Sutinen S, Karlsson PE, Sellden G (2003) Cell structural changes in the needles of Norway spruce exposed to long-term ozone and drought. Ann Bot 92:779–793

Klingberg J, Engardt M, Uddling J, Karlsson PE, Pleijel H (2010) Ozone risk for vegetation in the future climate of Europe based on stomatal ozone uptake calculations. Tellus A 63:174–187

Kolb TE, Matyssek R (2001) Limitations and perspectives about scaling ozone impact in trees. Environ Pollut 115:373–393

Kolb TE, Fredericksen TS, Steiner KC, Skelly JM (1997) Issues in scaling tree size and age responses to ozone: a review. Environ Pollut 98:195–208

Kopper BJ, Lindroth RL (2002) Effects of elevated carbon dioxide and ozone on the phytochemistry of aspen and performance of an herbivore. Oecologia 134:95–103

Körner C (2003) Carbon limitation in trees. J Ecol 91:4–17

Körner C (2006) Plant CO_2 responses: an issue of definition, time and resource supply. New Phytol 172:393–411

Kozovits AR, Matyssek R, Blaschke H, Göttlein A, Grams TEE (2005a) Competition increasingly dominates the responsiveness of juvenile beech and spruce to elevated CO_2 and/or O_3 concentrations throughout two subsequent growing seasons. Glob Change Biol 11:1387–1401

Kozovits AR, Matyssek R, Winkler B, Göttlein A, Blaschke H, Grams TEE (2005b) Aboveground space sequestration determines competitive success in juvenile beech and spruce trees. New Phytol 167:181–196

Kreuzwieser J, Gessler A (2010) Global climate change and tree nutrition: influence of water availability. Tree Physiol 30:1221–1234

Kronfuß G, Polle A, Tausz M, Havranek WM, Wieser G (1998) Effects of ozone and mild drought stress on gas exchange, antioxidants and chloroplast pigments in current-year needles of young Norway spruce [*Picea abies* (L.) Karst.]. Trees 12:482–489

Kubiske ME, Quinn VS, Heilman WE, McDonald EP, Marquardt PE, Teclaw RM, Friend AL, Karnosky DF (2006) Interannual climatic variation mediates elevated CO_2 and O_3 effects on forest growth. Glob Change Biol 12:1054–1068

Kubiske ME, Quinn VS, Marquardt PE, Karnosky DF (2007) Effects of elevated atmospheric CO_2 and/or O_3 on intra- and interspecific competitive ability of aspen. Plant Biol 9:342–355

Kytöviita M-M, Pelloux J, Fontaine V, Botton B, Dizengremel P (1999) Elevated CO_2 does not amerliorate effects of ozone on carbon allocation in *Pinus halepensis* and *Betula pendula* in symbiosis with *Paxillus involutus*. Physiol Plant 106:370–377

Lal R (1999) World soils and the greenhouse effect. IGBP Newslett 37:4–5

Landolt W, Pfenninger I, Lüthy-Krause B (1989) The effect of ozone and season on the pool sizes of cyclitols in Scots pine (*Pinus sylvestris*). Trees 3:85–88

Langebartels C, Ernst D, Heller W, Lütz C, Payer H-D, Sandermann H Jr (1997) Ozone responses of trees: results from controlled chamber exposures at the GSF phytotron. In: Sandermann H Jr, Wellburn AR, Heath RL (eds) Forest decline and ozone. Springer, Berlin

Langmann B, Duncan B, Textor C, Trentmann J, van der Werf GR (2009) Vegetation fire emissions and their impact on air pollution and climate. Atmos Environ 43:107–116

Laothawornkitkul J, Taylor JE, Paul ND, Hewitt CN (2009) Biogenic volatile organic compounds in the Earth system. New Phytol 183:27–51

Laurence JA, Amundson RG, Friend AL, Pell EJ, Temple PJ (1994) Allocation of carbon in plants under stress: an analysis of the ROPIS experiment. J Environ Qual 23:412–417

Lautner S, Ehlting B, Windeisen E, Rennenberg H, Matyssek R, Fromm J (2007) Calcium nutrition has a significant influence on wood formation in poplar. New Phytol 173:743–752

Lindroth RL (2010) Impacts of elevated atmospheric CO_2 and O_3 on forests: phytochemistry, trophic interactions, and ecosystem dynamics. J Chem Ecol 36:2–21

Lippert M, Steiner K, Payer H-D, Simons S, Langebartels C, Sandermann H Jr (1996) Assessing the impact of ozone on photosynthesis of European beech (*Fagus sylvatica* L.) in environmental chambers. Trees 10:268–275

Loreto F, Schnitzler JP (2010) Abiotic stresses and induced BVOCs. Trends Plant Sci 15:154–166

Loreto F, Velikova V (2001) Isoprene produced by leaves protects the photosynthetic apparatus against ozone damage, quenches ozone products, and reduces lipid peroxidation of cellular membranes. Plant Physiol 127:1781–1787

Löw M, Herbinger K, Nunn AJ, Häberle K-H, Leuchner M, Heerdt C, Werner H, Wipfler P, Pretzsch H, Tausz M, Matyssek R (2006) Extraordinary drought of 2003 overrules ozone impact on adult beech trees (*Fagus sylvatica*). Trees 20:539–548

Löw M, Häberle K-H, Warren CR, Matyssek R (2007) O_3 flux-related responsiveness of photosynthesis, respiration, and stomatal conductance of adult *Fagus sylvatica* to experimentally enhanced free-air O_3 exposure. Plant Biol 9:197–206

Luedemann G, Matyssek R, Fleischmann F, Grams TEE (2005) Acclimation to ozone affects host/pathogen interaction and competitiveness for nitrogen in juvenile *Fagus sylvatica* and *Picea abies* trees infected with *Phytophthora citricola*. Plant Biol 7:640–649

Luedemann G, Winkler JB, Matyssek R, Grams TEE (2009) Contrasting ozone x pathogen interaction as mediated through competition between juvenile European beech (*Fagus sylvatica*) and Norway spruce (*Picea abies*). Plant Soil 323:47–60

Lütz C, Anegg S, Gerant D, Alaoui-Sosse B, Gerard J, Dizengremel P (2000) Beech trees exposed to high CO_2 and to simulated summer ozone levels: effects on photosynthesis, chloroplast components and leaf enzyme activity. Physiol Plant 109:252–259

Luyssaert S, Schulze ED, Börner A, Knohl A, Hessenmöller D, Law BE, Grace J, Ciais P (2008) Old-growth forests as global carbon sinks. Nature 455:213–215

Maier-Maercker U (1998) Image analysis of the stomatal cell walls of *Picea abies* (L.) Karst. in pure and ozone-enriched air. Trees 12:181–185

Manes F, Vitale M, Donato E, Paoletti E (1998) O_3 and $O_3 + CO_2$ effects on a Mediterranean evergreen broadleaf tree, Holm Oak (*Quercus ilex* L.). Chemosphere 36:801–806

Mankovska B, Percy KE, Karnosky DF (2005) Impacts of greenhouse gases on epicuticular waxes of *Populus tremuloides* Michx.: results from an open-air exposure and a natural O_3 gradient. Environ Pollut 137:580–586

Manning WJ, v Tiedemann A (1995) Climate change: potential effects of increase atmospheric carbon dioxide (CO_2), ozone (O_3), and ultraviolet-B (UV-B) radiation on plant diseases. Environ Pollut 88:219–225

Marscher H (1991) Mechanisms of adaptation of plants to acid soils. Plant Soil 134:1–20

Matyssek R (2001) How sensitive is birch to ozone? Responses in structure and function. J For Sci 47:8–20

Matyssek R, Innes JL (1999) Ozone – a risk factor for trees and forests in Europe? Water Air Soil Pollut 116:199–226

Matyssek R, Sandermann H (2003) Impact of ozone on trees: an ecophysiological perspective. Prog Bot 64:349–404

Matyssek R, Reich PB, Oren R, Winner WE (1995) Response mechanisms of conifers to air pollutants. In: Smith WK, Hinckley TH (eds) Physiological ecology of coniferous forests. Physiological Ecology Series, Academic Press, New York

Matyssek R, Havranek WM, Wieser G, Innes JL (1997a) Ozone and the forests in Austria and Switzerland. In: Sandermann H Jr, Wellburn AR, Heath RL (eds) Forest decline and ozone: a comparison of controlled chamber and field experiments, Ecological Studies 127. Springer, Berlin

Matyssek R, Maurer S, Günthardt-Goerg MS, Landolt W, Saurer M, Polle A (1997b) Nutrition determines the 'strategy' of *Betula pendula* for coping with ozone stress. Phyton Ann Rei Bot 37:157–167

Matyssek R, Agerer R, Ernst D, Munch J-C, Osswald W, Pretzsch H, Priesack E, Schnyder H, Treutter D (2005) The plant's capacity in regulating resource demand. Plant Biol 7:560–580

Matyssek R, Le Thiec D, Löw M, Dizengremel P, Nunn AJ, Häberle K-H (2006) Interaction between drought stress and O_3 stress in forest trees. Plant Biol 8:11–17

Matyssek R, Sandermann H, Wieser G, Booker F, Cieslik S, Musselman R, Ernst D (2008) The challenge of making ozone risk assessment for forest trees more mechanistic. Environ Pollut 156:567–582

Matyssek R, Karnosky DF, Wieser G, Percy K, Oksanen E, Grams TEE, Kubiske M, Hanke D, Pretzsch H (2010a) Advances in understanding ozone impact on forest trees: messages from novel phytotron and free-air fumigation studies. Environ Pollut 158:1990–2006

Matyssek R, Wieser G, Ceulemans R, Rennenberg H, Pretzsch H, Haberer K, Löw M, Nunn AJ, Werner H, Wipfler P, Oßwald W, Nikolova P, Hanke DE, Kraigher H, Tausz M, Bahnweg G, Kitao M, Dieler J, Sandermann H, Herbinger K, Grebenc T, Blumenröther M, Deckmyn G, Grams TEE, Heerdt C, Leuchner M, Fabian P, Häberle K-H (2010b) Enhanced ozone strongly reduces carbon sink strength of adult beech (*Fagus sylvatica*) – resume from the free-air fumigation study at Kranzberg Forest. Environ Pollut 158:2527–2532

Maurer S, Matyssek R (1997) Nutrition and the ozone sensitivity of birch (*Betula pendula*), II. Carbon balance, water-use efficiency and nutritional status of the whole plant. Trees 12:11–20

Maurer S, Matyssek R, Günthhardt-Goerg MS, Landolt W, Einig W (1997) Nutrition and the ozone sensitivity of birch (*Betula pendula*). I Responses at the leaf level. Trees 12:1–10

McDonald EP, Kruger EL, Riemenschneider DE, Isebrands JG (2002) Competitive status influences tree-growth responses to elevated CO_2 and O_3 in aggrading aspen stands. Funct Ecol 16:792–801

Mieville A, Granier C, Liousse C, Guillaume B, Mouillot F, Lamarque JF, Grégoire JM, Pétron G (2010) Emissions of gases and particles from biomass burning during the 20th century using satellite data and an historical reconstruction. Atmos Environ 44:1469–1477

Mikkelsen TN, Ro-Poulsen H (1995) Exposure of Norway spruce to ozone increases the sensitivity of current-year needles to photoinhibitions and desiccation. New Phytol 128:153–163

Millenium Ecosystem Assessment (2005) Ecosystems and human well-being: synthesis. Island Press, Washington

Miller PR, McBride JM (1999) Oxidant air pollution impacts in the montane forest of Southern California, Ecological Studies 134. Springer, Berlin

Monks PS, Granier C, Fuzzi S, Stohl A et al (2009) Atmospheric composition change – global and regional air quality. Atmos Environ 43:5268–5350

Monson RK, Jaeger CH, Adams WW, Driggers EM, Silver GM, Fall R (1992) Relationship among isoprene emission rate, photosynthesis, and isoprene synthase activity as influenced by temperature. Plant Physiol 98:1175–1180

Mooney HA, Winner WE (1991) Partitioning response of plants to stress. In: Mooney HA, Winner WE, Pell EJ (eds) Response of plants to multiple stresses. Academic, San Diego

Moraes RM, Bulbovas P, Furlan CM, Domingos M, Meirelles ST, Delitti WBC, Sanz MJ (2006) Physiological responses of saplings of *Caesalpinia echinata* Lam., a Brazilian tree species, under ozone fumigation. Ecotox Environ Safe 63:306–312

Morris GA, Hersey S, Thompson AM, Pawson S, Nielsen JE, Colarco PR, McMillan WW, Stohl A, Turquety S, Warner J, Johnson BJ, Kucsera TL, Larko DE, Olrmans SJ, Witte JC (2006) Alaskan and Canadian forest fires exacerbate ozone pollution over Houston, Texas, on 19 and 20 July 2004. J Geophys Res 111:D24S03

Muramoto CA, Lopes CFF, Lacava CIV (2003) Study of tropospheric ozone in São Paulo metropolitan region. In: A&WMA's 96th annual conference & exhibition "Energy, economic and global challenges", San Diego

Musselman RC, Lefohn AS, Massman WJ, Heath RL (2006) A critical review and analysis of the use of exposure- and flux-based ozone indices for predicting vegetation effects. Atmos Environ 40:1869–1888

Myneni RB, Keeling CD, Tucker CJ, Asrar G, Nemani RR (1997) Increased plant growth in the northern high latitudes from 1981 to 1991. Nature 386:698–702

Nadelhoffer KJ, Emmett BA, Gundersen P, Kjønaas OJ, Koopmans CJ, Schleppi P, Tietema A, Wright RF (1999) Nitrogen deposition makes a minor contribution to carbon sequestration in temperate forests. Nature 398:145–148

Nali C, Paoletti E, Marabottini R, Della Rocca G, Lorenzini G, Paolacci AR, Ciaffi M, Badiani M (2004) Ecophysiological and biochemical, strategies of response to ozone in Mediterranean evergreen broadleaf species. Atmos Environ 38:2247–2257

Newell RE, Evans MJ (2000) Seasonal changes in pollutant transport to the north Pacific: the relative importance of Asian and European sources. Geophys Res Lett 27:2509–2512

Niinemets U (2010) Responses of forest trees to single and multiple environmental stresses from seedlings to mature plants: past stress history, stress interactions, tolerance and acclimation. Forest Ecol Manag 260:1623–1639

Nikolova P, Raspe S, Andersen C, Mainiero R, Blaschke H, Matyssek R, Häberle K-H (2009) Effects of the extreme drought in 2003 on soil respiration in a mixed Forest. Eur J For Res 128:87–98

Nikolova PS, Andersen CP, Blaschke H, Matyssek R, Häberle K-H (2010) Below-ground effects of enhanced tropospheric ozone and drought in a beech/spruce forest (Fagus sylvatica L./ Picea abies [L.] Karst). Environ Pollut 158:1071–1078

Norby RJ, Wullschleger SD, Gunderson CA, Johnson DW, Ceulemans R (1999) Tree response to rising CO_2 in experiments field: implications for the future forests. Plant Cell Environ 22:683–714

Nunn AJ, Reiter IM, Häberle K-H, Werner H, Langebartels C, Sandermann H, Heerdt C, Fabian P, Matyssek R (2002) "Free-air" ozone canopy fumigation in an old-growth mixed forest: concept and observations in beech. Phyton Ann Rei Bot 42:105–119

Nunn AJ, Kozovits AR, Reiter IM, Heerdt C, Leuchner M, Lütz C, Liu X, Winkler JB, Grams TEE, Häberle K-H, Werner H, Fabian P, Rennenberg H, Matyssek R (2005a) Comparison of ozone uptake and responsiveness between a phytotron study with young and a field experiment with adult beech (Fagus sylvatica). Environ Pollut 137:494–506

Nunn AJ, Reiter IM, Haberle K-H, Langebartels C, Bahnweg G, Pretzsch H, Sandermann H, Matyssek R (2005b) Response patterns in adult forest trees to chronic ozone stress: identification of variations and consistencies. Environ Pollut 136:365–369

Oksanen E (2003) Responses of selected birch (Betula pendula Roth) clones to ozone change over time. Plant Cell Environ 26:875–886

Olbrich M, Gerstner E, Bahnweg G, Häberle K-H, Matyssek R, Welzl G, Ernst D (2010) Transcriptional signatures in leaves of adult European beech trees (Fagus sylvatica L.) in an experimentally enhanced free air ozone setting. Environ Pollut 158:977–982

Olcese LE, Toselli BM (1998) Unexpected high levels of ozone measured in Córdoba, Argentina. J Atmos Chem 31:269–279

Pääkkönen E, Holopainen T (1995) Influence of nitrogen supply on the response of clones of birch (Betula pendula Roth.) to ozone. New Phytol 129:595–603

Pääkkönen E, Vahala J, Pohjolai M, Holopainen T, Kärenlampi L (1998a) Physiological, stomatal and ultrastructural ozone responses in birch (Betula pendula Roth) are modified by water stress. Plant Cell Environ 21:671–684

Pääkkönen E, Seppänen S, Holopainen T, Kokko H, Kärenlampi S, Kärenlampi L, Kangasjärvi J (1998b) Induction of genes for the stress proteins PR-10 and PAL in relation to growth, visible injuries and stomatal conductance in birch (Betula pendula) clones exposed to ozone and/or drought. New Phytol 138:295–305

Panek JA, Goldstein AH (2001) Response of stomatal conductance to drought in ponderosa pine: implications for carbon and ozone uptake. Tree Physiol 21:337–344

Panek JA, Kurpius MR, Goldstein AH (2002) An evaluation of ozone exposure metrics for a seasonally drought-stressed ponderosa pine ecosystem. Environ Pollut 117:93–100

Paoletti E (2006) Impact of ozone on Mediterranean forests: a review. Environ Pollut 144:463–474

Pearson M, Mansfield TA (1993) Interacting effects of ozone and water-stress on the stomatal-resistance of Beech (Fagus sylvatica L.). New Phytol 123:351–358

Pelloux J, Jolivet Y, Fontaine V, Banvoy J, Dizengremel P (2001) Changes in Rubisco and Rubisco activate gene expression and polypeptide content in *Pinus halepensis* Mill. subjected to ozone and drought. Plant Cell Environ 24:123–132

Peñuelas J, Staudt M (2010) BVOCs and global change. Trends Plant Sci 15:133–144

Percy KE, Awmack CS, Lindroth RL, Kubiske ME, Kopper BJ, Isebrands JG, Pregitzer KS, Hendrey GR, Dickson RE, Zak DR, Oksanen E, Sober J, Harrington R, Karnosky DF (2002) Altered performance of forest pests under CO_2- and O_3-enriched atmospheres. Nature 420:403–407

Pina JM, Moraes RM (2010) Gas exchange, antioxidants and foliar injuries in saplings of a tropical woody species exposed to ozone. Ecotox Environ Safe 73:685–691

Pinto DM, Blande JD, Souza SR, Nerg A-M, Holopainen JK (2010) Plant volatile organic compounds (VOCs) in ozone (O_3) polluted atmospheres: the ecological effects. J Chem Ecol 36:22–34

Poisson N, Kanakidou M, Crutzen PJ (2000) Impact of non-methane hydrocarbons on tropospheric chemistry and the oxidizing power of the global troposphere: 3-dimensional modelling results. J Atmos Chem 36:157–230

Polle A, Matyssek R, Günthardt-Goerg MS, Maurer S (2000) Defense strategies against ozone in trees: the role of nutrition. In: Agrawal SB, Agrawal M (eds) Environmental pollution and plant responses. Lewis Publishers, New York

Poorter H, Navas ML (2003) Plant growth and competition at elevated CO_2: on winners, losers and functional groups. New Phytol 157:175–198

Pretzsch H (1996) Growth trends in forests in Southern Germany. In: Spiecker H, Mielikäinen K, Köhl M, Skovsgaard JP (eds) Growth trends in european forests. Springer, Berlin

Pretzsch H, Dieler J (2010) The dependency of the size-growth relationship of Norway spruce (*Picea abies* [L.] Karst.) and European beech (*Fagus sylvatica* [L.]) in forest stands on long-term site conditions, drought events, and ozone stress. Trees 25:355–369

Pretzsch H, Dieler J, Matyssek R, Wipfler P (2010) Tree and stand growth of mature Norway spruce and European beech under long-term ozone fumigation. Environ Pollut 158:1061–1070

Pronos J, Merril L, Dahsten D (1999) Insects and pathogens in a pollution-stressed forest. In: Miller PR, McBride JM (eds) Oxidant air pollution impacts in the montane forest of southern California, Ecological Studies 134. Springer-Verlag, Berlin

Real E, Law KS, Weinzierl B, Fiebig M, Petzold A, Wild O, Methven J, Arnold S, Stohl A, Huntrieser H, Roiger A, Schlager H, Stewart D, Avery M, Sachse G, Browell E, Ferrare R, Blake D (2007) Processes influencing ozone levels in Alaskan forest fire plumes during long-range transport over the North Atlantic. J Geophys Res 112:D10S41

Reich PB (1987) Quantifying plant response to ozone: a unifying theory. Tree Physiol 3:63–91

Reilly J, Paltsev S, Felzer B, Wang X, Kicklighter D, Melillo J, Prinn R, Sarofim M, Sokolov A, Wang C (2007) Global economic effects of changes in crops, pasture, and forests due to changing climate, carbon dioxide, and ozone. Ergon Policy 36:5370–5383

Reiter IM, Haberle K-H, Nunn AJ, Heerdt C, Reitmayer H, Grote R, Matyssek R (2005) Competitive strategies in adult beech and spruce: space-related foliar carbon investment versus carbon gain. Oecologia 146:337–349

Retzlaff WA, Arthur MA, Grulke NE, Weinstein DA, Gollands B (2000) Use of a single-tree simulation model to predict effects of ozone and drought on growth of a white fir tree. Tree Physiol 20:195–202

Rezende FM, Furlan CM (2009) Anthocyanins and tannins in ozone-fumigated guava trees. Chemosphere 76:1445–1450

Riefler M, Novak O, Strnad M, Schmülling T (2006) *Arabidopsis* cytokinin receptor mutants reveal functions in shoot growth, leaf senescence, seed size, germination, root development, and cytokinin metabolism. Plant Cell 18:40–54

Rosenfeld D (1999) TRMM observed first direct evidence of smoke from forest fires inhibiting rainfall. Geophy Res Lett 26:3105–3108

Rosenstiel TN, Potosnak MJ, Griffin KL, Fall R, Monson RK (2003) Increased CO_2 uncouples growth from isoprene emission in an agriforest ecosystem. Nature 421:256–259

Sánchez-Ccoyllo OR, Ynoue YR, Martins DL, Andrade FM (2006) Impacts of ozone precursor limitation and meteorological variables on ozone concentration in Sao Paulo, Brazil. Atmos Environ 40:S552–S562

Sandermann H Jr, Wellburn AR, Heath RL (1997) Forest decline and ozone: synopsis. In: Sandermann H Jr, Wellburn AR, Heath RL (eds) Forest decline and ozone: a comparison of controlled chamber and field experiments, Ecological Studies 127. Springer, Berlin

Saugier B, Roy J, Mooney HA (2001) Terrestrial global productivity. Academic, San Diego

Saxe H, Ellsworth DS, Heath J (1998) Tree and forest functioning in an enriched CO_2 atmosphere. New Phytol 139:395–436

Scarascia-Mugnozza GE, Karnosky DF, Ceulemans R, Innes JL (2001) The impact of CO_2 and other greenhouse gases on forest ecosystems: an introduction. In: Karnosky DF, Scarascia-Mugnozza G, Ceulemans R, Innes JL (eds) The impacts of carbon dioxide and other greenhouse gases on forest ecosystems. CABI Press, Washington

Schulze E-D, Lange OL, Oren R (1989) Forest decline and air pollution – a study of spruce (*Picea abies*) on acid soils, Ecological studies 77. Springer, New York, p 475

Schulze E-D (1994) The regulation of plant transpiration: interactions of feedforward, feedback, and futile cycles. In: Schulze E-D (ed) Flux control in biological systems. Academic, New York

Schulze E-D, Beck E, Müller-Hohenstein K (2005) Plant ecology. Springer, Berlin

Schulze E-D, Luyssaert S, Ciais P, Freibauer A, Janssens IA, Soussana JF, Smith P, Grace J, Levin I, Thiruchittampalam B, Heimann M, Dolman AJ, Valentini R, Bousquet P, Peylin P, Peters W, Roedenbeck C, Etiope G, Vuichard N, Wattenbach M, Nabuurs GJ, Poussi Z, Nieschulze J et al (2009) Importance of methane and nitrous oxide for Europe's terrestrial greenhouse-gas balance. Nat Geosci 2:842–850

Schupp R, Rennenberg H (1988) Diurnal changes in the glutathione content of spruce needles (*Picea abies* L.). Plant Sci 57:113–117

Sharkey TD, Chen X, Yeh S (2001) Isoprene increases thermotolerance of fosmidomycin-fed leaves. Plant Physiol 125:2001–2006

Sheppard LJ (1994) Causal mechanisms by which sulphate, nitrate and acidity influence frost hardiness in red spruce. Review and hypothesis. New Phytol 127:69–82

Siegwolf RTW, Matyssek R, Saurer M, Maurer S, Günthardt-Goerg MS, Schmutz P, Bucher JB (2001) Stable isotope analysis reveals differential effects of soil nitrogen and nitrogen dioxide on the water-use efficiency in hybrid poplar leaves. New Phytol 149:233–246

Singh HB, Herlth D, Kolyer R, Chatfield R, Viezee W, Salas LJ, Chen Y, Bradshaw JD, Sandholm ST, Talbot R, Gregory GL, Anderson B, Sachse GW, Browell E, Bachmeier AS, Blake DR, Heikes B, Jacob D, Fuelberg HE (1996) Impact of biomass burning emissions on the composition of the South Atlantic troposphere: reactive nitrogen and ozone. J Geophys Res 101:24203–24219

Sitch S, Cox PM, Collins WJ, Huntingford C (2007) Indirect radiative forcing of climate change through ozone effects on the land-carbon sink. Nature 448:791–795

Skärby L, Ro-Poulsen H, Wellburn FAM, Sheppard LJ (1998) Impacts of ozone on forests: a European perspective. New Phytol 139:109–122

Sowden M, Zunckel M, van Tienhoven AM (2007) Assessment of the status of biogenic organic emissions and impacts on air quality in southern Africa. Tellus 59B:535–541

Spinnler D, Egh P, Körner C (2002) Four-year growth dynamics of beech-spruce model ecosystems under CO_2 enrichment on two different forest soils. Trees 16:423–436

Stockwell WR, Kramm G, Scheel HE, Mohnen VA, Seiler W (1997) Ozone formation, destruction and exposure in Europe and the United States. In: Sandermann H Jr, Wellburn AR, Heath RL (eds) Forest decline and ozone, a comparison of controlled chamber and field experiments, Ecological Studies 127. Springer, Berlin

Stohl A, Eckhardt S, Forster C, James P, Spichtinger N (2002) On the pathways and timescales of intercontinental air pollution transport. J Geophys Res 107:D4684

Takemoto BK, Bytnerowicz A, Fenn ME (2001) Current and future effects of ozone and atmospheric nitrogen deposition on California's mixed conifer forests. Forest Ecol Manag 144:159–173

Teixeir EC, de Santana ER, Wiegand F, Fachel J (2009) Measurement of surface ozone and its precursors in an urban area in South Brazil. Atmos Environ 43:2213–2220

Thompson AM (1992) The oxidizing capacity of the earth's atmosphere: probable past and future change. Science 256:1157–1165

Tjoelker MG, Luxmoore RJ (1991) Soil nitrogen and chronic ozone stress influence physiology, growth and nutrient status of *Pinus taeda* L. and *Liriodendron tulipifera* L. seedlings. New Phytol 119:69–81

Tjoelker MG, Volin JC, Oleksyn J, Reich PB (1995) Interaction of ozone pollution and light effects on photosynthesis in a forest canopy experiment. Plant Cell Environ 18:895–905

Torsethaugen G, Pell EJ, Assmann SM (1999) Ozone inhibits guard cell K^+ channels implicated in stomatal opening. Proc Natl Acad Sci U S A 96:13577–13582

Tunved P, Hansson HC, Kerminen VM, Ström J, Dal Maso M, Lihavainen H, Viisanen Y, Aalto PP, Komppula M, Kulmala M (2006) High natural aerosol loading over boreal forests. Science 312:261–263

Unsicker SB, Kunert G, Gershenzon J (2009) Protective perfumes: the role of vegetative volatiles in plant defense against herbivores. Curr Opin Plant Biol 12:479–485

Van Dingenen R, Dentener FJ, Raes F, Krol MC, Emberson L, Cofala J (2009) The global impact of ozone on agricultural crop yields under current and future air quality legislation. Atmos Environ 43:604–618

Van Tienhoven AM, Scholes MC (2003) Air pollution impacts on vegetation in South Africa. In: Emberson LD, Ashmore MR, Murray F (eds) Air pollution impacts on crops and forests: a global assessment. Imperial College Press, London

Vanderheyden D, Skelly J, Innes J, Hug C, Zhang J, Landolt W, Bleuler P (2001) Ozone exposure thresholds and foliar injury on forest plants in Switzerland. Environ Pollut 111:321–331

Vingarzan R (2004) A review of surface O_3 background levels and trends. Atmos Environ 38:3431–3442

Volin JC, Tjoelker MG, Oleksyn J, Reich PB (1993) Light environment alters response to ozone stress in *Acer saccharum* Marsh. and hybrid *Populus* L. seedlings. II. Diagnostic gas exchange and leaf chemistry. New Phytol 124:637–646

Volin JC, Reich PB, Givnish T (1998) Elevated carbon dioxide ameliorates the effects of ozone on photosynthesis and growth: species respond similarly regardless of photosynthetic pathway or plant functional group. New Phytol 138:315–325

Wamelink GWW, Wieggers HJJ, Reinds GJ, Kros J, Mol-Dijkstra JP, van Oijen M, de Vries W (2009) Modelling impacts of changes in carbon dioxide concentration, climate and nitrogen deposition on carbon sequestration by European forests and forest soils. Forest Ecol Manag 258:1794–1805

Wang H, Jacob DJ, Le Sager P, Streets DG, Park RJ, Gilliland AB, van Donkelaar A (2009) Surface ozone background in the United States: Canadian and Mexican pollution influences. Atmos Environ 43:1310–1319

Werner H, Fabian P (2002) Free-air fumigation of mature trees. Environ Sci Pollut Res 9:117–121

Westerling AL, Hidalgo HG, Cayan DR, Swetnam TW (2006) Warming and earlier spring increase western U.S. forest wildfire activity. Science 313:940–943

Wiedinmyer C, Tie XX, Guenther A, Neilson R, Granier C (2006) Future changes in biogenic isoprene emissions: how might they affect regional and global atmospheric chemistry? Earth Interact. doi:10.1175/EI174.1, 10

Wieser G (1999) Evaluation of the impact of ozone on conifers in the Alps: a case study on spruce, pine and larch in the Austrian Alps. Phyton Ann Rei Bot 39:241–252

Wieser G, Havranek WM (1993) Ozone uptake in the sun and shade crown of spruce: quantifying the physiological effects of ozone exposure. Trees 7:227–232

Wieser G, Havranek WM (1995) Environmental control of ozone uptake in Larix decidua Mill : a comparison between different altitudes. Tree Physiol 15:253–258

Wieser G, Matyssek R (2007) Linking ozone uptake and defense towards a mechanistic risk assessment for forest trees. New Phytol 174:7–9

Wieser G, Tausz M (eds) (2007) Trees at their upper limit. Tree life limitation at the alpine timberline, Springer Series Plant Ecophysiology. Springer, Dordrecht

Wieser G, Tegischer K, Tausz M, Häberle K-H, Grams TEE, Matyssek R (2002) Age effects on Norway spruce (*Picea abies*) susceptibility to ozone uptake: a novel approach relating stress avoidance to defense. Tree Physiol 22:583–590

Wilkinson S, Davies WJ (2010) Drought, ozone, ABA and ethylene: new insights from cell to plant to community. Plant Cell Environ 33:510–525

Winwood J, Pate AE, Price J, Hanke DE (2007) Effects of long-term, free-air ozone fumigation on the cytokinin content of mature beech trees. Plant Biol 9:265–278

Wipfler P, Seifert T, Heerdt C, Werner H, Pretzsch H (2005) Growth of adult Norway spruce (*Picea abies* [L.] Karst.) and European beech (*Fagus sylvatica* L.) under free-air ozone fumigation. Plant Biol 7:611–618

Wittig VE, Ainsworth EA, Long SP (2007) To what extent do current and projected increases in surface ozone affect photosynthesis and stomatal conductance of trees? A meta-analytic review of the last 3 decades of experiments. Plant Cell Environ 30:1150–1162

Wittig VE, Ainsworth EA, Naidu SL, Karnosky DF, Long SP (2009) Quantifying the impact of current and future tropospheric ozone on tree biomass, growth, physiology and biochemistry: a quantitative meta-analysis. Glob Change Biol 15:396–424

Wolfenden J, Mansfield TA (1991) Physiological disturbances in plants caused by air pollutants. Proc Roy Soc Edinburgh 97B:117–138

Yamaguchi M, Watanabe M, Matsumura H, Kohno Y, Izuta T (2010) Effects of ozone on nitrogen metabolism in the leaves of *Fagus crenata* seedlings under different soil nitrogen loads. Trees 24:175–184

Zunckel M, Venjonoka K, Pienaar JJ, Brunke E-G, Pretorius O, Koosailee A, Raghunandan A, van Tienhoven AM (2004) Surface ozone over southern Africa: synthesis of monitoring results during the Cross Border Air Pollution Impact Assessment Project. Atmos Environ 38:6139–6147

Chapter 8
Influence of Atmospheric and Climate Change on Tree Defence Chemicals

Jason Q.D. Goodger and Ian E. Woodrow

Abstract Environmental factors associated with atmospheric and climate change can potentially modify the structure and function of the world's forests. An important indirect effect of environmental variables such as elevated carbon dioxide (CO_2), air temperature, ozone (O_3), UV radiation, and water-related stress on forests results from the response of tree secondary metabolism. In particular, the concentrations of defence chemicals displayed by trees can change in response to certain climate change factors, and this may influence interactions with herbivores and pathogens, and the broader forest community. An evaluation of the literature relating to climate change effects on tree defence chemicals shows variable results in both direction and magnitude of concentration changes and a dearth of studies on chemicals other than carbon-based phenolics and terpenes. Nevertheless, some generalities are evident. Elevated CO_2, O_3, and UV-B tend to increase tree phenolics, while mono- and sesquiterpenes remain unchanged. Elevated temperature increases volatile terpene emissions and often foliar terpene concentrations, whereas phenolics are largely unaffected. Water stress tends to increase phenolic concentrations and mild stress can also increase terpene emissions, but the effect of excess water availability remains largely unknown. A greater understanding of the implications of global climate change factors on the defence chemistry of the world's forest trees would benefit from increasing the classes of defence chemicals examined, expanding the diversity of tree species and biomes studied, and incorporating long-term, multi-factor experiments. Clearly much more work is required to fully understand how the complexity of factors involved in global climate change influence defence chemistry in the world's forest trees, and how this in turn will influence future tree growth and fitness and forest ecosystem functioning.

J.Q.D. Goodger (✉) • I.E. Woodrow
School of Botany, The University of Melbourne, Parkville, Victoria, Australia
e-mail: jgoodger@unimelb.edu.au

M. Tausz and N. Grulke (eds.), *Trees in a Changing Environment*,
Plant Ecophysiology 9, DOI 10.1007/978-94-017-9100-7_8,
© Springer Science+Business Media Dordrecht 2014

8.1   Introduction

Forests of one form or another extend over some 30 % of the earth's 14.9 billion ha land area. In terms of cover, tropical forests are the most abundant (49 %) followed by boreal forests (24 %), temperate forests (13 %), sub-tropical forests (8 %) and plantations (5 %). These forests are an important and relatively stable pool in the global carbon cycle, storing some 45 % of terrestrial carbon. They are also important for a range of other ecological, economic and social services, including the conservation of biological diversity. Not surprisingly, primary (old growth) forests, which make up some 36 % of all forests, generally contain the greatest biological diversity. However, these forests are being either lost or transformed, for example, by selective logging at a rate of six million ha per year. Nevertheless, it is heartening that in 2005 some 400 million ha (11 % of total forested land) was designated worldwide for the conservation of biological diversity as the primary function (FAO 2006).

Conservation of biological diversity and a range of other services may, however, be under increasing threat from a set of forces that can potentially modify the structure and function of forests. These include factors such as land use changes and invasive species and pest outbreaks, which almost always have negative effects on forests. They also include highly pervasive factors associated with atmospheric and climate change, including elevated carbon dioxide (CO_2), air temperature, ozone (O_3), UV radiation, and water-related stress. Not all of these latter factors have negative effects on forests (e. g. CO_2), but they do all have a series of direct and indirect effects on trees which, taken together, are not straightforward to predict. For example, elevated CO_2 is known to promote plant growth by stimulating photosynthesis, but it also causes a dilution of nitrogen in foliage, which can in turn lead to at least two effects that can retard growth (Leakey et al. 2009). The first involves a slowing of the litter decay rate and thus the rate of return of nitrogen and other minerals to the soil and eventually to growing plants (see review by Lukac et al. 2010). The second involves an increase in the amount of foliage consumed by herbivory, which often consume sufficient plant material to satisfy their nitrogen demands. The balance between growth promotion on the one hand and retardation on the other is not easy to predict, and will certainly vary between species and environment.

Another important indirect effect of these environmental variables on forest structure and function results from the response of secondary metabolism and in particular, the concentrations of compounds that moderate interactions with herbivores and pathogens (i.e. defence chemicals). It is clear that maximisation of growth and reproductive fitness in trees requires an astute investment of limited energy and resources in chemical defence versus primary growth processes. Despite recent evidence suggesting that plant defence chemicals can share both primary and defensive roles (see Neilson et al. 2013), it remains logical that the optimal balance between these two investments will vary under different environmental conditions, both stressful and favourable to plant growth. Many theories have sought to predict such responses (e.g. Bryant et al. 1983; Herms and Mattson 1992). It is not our intention here to examine these theories. In this chapter, we will review recent work examining how changes in CO_2, air temperature, ozone, UV radiation, and water-related stress affect defence chemicals in trees. Trees can legitimately be treated separately from annuals, such as the model species *Arabidopsis*, because it is likely that selection has favoured quite different responses in long-lived species that cannot easily escape herbivores by quickly completing their lifecycle.

8.2 Carbon Dioxide

Forests play an important role in the global carbon cycle especially through their capacity for net CO_2 assimilation and for storage of large amounts of carbon in a relatively inert form (i.e. lignified tissues that are persistent and relatively resistant to decomposition). Indeed, it is the return to the atmosphere of carbon that has been sequestered by forests, both living (e.g. deforestation) and dead (e.g. coal), that account for much of the rise in atmospheric CO_2 from a pre-industrial level of some 280 ppm to the current annual average of close to 400 ppm. Such a rise in CO_2 has two primary effects on trees, which in turn effect changes in the concentration of defence chemicals. The first effect is a reduction in stomatal conductance (aperture). Interestingly, despite its importance for plant water and energy balances, and for CO_2 uptake, the mechanism by which guard cells respond to CO_2 is still unknown, but it is apparently a ubiquitous response in higher plants (Mott 2009). The second effect of CO_2 enhancement on trees, and C_3 plants generally (n.b. almost all trees have a C_3 photosynthetic system), is on the CO_2 fixing enzyme of photosynthesis, ribulose bisphosphate carboxylase/oxygenase (Rubisco). The ratio of carboxylation (photosynthesis) to oxygenation (photorespiration) catalysed by this enzyme is directly proportional to the CO_2 to O_2 ratio (Woodrow and Berry 1988), so any increase in CO_2 will increase the rate of net CO_2 assimilation. It will also increase the nitrogen use efficiency of CO_2 fixation because more CO_2 can be fixed per unit Rubisco, which itself comprises an appreciable proportion of leaf nitrogen in trees.

8.2.1 Effects of CO₂ Enrichment

The primary and secondary effects of CO_2 enrichment on the carbon and nitrogen metabolism of trees have been quantified in many experiments in glasshouses, open-topped chambers, and FACE (free air CO_2 enrichment) rings. A recent review of FACE studies (Leakey et al. 2009) made six general observations, including that in C_3 plants carbon uptake is enhanced by high CO_2 despite a reduction in Rubisco concentration, and that photosynthetic nitrogen and water use efficiency increases. Effects are, however, quite variable. For example, (Körner et al. 2005), reported little if any affect of high CO_2 exposure on growth in a mature deciduous forest in Europe, whereas Cole et al. (2010) estimated that natural increases in atmospheric CO_2 have caused a 53 % increase in the growth of quaking aspen (*Populus tremuloides*). It is not surprising, therefore, that chemical defence metabolism, which is influenced both by carbon and nitrogen supply (Fritz et al. 2006), responds to high CO_2 to establish a new balance between growth and chemical defence. It is also not surprising that such a balance will vary with many factors, including plant type, chemical defence compound (type and location), plant age, and the prevailing environmental conditions. With regard to plant type, research on high CO_2 effects on trees has focussed on representatives of very few genera (namely, *Acer, Betula, Citrus, Eucalyptus, Fagus, Picea, Pinus, Populus, Pseudotsuga, Quercus* and *Salix*). Importantly, four of these genera make up a sizeable proportion of the species reported as being the most common in a recent inventory of the world's forest (FAO 2006). Notably, species from some of the world's largest and most species-rich forests (i.e. wet tropical forests of Africa and South America) were not well represented in the survey.

One of the most important frameworks for interpreting the results of studies of the response of defence chemicals to high CO_2 involves the assumption that the concentrations of these chemicals are sensitive to substrate (i.e. carbon) supply. In other words, if carbon supply rises through rising photosynthesis, as is the case in most high CO_2 studies, then carbon-based defence chemicals (i.e. those that do not contain nitrogen) should also rise (Bryant et al. 1983; Herms and Mattson 1992). Implicit in this thinking is the assumption that metabolic pathways involved in the synthesis of carbon-based defence chemicals have extra capacity – i.e. that feedback inhibition deriving from a lack of biosynthetic capacity or a lack of capacity to store additional defence chemicals does not negate completely the effects of increased carbon supply. If there is indeed extra capacity, then it is unlikely that it will be even across all of the defence chemicals, and it is unlikely that it will be the same across all species and conditions. Put another way, different species or groups of species will likely have different evolutionary strategies for balancing chemical defence, growth and reproduction (Lindroth 2010).

8.2.2   Carbon-Based Defence Chemicals

Measurements of numerous carbon-based defence chemicals from a range of tree species have largely agreed with the aforementioned ideas, and several generalisations can be made. These relate mainly to studies of photosynthetic tissues. First, mass-based measurements of total carbon-based defence chemicals generally show an increase in concentration (e.g. see meta-analysis by (Koricheva et al. 1998). This increase has generally been found to mirror increases in the carbon to nitrogen ratio, which is driven largely by a reduction in photosynthetic nitrogen and an increase in carbohydrates. Second, different classes of compounds (i.e. those that are synthesised by different metabolic pathways; see Fig. 8.1) often respond

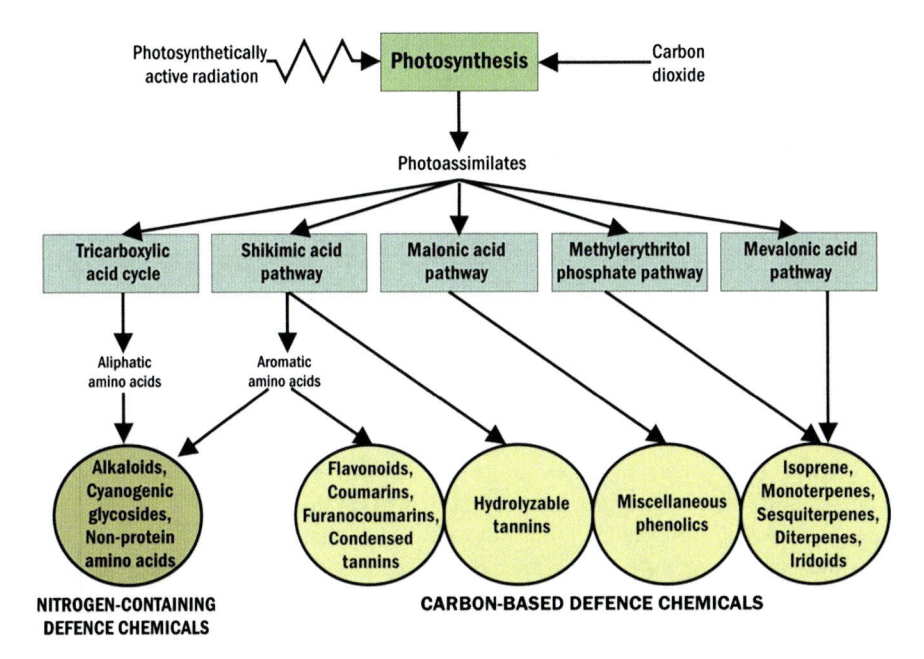

Fig. 8.1 Generalized scheme representing the flow of carbon from photosynthesis through to the major pathways of chemical defence biosynthesis in plants

quantitatively differently to high CO_2. For example, the meta-analyses of trees by Stiling and Cornellisen (2007) and Koricheva et al. (1998) found significant increases in both phenolics and tannins across a range of species. Furthermore, Koricheva et al. (1998) separated measurements of hydrolysable tannins from those of other phenolic compounds derived from phenylpropanoid metabolism, and found both groups to rise significantly under high CO_2. It should be noted that phenolics are a chemically diverse group of compounds involved in both primary and secondary metabolism, and many do not have a clear link to chemical defence.

Terpenoids, on the other hand, have proved much less responsive to high CO_2, with several studies recording a decrease in concentration under high CO_2 (see meta-analysis by Zvereva and Kozlov (2006). This is despite a recent FACE study on a 27-year-old loblolly pine (*Pinus taeda*) plantation finding mono- and diterpene resin mass flow increased by an average of 140 % under elevated CO_2 in trees growing on low-nitrogen soils (Novick et al. 2012). In general, the results of meta-analyses across tree species suggest terpenoids do not show significant responses to elevated CO_2 (Koricheva et al. 1998; Lindroth 2010; Stiling and Cornellisen 2007). It is noteworthy that Novick and co-workers did not detect a significant increase in loblolly pine resin flow under elevated CO_2 for fertilized or carbohydrate-limited trees, supporting the aforementioned importance of the carbon to nitrogen ratio in carbon-based defence responses.

Finally, as noted above, considerable variation has been found in the responses of carbon-based defence chemicals of different species and groupings of species to elevated CO_2. For example, in their meta-analysis, Zvereva and Kozlov (2006) examined studies on the effect of high CO_2 on carbon-based defence chemicals and found a significant rise in angiosperms but a decline in gymnosperms. In a similar manner, Cseke et al. (2009) measured gene expression in two clones of quaking aspen, one being more responsive to high CO_2 in terms of growth than the other. They found evidence of clear differences in the way in which a higher carbon flux is allocated amongst the pathways of secondary metabolism.

8.2.3   Nitrogen-Containing Defence Chemicals

In contrast to the plethora of studies of carbon-based defence chemicals, there has been remarkably little research on those chemicals that contain nitrogen, such as alkaloids and cyanogenic glycosides (see Fig. 8.2). In fact, there has been only one study of this type, on saplings of sugar gum (*Eucalyptus cladocalyx*), which contains relatively high amounts of the cyanogenic glycoside prunasin (Gleadow et al. 1998). It was found that both leaf nitrogen and prunasin concentration declined under high CO_2, but the proportion of nitrogen allocated to prunasin actually increased. Thus, it could be argued that nitrogen was effectively reallocated from photosynthesis to defence.

8.3   Temperature

Global temperatures are predicted to rise during the course of the twenty-first century and beyond. The average global temperature increased by 0.76 °C during the twentieth century, and a further increase of up to 4.0 °C is projected during the twenty-first century (Solomon et al. 2007). Temperature increases are expected to be greatest over land and at most high northern latitudes, thereby influencing the world's forests and those in the Northern hemisphere in particular. Although this section focuses on the direct effects of temperature on tree defence chemicals, it should be noted that temperature changes can also influence herbivore ranges and expansions, and may have large consequences for tree species that lack coevolved defences. For instance, a recent study suggests that global warming has increased the range of mountain pine beetles (*Dendroctonus ponderosae*) so that they now encounter the high-elevation whitebark pine (*Pinus albicaulis*), which is more susceptible to beetle attack than the lower-elevation host species lodgepole pine (*Pinus contorta*; Raffa 2013).

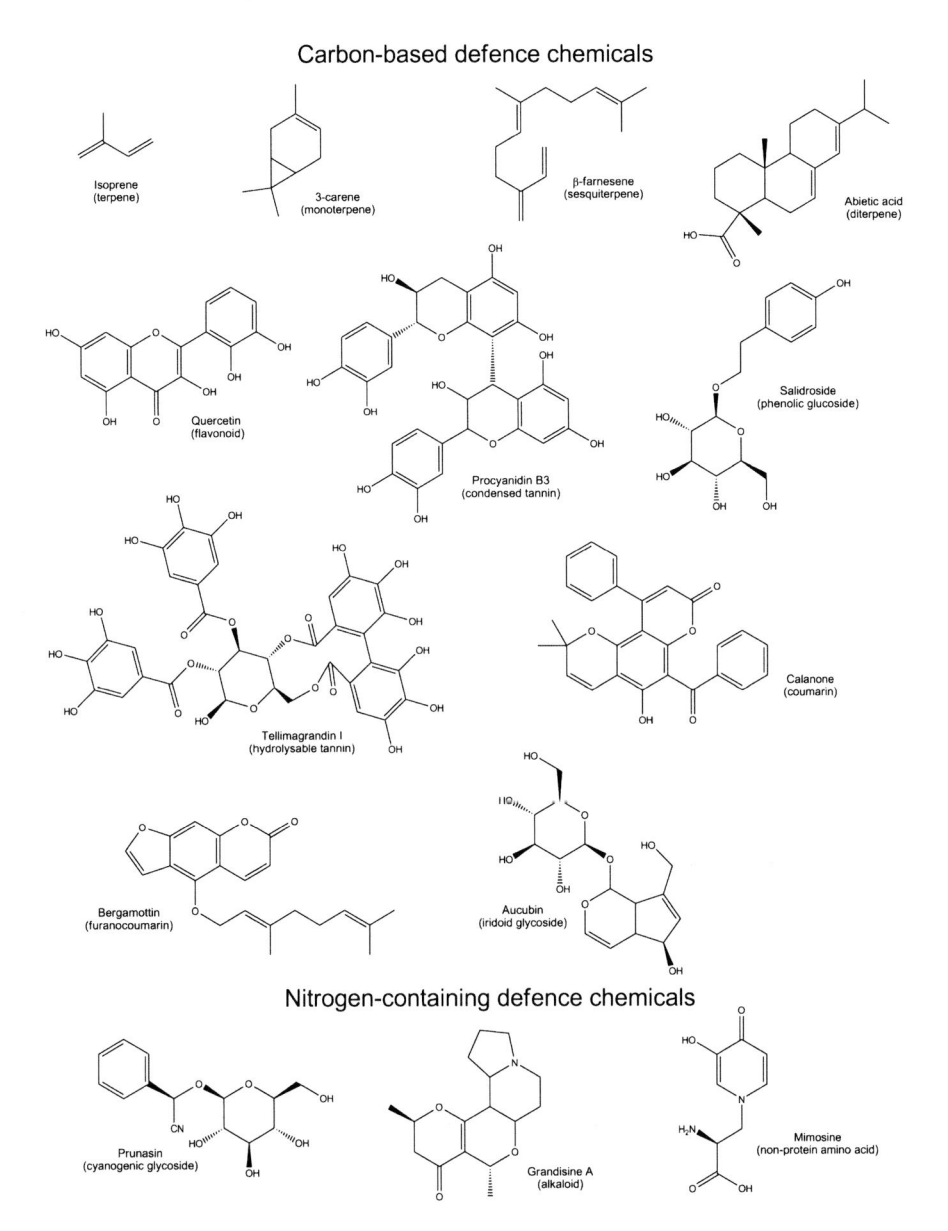

Carbon-based defence chemicals

Isoprene (terpene)

3-carene (monoterpene)

β-farnesene (sesquiterpene)

Abietic acid (diterpene)

Quercetin (flavonoid)

Procyanidin B3 (condensed tannin)

Salidroside (phenolic glucoside)

Tellimagrandin I (hydrolysable tannin)

Calanone (coumarin)

Bergamottin (furanocoumarin)

Aucubin (iridoid glycoside)

Nitrogen-containing defence chemicals

Prunasin (cyanogenic glycoside)

Grandisine A (alkaloid)

Mimosine (non-protein amino acid)

Fig. 8.2 Examples of some of the chemical defences found in trees. The general chemical class of each compound is given in *parentheses*

8.3.1   Temperature Effects on Volatile Defence Chemicals

The key effect of rising temperatures on tree defence chemicals is likely to relate to their volatile emissions. Trees emit a vast array and substantial quantities of volatile organic compounds, including terpenoids (e.g. isoprene, monoterpenes and sesquiterpenes; see Fig. 8.2) and oxygenated hydrocarbons (e.g. alcohols, aldehydes and ketones; Kesselmeier and Saudt 1999). The rate of emission of such volatiles is generally highly correlated with, and apparently dependent on, air temperature. This relationship was exemplified by early work on slash pine (*Pinus elliottii*) that showed that a linear increase in temperature could result in an exponential increase in monoterpene emissions (Tingey et al. 1980). Since then, similar increases in emissions with temperature have been observed for a variety of different terpenoids from numerous tree species, including isoprene from quaking aspen (Fuentes et al. 1999) and white oak (*Quercus alba*; Sharkey et al. 1996), monoterpenes from holm oak (*Quercus ilex*; Loreto et al. 1996) and sesquiterpenes from loblolly pine (Helmig et al. 2006).

The simplest terpenoid is isoprene and it is produced in leaf chloroplasts from photosynthetic intermediates. Although not generally considered a plant defence chemical, isoprene has recently been shown to have deterrent properties towards insect herbivores (Laothawornkitkul et al. 2008). The proportion of fixed carbon emitted as isoprene rapidly increases with temperature via *de novo* biosynthesis. For example, a study on holm oak found that at 30 °C, 2 % of carbon fixed by photosynthesis was emitted as isoprene, but at 40 °C, 15 % was emitted (Loreto et al. 1996). This large proportional increase is attributed to the fact that photosynthesis remains constant or declines at temperatures above 30 °C while isoprene biosynthesis and emission increases. Thus it appears likely that isoprene biosynthesis and emission will increase as global temperatures rise. Given the high proportion of fixed carbon lost as isoprene from the world's forests under global warming, the process of isoprene emission may offset some of the photosynthetic advantages of increased atmospheric CO_2. Moreover, higher isoprene emissions from trees will play an increasingly important role in the atmospheric hydrocarbon budget, influencing air quality and, in particular, contributing to tropospheric ozone depletion episodes (Sharkey et al. 1996).

Monoterpenes and sesquiterpenes are polymers of isoprene and are relatively well characterised as defence chemicals. They are stored predominantly in specialized secretory structures such as resin ducts, secretory cavities, or trichomes which are commonly housed in photosynthetic tissues of trees. The emission of these terpenes is largely dependent on their volatility or on damage to the secretory structures. Generally, terpene emission from trees is assumed to originate via evaporative processes acting on the terpene pools of secretory structures (Grote and Niinemets 2008) and may protect membranes from denaturation during heat stress, thereby increasing the thermotolerance of photosynthesis (Sallas et al. 2003). Evaporation depends mostly on terpene saturation vapour pressure, which is directly related to leaf or needle temperature. Consequently, terpene emissions from trees are commonly calculated by temperature-dependent algorithms (Guenther et al. 1993), and are predicted to increase as global temperature rises. Unlike the situation for isoprene, increases in evaporative emission rates of terpenes have generally been assumed to be independent of de novo biosynthesis. If this assumption is correct, increased emissions may not be balanced by increased biosynthesis. This could potentially lead to decreased foliar terpene concentrations, with subsequent implications for herbivory given their role in chemical defence. Nonetheless, recent work has cast doubt on the assumption that tree monoterpene emission is independent of de novo biosynthesis. The results of a study on four common boreal and alpine forest species showed that at least some monoterpene emissions do result from de novo biosynthesis in all studied species (Ghirardo et al. 2010). In fact, the study showed the proportion of monoterpene emissions that originate directly from de novo biosynthesis to range from 10 % in European larch (*Larix decidua*), to 34 % in Norway spruce (*Picea abies*), 58 % in Scots pine (*Pinus sylvestris*), and to a remarkable 100 % in silver birch (*Betula pendula*; (Ghirardo et al. 2010). If the relationship between terpene emission and concomitant biosynthesis proves to be more general, then increased emissions of mono- and sesquiterpenes with global warming may further offset photosynthetic advantages of increased atmospheric CO_2, especially in species such as silver birch.

Research that has focussed on the concentration of terpenes in foliage of trees subjected to elevated temperatures has produced mixed results. For example, a study on Scots pine seedlings found monoterpene concentrations in needles were not affected by elevated temperature (Holopainen and Kainulainen 2004). In contrast, a study involving both Scots pine and Norway spruce seedlings found the concentrations of almost all examined monoterpenes to increase in both needles and stems (Sallas et al. 2003). In the latter study, the authors suggested that the increase in terpene production might compensate potential terpene emission increases at elevated temperatures. Despite such variable results, a meta-analysis of studies on herbaceous and woody plant species, including a number of tree species, found concentrations of terpenes to generally increase in both green and woody plant parts when plants were subjected to elevated temperature treatments (Zvereva and Kozlov 2006). If the general response of trees to global warming is an increase in terpene concentrations, and if this increase is not counteracted by increased emission rates, then the increased herbivore deterrence and/or toxicity of terpenes may have serious implications not only for herbivores, but also for the wider forest communities that depend on them.

8.3.2 Temperature Effects on Non-volatile Defence Chemicals

The influence of global warming on the numerous non-volatile defence chemicals that trees produce is less clear. Most of the research related to this topic has been conducted on foliar phenolics. By definition, phenolics possess one or more phenol groups (aromatic rings with at least one hydroxyl functional group attached), and common examples include condensed tannins and flavonoids (see Fig. 8.2). Most work has been conducted on the phenolics of northern hemisphere forest species subjected to elevated temperature treatments, but even within this relatively restricted group, few consistent trends have been observed. The aforementioned meta-analysis of studies on herbaceous and woody plant species found a general trend of decreased phenolic concentrations in both green and woody plant parts under elevated temperature (Zvereva and Kozlov 2006). Nevertheless, some studies have observed no change in total phenolics with elevated temperature, such as those on red maple (*Acer rubrum*) saplings (Williams et al. 2003) and Scots pine seedlings (Holopainen and Kainulainen 2004).

Despite inconsistency in the direction and magnitude of changes in phenolic concentration with temperature treatments, the most common observation reported is a reduction in total foliar phenolic concentrations with increasing temperature, although this is not necessarily representative of all individual phenolics in such studies. For example, a study on seedlings of white birch (*Betula pubescens*), silver birch, and dark-leaved willow (*Salix myrsinifolia*) found elevated temperature generally resulted in decreased concentrations of phenolic acids, salicylates and flavonoids, but only condensed tannin concentration decreased in silver birch (Veteli et al. 2007). Similarly, a study on silver birch seedlings at elevated temperature found decreased total phenolics in the leaves, due largely to a decrease in flavonol glycosides and cinnamoylquinic acids, and a decrease in condensed tannins due to a decrease in their precursor (+)-catechin (Kuokkanen et al. 2001). Nevertheless, in that study, the concentration of phenolic glucosides such as salidroside was observed to increase. Furthermore, a study using dark-leaved willow cuttings found the concentration of total phenolics significantly decreased by 25 % under elevated temperatures, and in particular the concentration of chlorogenic acid decreased by 25 %, salicylic alcohol by 27 %, salicortin by 26 % and eriodikytol-di-glucoside by 100 %. However, quercetin and condensed tannin concentrations were unaffected by the elevated temperature treatment (Veteli et al. 2002).

It remains unclear whether temperature has a direct influence on the biosynthesis of tree phenolics or if confounding variables can explain some of the observed changes in phenolic concentrations at elevated temperatures. For instance, a study using saplings of Scots pine found a slight decrease in the concentration of 20 different phenolics at elevated temperature, but the authors concluded that this could be due to mass-based dilution given the greater needle biomass observed in the elevated temperature treated plants (Raisanen et al. 2008). Furthermore, a study on 4-year-old English oak (*Quercus robur*) found an increase in condensed tannins with temperature treatment, but the authors suggested that this result may be confounded by the fact that elevated temperature induced earlier bud burst and that condensed tannins increase with age (Dury et al. 1998). Similarly, the apparent decline in phenolic glucosides in elevated temperature-treated willows could also be an ontogenetic response due to accelerated physiological ageing in the treated trees, and the fact that phenolic glucosides decrease with ontogeny (Veteli et al. 2002). Thus the inherent difficulty of separating biomass and ontogenetic effects from those of elevated temperature may confound interpretation of results in many studies.

Nevertheless, a number of different mechanistic explanations have been proposed to explain a decrease in the foliar concentration of particular phenolics with predicted global temperature rise. First, a decrease in phenolic biosynthesis could be due to differences in activity of enzymes within the biosynthetic pathway at different temperatures (Veteli et al. 2002). Second, elevated temperatures may result in greater losses of carbon via maintenance respiration and, consequently, the biosynthesis of phenolic compounds, especially the glycosylation of flavonoids, would be retarded (Kuokkanen et al. 2001). Third, elevated temperatures may enhance enzymatic breakdown of certain phenolics, especially phenolic glucosides (Veteli et al. 2002). Whether due to one or a combination of these mechanisms, a general reduction in phenolic concentrations is likely due to impact on tree fitness, because of the many and varied primary and secondary roles of phenolics in trees.

Only a few other non-volatile defence chemicals of trees have been examined in the context of global warming. They include the triterpenoid, papyriferic acid, which has been observed to increase in stems of silver birch seedlings subjected to elevated temperature (Kuokkanen et al. 2001), and resin acids which have shown variable responses to temperature treatments. With respect to the latter, a study on Scots pine seedlings found the concentrations of pimaric and dehydroabietic acids in needles, and sandaracopimaric and isopimaric acids in stems to vary significantly across five different temperature treatments (Holopainen and Kainulainen 2004). Despite the significant variation in that study, there was no clear trend of an increase or decrease with increasing temperature. Directionally more consistent results were found in a study on seedlings of two boreal tree species, where elevated temperature increased concentrations of some resin acids in Scots pine needles and total resin acids in Norway spruce needles (Sallas et al. 2003).

8.3.3 Interactive Effects of Elevated CO_2 and Temperature

Potential interactions between rising temperatures and increasing atmospheric CO_2 may have important effects on tree secondary metabolism. Nonetheless, relatively few studies have explored the effects of both parameters in such a way that both their individual and combined effects on tree defence chemicals can be evaluated (Zvereva and Kozlov 2006). A meta-analysis of the studies that have examined the effect of both parameters on plant quality, including terpenes and phenolics, found the results to be taxon-specific and variable depending on the compound analysed (Zvereva and Kozlov 2006). Despite this, the authors generalised that responses to enriched CO_2 can either be independent of temperature, offset by elevated temperature, or apparent only with elevated temperature. These ostensibly contrasting generalisations were based on the fact that several studies concluded temperature does not influence tree secondary metabolism under CO_2 elevation, while others demonstrated clear interactive effects of elevated temperature and CO_2. For example, no significant interaction was found between temperature and CO_2 for phenolic concentrations in leaves of red maple saplings (Williams et al. 2003) or in leaves or stems of silver birch seedlings (Kuokkanen et al. 2001). Indeed, the authors of the latter study suggested that increasing either temperature or CO_2 alone may affect tree secondary metabolism more than a combination of the two factors (Kuokkanen et al. 2001). Moreover, a study on seedlings of three boreal tree species found that phenolic concentration decreased under elevated temperature, increased under elevated CO_2, but was unchanged under a combination of the two factors (Veteli et al. 2007). The study revealed the combined effects of elevated temperature and CO_2 were additive, thereby cancelling out their individual effects (Veteli et al. 2007). A similar trend was found for terpenes and resin acids in a study on Scots pine and Norway spruce seedlings, where significant interactive effects of elevated temperature and CO_2 on needle and stem concentrations of some individual compounds were observed. In that study, the effect of elevated temperature alone was opposite to and much greater than elevated CO_2 (Sallas et al. 2003). Nevertheless, the effect of the combined treatments was generally intermediate between the two factors, again suggesting that temperature and CO_2 have an additive effect on tree defence chemicals.

8.3.4   Phenological Effects of Elevated CO_2 and Temperature

Arguably one of the greatest effects of global warming on forest-herbivore interactions may relate to phenology rather than defence chemicals. Elevated temperatures can lead to phenological mismatches between trees and their insect herbivores (Bidard-Bouzat and Imeh-Nathaniel 2008). Many insects have evolved life histories in which their maximum nutritional demands coincide with the season when developing tree tissue is maximally available. For example, approximately 50 % of forest insect pest species are early-season specialists consuming immature plant tissue when it is generally easier to digest, more nutritious and lower in defence chemicals (Ayres 1993). Increased global temperatures may speed up leaf development rates thus reducing the time that immature foliage is available to these herbivores. Such a phenological mismatch, even on the scale of a few days, can halve herbivore fecundity and therefore greatly impact forest-herbivore communities (Ayres 1993). Rising temperatures, however, are likely to affect insect herbivores similarly by speeding up development, thereby counteracting any such phenological mismatch. In fact, increased insect growth rates may result in increased rates of feeding over shorter time periods, potentially tipping the balance in favour of the herbivores with dramatic impacts for the forest trees.

As with tree defence chemicals, elevated CO_2 is likely to interact with elevated temperature to influence any potential phenological effects. For example, elevated temperatures generally result in accelerated development and enhanced growth and reproductive performance of insects, but elevated CO_2 tends to result in the opposite (Lindroth 2010). Indeed, it has been suggested that temperature increases are likely to mitigate predicted negative effects of enriched CO_2 on insects (Zvereva and Kozlov 2006). Hence, in a similar manner to the combined effects of these factors on tree defence chemicals, elevated temperature and CO_2 may have additive phenological effects.

8.4 Ozone

Unlike CO_2, variations in the O_3 concentration of the lower atmosphere are relatively regional, although there has been an increase in the global background level of some 40 % since the industrial revolution (Solomon et al. 2007). The reason for regional increases in O_3 relates to the mechanism of production of this gas. O_3 is largely produced through the oxidation of atmospheric hydrocarbons in the presence of light and nitrogen oxides. All three of the components of this reaction are highly variable. Hydrocarbon and nitrogen oxide concentrations depend on the intensity of industrial activities, and hydrocarbon concentrations can be further increased by plant emissions (see Sect. 8.3.1). Light intensity varies diurnally, seasonally and regionally. Not surprisingly, therefore, it is predicted that the largest increases in O_3 concentration will occur in the Northern hemisphere, and that globally O_3 will rise to an average concentration of just under 70 ppb by 2050 (Karnosky et al. 2005). Moreover, it is estimated that by the end of the century some 60 % of the world's forests will be significantly affected by O_3 (Vingarzan 2004).

8.4.1 Effects of Enhanced O_3

The main mechanism by which O_3 affects plants involves diffusion of the gas through the stomata and into the liquid phase of cells around the intercellular air spaces. O_3 readily reacts with especially lipids and proteins in both the cell wall and plasma membrane, resulting in damage and the release of products such as reactive oxygen species and hydrogen peroxide (Kangasjärvi et al. 2005). This leads to a number of responses, including stomatal closure, production of antioxidants, programmed cell death and other defence responses. Overall, enhanced O_3 causes reductions in photosynthesis and leaf area, and accelerated senescence of leaves, which together result in lower productivity. Wittig et al. (2009), in their meta-analysis of a large number of studies, estimated that the current average ambient level of O_3 of 40 ppb has resulted in a 7 % reduction in the biomass (compared to pre-industrial conditions) of the temperate and boreal forests of the Northern hemisphere. Interestingly, they found that like some CO_2 effects (see Sect. 8.2), gymnosperms were significantly less sensitive to O_3 than angiosperms.

Given that O_3 elicits defence and stress responses involving, amongst other things, the production of reactive oxygen species; it is not surprising that changes in the concentrations of secondary metabolites occur. Indeed, recent work on European beech (*Fagus sylvatica*) saplings showed that elevated O_3 effects an increase in the transcript abundance of genes that are involved in the shikimate pathway (Betz et al. 2009a), which in turn causes an increase in a range of flavonoids and simple phenolics with antioxidant properties. Interestingly, changes in the monomeric composition of lignin were also measured under the enhanced O_3 conditions (Betz et al. 2009b). Most studies of the effects of high O_3 on defence chemicals have, however, not focussed on the mechanisms of the response; rather they have determined the concentration of these chemicals. Like the research on CO_2 effects (see Sect. 8.2), studies have involved a range of tree species, conditions and experimental designs, and the defence chemicals measured have come almost exclusively from the phenolic and terpenoid groups. No studies of, for example, nitrogen-based defence chemicals have apparently been undertaken. Also, like the CO_2 studies, considerable variability – both qualitative and quantitative – has been recorded, depending on species, conditions, and chemical type examined.

In order to make generalisations about these studies of defence chemicals, (Valkama et al. 2007) undertook a meta-analysis of 63 studies of enhanced O_3 effects involving 22 tree species and a range of phenolic and terpenoid compounds. They found that medium- and long-term exposure (i.e. >6 months) to enhanced O_3 (i.e. >1.5 times ambient) caused an increase in total levels of terpenes and phenolics of 8 % and 16 %, respectively. However, there was considerable variability within these groups of compounds. For example, consistent with the findings of Betz et al. (2009), flavanoids and phenolic acids were found to increase significantly across species, whereas tannins were not. Of the terpenes, a significant increase in concentration was detected for diterpene resin acids, but not for mono- or sesquiterpenes. The analysis of terpene studies also showed that the tree's ontogeny can determine the magnitude of the O_3 response. It was found that mature trees were more responsive than both saplings and seedlings. The authors also drew attention to the considerable variability between species and groups of species. For example, angiosperms were more responsive overall than gymnosperms with respect to the total level of phenolics. There was, however, considerable variability in the responses of individuals within these groups, as summarised in several recent reviews (Bidard-Bouzat and Imeh-Nathaniel 2008; Lindroth 2010; Valkama et al. 2007).

8.4.2 Interactive Effects of Enhanced CO_2 and O_3

While both O_3 and CO_2 enhancement have been shown to be able to increase the concentrations of defence chemicals in many tree species, there is reason to propose that concurrent increases in these gas concentrations may show a more complex interactive effect. First, stomatal conductance is reduced under high CO_2, which may decrease the amount of O_3 that can diffuse into leaves. Second, the higher rates of photosynthesis and carbohydrate production under high CO_2 may also increase the capacity of plants to cope with the adverse effects of O_3. Studies of the interactive effects of O_3 and CO_2 were recently reviewed by Valkama et al. (2007) and Lindroth (2010). They concluded that there is evidence of both additive and interactive effects, depending on both the species and the type of defence chemicals examined. In some cases, CO_2 was found to be able to negate the positive effects on a defence chemical of O_3 (e.g. some phenolics; Peltonen et al. 2005), while in other cases there was synergy, with the effect of both gases being greater than the sum of each treatment (e.g. terpenes; Valkama et al. 2007).

8.5 Ultraviolet-B Radiation

In contrast to tropospheric O_3, which is gradually increasing worldwide (see Sect. 8.4.1), stratospheric O_3 has been decreasing, particularly over the last few decades due largely to a marked rise in the concentration of chemicals such as chlorofluorocarbons (CFCs). While trees are not directly affected by events in the stratosphere, the depletion of O_3 has indirect effects, which derive from the fact that O_3 is a strong absorber of UV-B radiation. Modeling has recently shown that there has been a significant rise in the intensity of UV-B at the earth's surface over the last three decades at all latitudes other than the equatorial region (McKenzie et al. 2011). This rise is, moreover, predicted to stabilise, then decline due to reductions in the outputs of O_3-depleting chemicals, with full recovery to the 1980s level likely sometime in the middle of this century (Schrope 2000). Nevertheless, the kinetics of recovery of O_3 levels are dependent on other relatively unpredictable factors such as atmospheric temperature, and UV-B levels can be further influenced by other factors such as cloudiness and aerosol concentration (McKenzie et al. 2011). Clearly, forests will be subjected to enhanced UV-B radiation for some time to come.

8.5.1 Effects of UV-B Enhancement

UV-B radiation affects trees because it is strongly absorbed by macromolecules, including functionally important ones such as proteins and nucleic acids. Absorbance of UV-B radiation causes structural changes in these molecules, which can then lead to deleterious effects including genetic mutations, loss of function of proteins and oxidative stress involving production of active oxygen species. Many studies have sought to examine the effect of enhanced UV-B on overall growth and development of plants, including many tree species. For example, the recent meta-analysis by Li et al. (2010) of studies of enhanced UV-B effects on plants found that some 38 different woody species had been examined. Overall, most evidence suggests that when plants are grown under natural conditions with enhanced UV-B radiation, photosynthetic capacity (i.e. the area based rate of photosynthesis) is not significantly affected, but growth is retarded by small changes in leaf expansion (Ballaré et al. 2011). Trees are generally less sensitive than herbaceous plants, but there is still considerable variation between species (Li et al. 2010). Much of this variation probably relates to differences in UV-B tolerance under normal conditions and differences in the way in which plants acclimate to enhanced UV-B radiation. This latter process is complex and involves UV-B specific and non-specific signalling pathways that moderate a series of processes including morphological changes, accumulation of UV screening compounds (e.g. flavonoids), enhancement of DNA repair, and production of antioxidants (Jenkins 2009).

It is not surprising, therefore, that increases in UV-B radiation are associated with changes in plant defence chemicals, given that up-regulation of especially phenolic metabolism is required to sustain any increase in UV screening and antioxidant compounds, and that many of these compounds also function in defence (e.g. some flavonoids are feeding and oviposition deterrents). Such a diversion of carbon flux could also affect other pathways such as terpene metabolism. Studies of bulk extracts of UV screening compounds from a wide range of species have indeed shown an increase in ability to absorb UV-B radiation, with an average increase of about 10 % for field based studies (Searles et al. 2001). Consistent with this, Li et al. (2010) found that tree species showed a significant increase in UV-B absorbing compounds, especially phenolics, when subjected to both low (18–40 % of ambient UV-B) and high (>40 % of ambient UV-B) radiation treatments. They also noted considerable variation in the response level between species. Overall, tree species show a relatively consistent increase of phenolic metabolism under UV-B enhancement, with increases in a broad range of compounds reported, including flavonoids such as anthocyanins, isoflavonoids and flavonol glycosides, coumarins, tannins and phenolic acids (Bidard-Bouzat and Imeh-Nathaniel 2008). Responses can, however, be complex and few clear patterns have yet to emerge. For example, in a study of 11 phenolic compounds in silver birch leaves, it was found that three compounds increased in concentration while one decreased (Lavola et al. 1997). Similarly, in rough lemon (*Citrus jambhiri*), the concentration of furanocoumarins (which like the flavonoids also absorb strongly in the UV-B) was found to decrease under enhanced levels of UV-B radiation (Asthana et al. 1993).

Relatively few studies of the effect of enhanced UV-B radiation on other groups of defence compounds have been undertaken. For example, it has been reported that terpenoid levels in the bark of silver birch are unaffected by enhanced UV-B radiation (Tegelberg et al. 2002), whereas the level of terpene-derived carotenoids, which serve a photoprotection role, was higher in leaves of holm oak trees growing at high compared to low elevations (Filella and Penuelas 1999). There have been a few studies of nitrogen-based defence chemicals (alkaloid and cyanogenic glycosides) in herbaceous plants, but as yet no tree species have been examined. This is clearly an area that needs additional research because, like high CO_2 effects, UV-B alters both carbon and nitrogen fluxes so it is likely that changes across all groups of defence chemicals will occur (see Fig. 8.1).

8.5.2 Interactive Effects of UV-B and Climate Factors

Globally, increases in UV-B radiation are likely to be combined with rises in CO_2, O_3 and temperature, and changes in water availability. Thus, in order to understand how ecosystems will respond to these changes, it is important to undertake factorial experiments in which the effects of combinations of changes are measured. Studies of the interactive effects of UV-B enhancement and drought have generally shown that the overall growth effect is less than the additive effect of both factors – i.e. UV-B sensitivity is reduced under drought (Caldwell et al. 2007). In contrast, there is little evidence of interactions between UV-B enhancement and either temperature (Day et al. 1999) or elevated CO_2 (Zhao et al. 2003), although most of this work has involved herbaceous species. A recent study of four clones of dark-leaved willow examined the combined effects of UV-B, CO_2 and temperature on both growth and levels of phenolics (Paajanen et al. 2011). No clear results regarding interactive effects emerged from this study largely because the clones had high constitutive levels of phenolics and were thus quite resistant to the UV-B treatment.

8.6 Water-Related Stress

A warmer atmosphere is capable of carrying greater quantities of gaseous water. The subsequent convergence of moisture-laden air masses leads to air uplift, cloud formation and precipitation as rain, hail or snow (O'Gorman and Schneider 2009). Therefore global warming is predicted to result in increased precipitation. The predicted precipitation increases, however, are likely to be regionally specific and not uniform across the earth. Indeed, simulations based on predicted temperature increases in the twenty-first century suggest mean precipitation will generally increase in the deep tropics and extra-tropics, but decrease in the sub-tropics (Sun et al. 2007). Such regional contrasts are due to intense rainfall being an inherently local event that is a direct consequence of the supply of atmospheric moisture from farther afield, where it may otherwise have contributed to more moderate rainfall (O'Gorman and Schneider 2009). Thus rapid increases in precipitation intensity in one region imply a decrease in intensity, duration and/or frequency in other regions (Trenberth et al. 2003). In addition, precipitation extremes are widely held to increase proportionately to the mean atmospheric water vapour content (Pall et al. 2007). Therefore, global warming is also predicted to increase the frequency and intensity of precipitation extremes such as flooding and drought, and this phenomenon is predicted to affect almost all regions of the earth (Kharin et al. 2007). Any changes in precipitation patterns directly influence soil water availability, which in turn affects tree water status and can impact on tree secondary metabolism.

8.6.1 Effects of Excess Water

The most recent models for the influence of global warming on precipitation predict increased intensity of rain and snow in much of the Northern hemisphere (Min et al. 2011), including an increased risk of severe flooding episodes (Pall et al. 2011). In fact, analysis of meteorological data suggests increases in greenhouse-gas concentrations have already resulted in intensification of heavy precipitation events over large swathes of land in the Northern hemisphere during the latter half of the twentieth century (Min et al. 2011). Despite the likelihood of increased soil moisture in such regions, surprisingly few studies have examined the influence of excess water availability on tree secondary metabolism, although its effect on tree nutrition, which may indirectly influence defence chemicals, has received more attention (for review see Kreuzwieser and Gessler 2010).

One of the few studies addressing this topic examined foliar phenolics in tea-leafed willow (*Salix phylicifolia*) trees growing in three different natural habitats: one with permanently water-logged soil, one subjected to regular periods of flooding annually and one growing on well-drained soil, never subjected to water-logging (Sipura et al. 2002). The study found concentrations of condensed tannins and their precursor (+)-catechin, together with ampelopsin and a myricetin derivative, were all highest in the water-logged trees, intermediate in the periodically flooded trees, and significantly lowest in the dry zone trees. These results suggest that the degree of excess water stress can increase the concentration of foliar phenolics, although it should be noted that leaf size (and possibly LMA which is well correlated with phenolics) decreased significantly as the habitats became wetter. In contrast, a study on silky willow (*Salix sericea*) seedlings grown in a common garden experiment found that flooding decreased the concentration of the phenolic glycoside salicortin, but had no effect on 2′-cinnamoylsalicortin concentration (Lower et al. 2003). Given that global warming is predicted to result in increased precipitation in many regions with an increased likelihood of extreme precipitation events such as flooding, research on the effect of excess water stress on tree defence chemicals is required.

8.6.2 Effects of Drought Stress

In contrast to excess water availability, the influence of limited water availability or drought on tree defence chemicals, particularly carbon-based chemicals, has received more attention. It has been predicted that secondary metabolism should increase under moderate drought stress due to tree growth being limited more than photosynthesis, resulting in the accumulation of carbohydrates and an increase in carbon-based defence chemicals (Ayres 1993). In support of this prediction, a study on black poplar (*Populus nigra*) saplings found that total phenolic glycoside concentrations were 89 % higher in the drought-stressed trees relative to the well-watered control (Hale et al. 2005). Nonetheless, the results of a recent study on apple (*Malus x domestica*) seedlings found that the soluble sugars sorbitol, glucose and fructose increase with low and moderate drought stress as predicted, but the concentrations of defence phenolics did not match predictions as phloridzin decreased with increasing stress while phloretin showed no changes (Gutbrodt et al. 2012). Similarly, in a study of quaking aspen and sugar maple (*Acer saccharum*) seedlings, the concentration of condensed tannins were not altered by drought treatment in either species, nor were those of the hydrolyzable tannins, gallotannin and ellagitannin, which are limited to maples (Roth et al. 1997). In fact, the only defence chemicals that were altered by drought treatment in that study were the aspen phenolic glycosides salicortin and tremulacin, which were significantly reduced by 21 and 14 %, respectively (Roth et al. 1997).

Roth and co-workers attributed their lack of parity with predictions to the following factors. First, drought may have suppressed activity of enzymes required for phenolic biosynthesis, preventing an increase in production. Second, the drought treatment may have been severe enough to retard photosynthesis more than nutrient uptake, leading ultimately to reductions in secondary metabolism due to carbon limitations. In support of this latter notion, a study on Douglas fir (*Pseudotsuga menziesii*) growing different distances from a water source found a non-linear relationship between water stress and condensed tannin concentration, whereby tannin concentrations initially increased under mild water stress, but then levelled off under moderate stress before decreasing under severe stress (Horner 1990). Such non-linear relationships may explain some of the variability observed in studies on drought stress and tree defence chemicals, as most studies have employed only one stress treatment compared to a well-watered control. Thus, to describe potential non-linear relationships between water stress and tree defence chemicals, studies may need to employ additional treatments covering a range of stress levels or multiple harvests during progressive drought.

The influence of drought stress on volatile emissions from trees has also received attention (Kesselmeier and Saudt 1999; Rennenberg et al. 2006) for reviews). In general, emission rates of monoterpenes are mostly unaffected under moderate water stress, when CO_2 and water gas exchange decline, but are reduced under severe water stress. Monoterpene emissions can then increase substantially after rewatering during stress recovery (Kesselmeier and Gessler 1999). With respect to isoprene emissions, a 3 month study on two Eastern cottonwood (*Populus deltoides*) plantations showed emissions were decreased under severe water stress, but stimulated under short term water stress (Pegoraro et al. 2005). These results suggest that moderate soil water stress has the potential to counteract the effect of elevated CO_2 on isoprene by increasing production while decreasing CO_2 assimilation. Global warming is expected to result in mean temperature increases and localised reductions in precipitation in some regions of the world. In these regions, increased isoprene emissions may have negative air-quality impacts on regional atmospheric chemistry (Rosenstiel et al. 2003).

As with the other climate change factors; there has been remarkably little research on the influence of drought stress on nitrogen-containing defence chemicals in trees. One study compared mean foliar cyanogenic glycoside (prunasin) concentration between two sugar gum populations growing in areas with contrasting rainfall (Woodrow et al. 2002). In that study, mean prunasin concentration was found to be higher in the population from the drier site, although the authors attributed this to a concomitant increase in foliar nitrogen. Nevertheless, in a subsequent greenhouse study using sugar gum seedlings, foliar prunasin concentration increased markedly under water stress, and this was independent of changes in leaf nitrogen or any other identifiable variable (Gleadow and Woodrow 2002).

8.7   Conclusions

Factors related to current and predicted climate change are likely to influence at least some of the many defence chemicals produced by the world's forest trees. Climate change factors can act directly and/or indirectly on tree defence chemicals in a number of ways. Direct effects on the concentration of a given defence chemical include changes in biosynthetic enzyme activity, substrate availability, catabolic activity (enzymatic or otherwise) and volatile emission rates. Indirect effects include changes in nutrient availability, which can have flow-on effects on the rate of biosynthesis of particular defence chemicals. The key climate change factors likely to influence tree foliar enzyme biosynthesis are atmospheric CO_2 concentrations, global temperatures, ozone stress, UV levels and water availability, and these factors may act independently or interactively. The direction and magnitude of the effect of these factors on tree defence chemicals appears to be species- and possibly genotype-specific, as well as dependent on the types of defence chemicals produced. Despite the complexity of climatic factors and tree responses, some general trends for defence chemicals under a changing climate are apparent from the research conducted to date:

- Elevated CO_2 consistently results in increased phenolic concentrations, whilst terpenes generally remain unchanged. Nitrogen-containing chemicals have rarely been examined, but in a single study, cyanogenic glycoside content was observed to decrease.
- Elevated temperature consistently results in increased isoprene emissions and may also increase mono- and sesquiterpene emissions. Foliar terpene concentrations may also increase, but phenolic concentrations often remain unchanged, or may decrease. Nitrogen-containing chemicals have not been examined.
- Drought stress can result in increased phenolic concentrations and mild stress can increase terpene emissions. Nitrogen-containing chemicals have rarely been examined, but the cyanogenic glycoside content has been observed to increase under drought stress. The influence of excess water availability is largely unknown, but increased phenolic concentrations have been observed.

- Enhanced O_3 tends to increase phenolics and diterpene resin acids, but appears to have no effect on mono- and sesquiterpenes. Nitrogen-containing chemicals have not been examined.
- UV-B consistently increases phenolics, whereas terpenes remain unchanged. Nitrogen-containing chemicals have not been examined.

8.7.1 Future Directions

A greater understanding of the implications of global climate change for defence chemistry in the world's forest trees would benefit from extending research in a number of clear directions. As outlined succinctly by Lindroth (2010), future research in this area should aim to increase the classes of defence chemicals quantified, expand the diversity of tree species and biomes studied, and incorporate long-term, multi-factor experiments.

The number and type of chemicals examined to date has been restricted to only a few classes of chemical compounds, primarily the carbon-based phenolics and terpenoids, and commonly only select compounds within each class. The influence of global climate change on other classes of carbon-based defence chemicals such as iridoid glycosides is unknown, despite their importance in some tree species. Moreover, nitrogen-containing compounds such as alkaloids, cyanogenic glycosides, and non-protein amino acids can be very important in certain forest ecosystems, but have received scant attention. Irrespective of chemical class, the influence of climate change on the mode of expression of chemical defence, i.e. whether inducible or constitutive, also requires greater attention in the future. For example, a recent study on a non-tree species, *Brassica napus*, found the inducibility of glucosinolates could be altered by both elevated CO_2 and changes in ozone levels (Himanen et al. 2008). Similar changes in tree species could have great implications for forest ecosystems.

The number of species and biomes studied to date has also been relatively restricted and has resulted in much of the world's forests being under-represented. The majority of research has focused on a select group of deciduous and coniferous tree species from particular Northern hemisphere biomes. Future studies should incorporate a greater number of boreal and tropical forests, as both are critically important to the global carbon cycle and a range of other ecosystem services, but neither has received much attention; a problem that is particularly acute for understanding effects on ecosystem services. Similarly, Southern hemisphere forests have rarely been investigated in the context of defence chemistry and climate change, despite the increasing importance and prevalence of plantation forests of genera such as *Eucalyptus* throughout the world.

Arguably the most important improvement to future experimentation in this field will relate to the simultaneous examination of multiple climatic factors. In a recent review, Bidart-Bouzat and Imeh-Nathaniel (2008) concluded that future studies should focus on simultaneously testing the effects of multiple climate change factors to gain a more realistic perspective of how climate change may impact defence chemical production. The evaluation of one or two factors at a time, which has been the general strategy to date, makes predicting the response of trees to changes in the complexity of climate factors very difficult. This is because climatic factors can induce differential responses in trees, in terms of both direction and magnitude of defence chemical changes, and future scenarios affecting the world's forests will be the result of these complex interactions, not single factor changes. According to Lindroth (2010), a better understanding of the ways the various atmospheric and climatic factors interact is needed to more realistically approximate future environments, and to provide a diversity of conditions under which to test the effects of individual factors on tree defence chemicals.

In addition to the inclusion of multiple factors, future experimental designs also need to take into consideration the developmental stage of trees used and the length of experiments. The majority of published studies have been performed on seedlings and saplings, but these immature stages may respond in a very different manner to adult trees and represent a relatively small proportion of the lifespan of tree species. Moreover, the majority of studies to date have been performed over relatively short time periods (months to several years), which are considerably less than the lifespan of forest trees, and do not provide knowledge on the perpetuation of effects over time. Furthermore, short-term experiments with few sampling times can fail to detect non-linear effects of climatic factors on tree defence chemicals. Future research should attempt to overcome logistic and funding constraints to focus more on long-term effects of global change factors on tree defence chemicals. Long-term experiments will also enable studies on indirect effects of climate change on tree defence chemistry such as nutrient cycling at the ecosystem level, potential evolutionary responses of herbivorous insects, as well as effects on mutualistic associations of trees with pollinators and mycorrhizae.

Clearly much more work is required to fully understand how the complexity of factors involved in global climate change will influence the defence chemistry of the world's forest trees and how this in turn will influence future tree growth and fitness in the face of predation.

Acknowledgments This work was supported by an Australian Research Council Discovery grant to JQDG & IEW (DP1094530).

References

Asthana A, McCloud ES, Berenbaum MR, Tuveson RW (1993) Phototoxicity of *Cirrus jambhiri* to fungi under enhanced UVB radiation: role of furanocoumarins. J Chem Ecol 19:2813–2830

Ayres MP (1993) Plant defense, herbivory, and climate change. In: Kareiva PM, Kingsolver JG, Huey RB (eds) Biotic interactions and global change. Sinauer Associates, Sunderland

Ballaré CL, Caldwell MM, Flint SD, Robinson SA, Bornman JF (2011) Effects of solar ultraviolet radiation on terrestrial ecosystems. Patterns, mechanisms, and interactions with climate change. Photochem Photobiol Sci 10:226–241

Betz GA, Gerstner E, Stich S, Winkler B, Welzl G, Kremmer E, Langebartels C, Heller W, Sandermann H, Ernst D (2009a) Ozone affects shikimate pathway genes and secondary metabolites in saplings of European beech (*Fagus sylvatica* L.) grown under greenhouse conditions. Trees 23:539–553

Betz GA, Knappe C, Lapierre C, Olbrich M, Welzl G, Langebartels C, Heller W, Sandermann H, Ernst D (2009b) Ozone affects shikimate pathway transcripts and monomeric lignin composition in European beech (*Fagus sylvatica* L.). Eur J For Res 128:109–116

Bidard-Bouzat MG, Imeh-Nathaniel A (2008) Global change effects on plant chemical defenses against insect herbivores. J Integr Plant Biol 50:1339–1354

Bryant JP, Chapin FSI, Klein DR (1983) Carbon/nutrient balance of boreal plants in relation to vertebrate herbivory. Oikos 40:357–368

Caldwell MM, Bornman JF, Ballaré CL, Flint SD, Kulandaivelu G (2007) Terrestrial ecosystems, increased solar ultraviolet radiation, and interactions with other climate change factors. Photochem Photobiol Sci 6:252–266

Cole CT, Anderson JE, Lindroth RL, Waller DM (2010) Rising concentrations of atmospheric CO_2 have increased growth in natural stands of quaking aspen (*Populus tremuloides*). Glob Chang Biol 16:2186–2197

Cseke LJ, Tsai CJ, Rogers A, Nelsen MP, White HL, Karnosky DF, Podila GK (2009) Transcriptomic comparison in the leaves of two aspen genotypes having similar carbon assimilation rates but different partitioning patterns under elevated [CO_2]. New Phytol 182:891–911

Day TA, Ruhland CT, Grobe CW, Xiong F (1999) Growth and reproduction of Antarctic vascular plants in response to warming and UV radiation reductions in the field. Oecologia 119:24–35

Dury SJ, Good JEG, Perrins CM, Buse A, Kaye T (1998) The effects of increasing CO_2 and temperature on oak leaf palatability and the implications for herbivorous insects. Glob Chang Biol 4:55–61

FAO (2006) Global forest resources assessment 2005. Progress towards sustainable forest management. FAO, Rome

Filella I, Penuelas J (1999) Altitudinal differences in UV absorbance, UV reflectance and related morphological traits of *Quercus ilex* and *Rhododendron ferrugineum* in the Mediterranean region. Plant Ecol 145:157–165

Fritz C, Palacios-Rojas N, Feil R, Stitt M (2006) Regulation of secondary metabolism by the carbon-nitrogen status in tobacco: nitrate inhibits large sectors of phenylpropanoid metabolism. Plant J 46:533–548

Fuentes JD, Wang D, Gu L (1999) Seasonal variations in isoprene emissions from a boreal aspen forest. J Appl Meteorol 38:855–869

Ghirardo A, Koch K, Taipale R, Zimmer I, Schnitzler J-P, Rinne J (2010) Determination of *de novo* and pool emissions of terpenes from four common boreal/alpine trees by $^{13}CO_2$ labelling and PTR-MS analysis. Plant Cell Environ 33:781–792

Gleadow RM, Woodrow IE (2002) Defense chemistry of cyanogenic *Eucalyptus cladocalyx* seedlings is affected by water supply. Tree Physiol 22:939–945

Gleadow RM, Foley WJ, Woodrow IE (1998) Enhanced CO_2 alters the relationship between photosynthesis and defence in cyanogenic *Eucalyptus cladocalyx* F. Muell. Plant Cell Environ 21:12–22

Grote R, Niinemets Ü (2008) Modeling volatile isoprenoid emissions – a story with split ends. Plant Biol 10:8–28

Guenther AB, Zimmermann PR, Harley PC, Monson RK, Fall R (1993) Isoprene and monoterpene emission rate variability: model evaluations and sensitivity analyses. J Geophys Res 98:12609–12617

Gutbrodt B, Dorn S, Mody K (2012) Drought stress affects constitutive but not induced herbivore resistance in apple plants. Arthropod-Plant Interact 6:171–179

Hale BK, Herms DA, Hansen RC, Clausen TP, Arnold D (2005) Effects of drought stress and nutrient availability on dry matter allocation, phenolic glycosides, and rapid induced resistance to two lymantriid defoliators. J Chem Ecol 31:2601–2620

Helmig D, Ortega J, Guenther A, Herrick J, Geron C (2006) Sesquiterpene emissions from loblolly pine and their potential contribution to biogenic aerosol formation in the southeastern US. Atmos Environ 40:4150–4157

Herms DA, Mattson WJ (1992) The dilemma of plants: to grow or defend. Q Rev Biol 67:283–335

Himanen SJ, Nissinen A, Auriola S, Poppy GM, Stewart CN, Holopainen JK, Nerg A (2008) Constitutive and herbivore-inducible glucosinolate concentrations in oilseed rape (*Brassica napus*) leaves are not affected by Bt Cry1Ac insertion but change under elevated atmospheric CO_2 and O_3. Planta 227:427–437

Holopainen JK, Kainulainen P (2004) Reproductive capacity of the grey pine aphid and allocation response of Scots pine seedlings across temperature gradients: a test of hypotheses predicting outcomes of global warming. Can J For Res 34:94–102

Horner JD (1990) Non-linear effects of water deficits on foliar tannin concentration. Biochem Syst Ecol 18:211–213

Jenkins GI (2009) Signal transduction in responses to UV-B radiation. Annu Rev Plant Biol 60:407–431

Kangasjärvi J, Jaspers P, Kollist H (2005) Signalling and cell death in ozone-exposed plants. Plant Cell Environ 28:1021–1036

Karnosky DF, Pregitzer KS, Zak DR, Kubiske ME, Hendrey GR, Weinstein D, Nosal M, Percy KE (2005) Scaling ozone responses of forest trees to the ecosystem level in a changing climate. Plant Cell Environ 28:965–981

Kesselmeier J, Saudt M (1999) Biogenic volatile organic compounds (VOC): an overview on emission, physiology, and ecology. J Atmos Chem 33:23–88

Kharin VV, Zwiers FW, Zhang X, Hegerl GC (2007) Changes in temperature and precipitation extremes in the IPCC ensemble of global coupled model simulations. J Climate 20:1419–1444

Koricheva J, Larsson S, Haukioja E, Keinanen M (1998) Regulation of woody plant secondary metabolism by resource availability: hypothesis testing by means of meta-analysis. Oikos 83:212–226

Körner C, Asshoff R, Bignucolo O, Hattenschwiller S, Keel SG, Pelaez-Riedl S, Pepin S, Siegwolf RTW, Zotz G (2005) Carbon flux and growth in mature deciduous forest trees exposed to elevated CO_2. Science 309:1360–1362

Kreuzwieser J, Gessler A (2010) Global climate change and tree nutrition: influence of water availability. Tree Physiol 30:1221–1234

Kuokkanen K, Julkunen-Tiitto R, Keinanen M, Niemela P, Tahvanainen J (2001) The effect of elevated CO2 and temperature on the secondary chemistry of *Betula pendula* seedlings. Trees 15:378–384

Laothawornkitkul J, Paul ND, Vickers CE, Possell M, Taylor JE, Mullineaux PM, Hewitt CN (2008) Isoprene emissions influence herbivore feeding decisions. Plant Cell Environ 31:1410–1415

Lavola A, Julkunen-Tiitto R, Aphalo P, de la Rosa T, Lehto T (1997) The effect of U.V.-B radiation on U.V.-absorbing secondary metabolites in birch seedlings grown under simulated forest soil conditions. New Phytol 137:617–621

Leakey ADB, Ainsworth EA, Bernacchi CJ, Rogers A, Long SP, Ort DR (2009) Elevated CO_2 effects on plant carbon, nitrogen, and water relations: six important lessons from FACE. J Exp Bot 60:2859–2876

Li FR, Peng SL, Chen BM, Hou YP (2010) A meta-analysis of the responses of woody and herbaceous plants to elevated ultraviolet-B radiation. Acta Oecol 36:1–9

Lindroth RL (2010) Impacts of elevated atmospheric CO_2 and O_3 on forests: phytochemistry, trophic interactions, and ecosystem dynamics. J Chem Ecol 36:2–21

Loreto F, Ciccioli P, Cecinato A, Brancaleoni E, Frattoni M, Tricoli D (1996) Influence of environmental factors and air composition on the emission of a-pinene from *Quercus ilex* leaves. Plant Physiol 110:267–275

Lower SS, Kirshenbaum S, Orians CM (2003) Preference and performance of a willow-feeding leaf beetle: soil nutrient and flooding effects on host quality. Oecologia 136:402–411

Lukac M, Calfapietra C, Lagomarsino A, Loreto F (2010) Global climate change and tree nutrition: effects of elevated CO_2 and temperature. Tree Physiol 30:1209–1220

McKenzie RL, Aucamp PJ, Bais AF, Bjorn LO, Ilyas M, Madronich S (2011) Ozone depletion and climate change: impacts on UV radiation. Photochem Photobiol Sci 10:182–198

Min S-K, Zhang X, Zwiers FW, Hegerl GC (2011) Human contribution to more-intense precipitation extremes. Nature 470:378–381

Mott KA (2009) Stomatal responses to light and CO_2 depend on the mesophyll. Plant Cell Environ 32:1479–1486

Neilson EH, Goodger JQD, Woodrow IE, Møller BL (2013) Plant chemical defense: at what cost? Trends Plant Sci 18:251–258

Novick KA, Katul GG, McCarthy HR, Oren R (2012) Increased resin flow in mature pine trees growing under elevated CO_2 and moderate soil fertility. Tree Physiol 32:752–763

O'Gorman PA, Schneider T (2009) The physical basis for increases in precipitation extremes in simulations of 21st-century climate change. Proc Natl Acad Sci U S A 106:14773–14777

Paajanen R, Julkunen-Tiitto R, Nybakken L, Petrelius M, Tegelberg R, Pusenius J, Rousi M, Kellomaki S (2011) Dark-leaved willow (*Salix myrsinifolia*) is resistant to three-factor (elevated CO_2, temperature and UV-B-radiation) climate change. New Phytol 190:161–168

Pall P, Allen MR, Stone DA (2007) Testing the Clausius-Clapeyron constraint on changes in extreme precipitation under CO_2 warming. Climate Dynam 28:351–363

Pall P, Aina T, Stone DA, Stott PA, Nozawa T, Hilberts AGJ, Lohmann D, Allen MR (2011) Anthropogenic greenhouse gas contribution to flood risk in England and Wales in autumn 2000. Nature 470:382–386

Pegoraro E, Rey A, Barron-Gafford G, Monson RK, Malhi Y, Murthy R (2005) The interacting effects of elevated atmospheric CO_2 concentration, drought and leaf-to-air vapour pressure deficit on ecosystem isoprene fluxes. Oecologia 146:120–129

Peltonen PA, Vapaavuori E, Julkunen-Tiitto R (2005) Accumulation of phenolic compounds in birch leaves is changed by elevated carbon dioxide and ozone. Glob Chang Biol 11:1305–1324

Raisanen T, Ryyppo A, Julkunen-Tiitto R, Kellomaki S (2008) Effects of elevated CO_2 and temperature on secondary compounds in the needles of Scots pine (*Pinus sylvestris* L.). Trees 22:121–135

Rennenberg H, Loreto F, Polle A, Brilli F, Fares S, Beniwal RS, Gessler A (2006) Physiological responses of forest trees to heat and drought. Plant Biol 8:556–571

Rosenstiel TN, Potosnak MJ, Griffin KL, Fall R, Monson RK (2003) Increased CO_2 uncouples growth from isoprene emission in an agriforest ecosystem. Nature 421:256–259

Roth S, McDonald EP, Lindroth RL (1997) Atmospheric CO_2 and soil water availability: consequences for tree insect interactions. Can J For Res 27:1281–1290

Sallas L, Luomala EM, Utriainen J, Kainulainen P, Holopainen JK (2003) Contrasting effects of elevated carbon dioxide concentration and temperature on rubisco activity, chlorophyll fluorescence, needle ultrastructure and secondary metabolites in conifer seedlings. Tree Physiol 23:97–108

Schrope M (2000) Successes in fight to save ozone layer could close holes by 2050. Nature 408:627

Searles PS, Flint SD, Caldwell MM (2001) A meta-analysis of plant field studies simulating stratospheric ozone depletion. Oecologia 127:1–10

Sharkey TD, Singsaas EL, Vanderveer PJ, Geron C (1996) Field measurements of isoprene emission from trees in response to temperature and light. Tree Physiol 16:649–654

Sipura M, Ikonen A, Tahvanainen J, Roininen H (2002) Why does the leaf beetle *Galerucella lineola* F. attack wetland willow? Ecology 83:3393–3407

Solomon S, Qin D, Manning M et al (eds) (2007) Climate change 2007: the physical science basis contribution of working group I to the fourth assessment report of the Intergovernmental Panel on Climate Change. Cambridge University Press, Cambridge

Stiling P, Cornellisen T (2007) How does elevated carbon dioxide (CO_2) affect plant-herbivore interactions? A field experiment and meta-analysis of CO_2-mediated changes on plant chemistry and herbivore performance. Glob Chang Biol 13:1823–1842

Sun Y, Solomon S, Dai A, Portmann RW (2007) How often will it rain? J Climate 20:4801–4818

Tegelberg R, Aphalo PJ, Julkunen-Tiitto R (2002) Effects of long-term, elevated ultraviolet-B radiation on phytochemicals in the bark of silver birch (Betula pendula). Tree Physiol 22:1257–1263

Tingey DT, Manning M, Grothaus LC, Burns WF (1980) Influence of light and temperature on monoterpene emission rates from slash pine. Plant Physiol 65:797–801

Trenberth KE, Dai A, Rasmussen RM, Parsons DB (2003) The changing character of precipitation. Bull Am Meteorol Soc 84:1205–1217

Valkama E, Koricheva J, Oksanen E (2007) Effects of elevated O_3, alone and in combination with elevated CO_2, on tree leaf chemistry and insect herbivore performance: a meta-analysis. Glob Chang Biol 13:184–201

Veteli TO, Kuokkanen K, Julkunen-Tiitto R, Roininen H, Tahvanainen J (2002) Effects of elevated CO_2 and temperature on plant growth and herbivore defensive chemistry. Glob Chang Biol 8:1240–1252

Veteli TO, Mattson WJ, Niemela P, Julkunen-Tiitto R, Kellomaki S, Kuokkanen K, Lavola A (2007) Do elevated temperature and CO_2 generally have counteracting effects on phenolic phytochemistry of boreal trees? J Chem Ecol 33:287–296

Vingarzan R (2004) A review of surface ozone background levels and trends. Atmos Environ 38:3431–3442

Williams RS, Lincoln DE, Norby RJ (2003) Development of gypsy moth larvae feeding on red maple saplings at elevated CO_2 and temperature. Oecologia 137:114–122

Wittig VE, Ainsworth EA, Naidu SL, Karnosky DF, Long SP (2009) Quantifying the impact of current and future tropospheric ozone on tree biomass, growth, physiology and biochemistry: a quantitative meta-analysis. Glob Chang Biol 15:396–424

Woodrow IE, Berry JA (1988) Enzymatic regulation of photosynthetic CO_2 fixation in C3 Plants. Annu Rev Plant Physiol Plant Mol Biol 39:533–594

Woodrow IE, Slocum DJ, Gleadow RM (2002) Influence of water stress on cyanogenic capacity in Eucalyptus cladocalyx. Funct Plant Biol 29:103–110

Zhao D, Reddy KR, Kakani VG, Read JJ, Sullivan JH (2003) Growth and physiological responses of cotton (Gossypium hirsutum L) to elevated carbon dioxide and ultraviolet-B radiation under controlled environmental conditions. Plant Cell Environ 26:771–782

Zvereva EL, Kozlov MV (2006) Consequences of simultaneous elevation of carbon dioxide and temperature for plant-herbivore interactions: a metaanalysis. Glob Chang Biol 12:27–41

Chapter 9
Control over Growth in Cold Climates

Sergio Rossi, Annie Deslauriers, Carlo Lupi, and Hubert Morin

Abstract Trees harmonize their growing cycles with the natural seasonal changes, and this is crucial in ecosystems with marked climatic differences between the periods favorable and unfavorable to the physiological activities. Wood formation, or xylogenesis, is a complex and fascinating example of an intermittent, temperature-sensitive growth process that can be investigated at several temporal scales, from daily to annual. The period in which wood formation occurs is the time window when xylem is differentiating and when environmental factors can act directly on the cells constituting the tree ring and, therefore, on wood characteristics and properties. In this chapter, the timings and dynamics of the different phases of xylem cell differentiation are described in detail. The role of some environmental factors affecting xylogenesis at short time scales, such as temperature, photoperiod, dates of snowmelt and soil nitrogen are also discussed for forest ecosystems of cold climates. Although many questions still remain unanswered, the recent findings from monitoring xylogenesis have provided valuable cues for improving the understanding of the physiology and ecology of secondary growth in trees.

9.1 Introduction

Living organisms must harmonize their cycles with the natural seasonal changes (Purnell 2003). This adaptation is essential in the temperate and boreal regions of the world, which experience alternating and distinct winter-summer successions, revealed by the growth layers of decay fungi fruiting bodies, the pseudo-concentric growths of some lichen thalli, the needles of different ages along branches of

S. Rossi (✉) • A. Deslauriers • C. Lupi • H. Morin
Département des Sciences Fondamentales, Université du Québec à Chicoutimi, 555 Boulevard de l'Université, Chicoutimi, QC G7H2B1, Canada
e-mail: sergio.rossi@uqac.ca

M. Tausz and N. Grulke (eds.), *Trees in a Changing Environment*,
Plant Ecophysiology 9, DOI 10.1007/978-94-017-9100-7_9,
© Springer Science+Business Media Dordrecht 2014

evergreen tree species, the modular lengthening of below- and above-ground tree corms and the cyclic growth layers of wood. In trees, the favorable and adverse climatic conditions occurring during the year irreversibly mark the wood with the annual rings.

Tree rings are the result of a gradual accumulation of cells that the plants produce to renew and increase their transport system, to store substances and assure mechanical support. At the basis of this process is the cambium, a meristematic tissue constituted by cells whose task is to divide incessantly without differentiating and to produce new cells. A few of the produced cells (initials) will maintain the meristematic aptitude, while the others (derivatives) will develop to produce the transport system (xylem and phloem). So, a tree ring is the product of a cycle of activity of cambium which, in the boreal and temperate regions, corresponds to the period between spring and autumn.

In dendrochronology, tree-ring width or density are considered as a whole and time series are analyzed at an annual scale (Fritts 1976). However, wood formation is a slow and complex process of cell division, growth and maturation, so the annual resolution is insufficient to perform suitable climate-growth relationships at the correct, fine temporal scale. Instead, the understanding of the mechanisms and dynamics of wood formation, timings of xylem production and differentiation and the assessing of the climatic and physiological factors influencing stem radial growth require a time resolution shorter than 1 year (Eckstein 2004; Rossi and Deslauriers 2007a). A number of reasons contribute to this need:

1. In temperate and boreal regions, growth is intermittent, with cell division and differentiation taking place only in a distinct period of the year and mainly within a few months. In conifers, xylogenesis was observed to last from 100 to 150 days (Rossi et al. 2008b; Gruber et al. 2010; Moser et al. 2010). An intra-annual resolution allows the observations or analyses to be matched with the time scale at which growth occurs. Nevertheless, the intra- and inter-specific variability and the overall period of xylogenesis still remain partially or even completely unknown in most species, thus preventing an effective experimental design of some investigations or a clear interpretation of tree growth as well as its relationship with climatic factors.

2. The period in which wood formation occurs is the time window when xylem is differentiating and when environmental factors can act directly on the cells constituting the tree ring and, therefore, on wood characteristics (e.g. cell lumen, area, wall thickness and percentage of earlywood-latewood). In *Abies balsamea* seedlings, reductions of up to 50 % in lumen area and cell diameter were observed in the cells produced during a period of water stress in June (Rossi et al. 2009b). On the contrary, no reduction in size was observed in the cells produced after the drought treatment had ceased and water supply had entirely recovered.

3. Plants routinely experience meteorological episodes that occur, as the name itself suggests, punctually, such as late frosts, extreme spring temperatures, intense rainfalls, early-autumn snowfalls. In other cases, contrasting events can

alternate during short periods, such as the widely fluctuating availability of water in sandy soils with irregular or sporadic rainfall regimes.

4. Phenology and climate-growth relationships are age- and/or size-dependent. Although neither anatomical change nor physiological abnormalities have been observed in the cambia of older trees, different patterns of cambial activity, durations of growth, sensitivity and periods of significant responses to climate have been observed in trees of different ages and sizes (Szeicz and MacDonald 1994, 1995; Carrer and Urbinati 2004; Rossi et al. 2008a). As a consequence, the need arises to verify the available growth-climate models with investigations at high resolution including short time scale dynamics of growth compared with weather variables directly measured on site.

5. It is an oversimplification to consider the tree-ring as a mere collection of cells accumulated one after another like bricks in a wall. The conducting system of vascular plants is a periodically-renewable tissue originated by complex evolutionary dynamics providing the production and complete differentiation of a population of cells forming the xylem. During xylogenesis, the phases of xylem differentiation of each cell are separated in space and time according to a partial and overlapping delay. The cells produced by cambium and disposed along the radial row gradually undergo maturation in the same order as they are produced, following the rule of "first in first out".

The above arguments imply that (i) each cell is produced and differentiated in a specific and exclusive period of time, and consequently, (ii) each cell incorporates a particular and unique climatic signal occurring during its development. Although just a few days could be required for cells to complete maturation, a latewood tracheid of *Picea abies* produced in mid-August can remain in differentiation for 40 days, not reaching the mature stage until the end of September, and more than 30 days are necessary to complete secondary wall formation in latewood of *A. balsamea* (Deslauriers et al. 2003a; Rossi et al. 2006a). Determination of the timings and rates of cell formation is therefore crucial when assessing the effects of climate on growth.

Because 50–60 % of structural carbohydrates are allocated in the branches and stems in young conifers (Bernoulli and Körner 1999), and greater amounts are expected in older trees, the dynamics of biomass accumulation are strictly related to the secondary growth of wood. Understanding how and when plants produce their annual structures during the growing season and react to environmental conditions in natural, fast growing and urban systems is therefore of great interest (Downes et al. 2002). The aim of this chapter is (i) to describe how wood production takes place at short time scales and to assess the different phases of xylem cell differentiation in the stem, and (ii) to discuss the most recent findings on the role of some important environmental factors affecting growth in forests of cold climates.

9.2   Historical Overview of the Methods

During the past decade, the ecological and economic importance of forest ecosystems has encouraged investigations on cambial activity using time scales shorter than 1 year (Wimmer et al. 2002; Deslauriers et al. 2003a, b; Downes et al. 2004; Rossi and Deslauriers 2007a). However, the cambium and its derivatives have been a stimulating subject of exploration by plant physiologists and ecologists for more than a century. The first observations on the duration of wood formation with histological examinations made with a light microscope date from the beginning of the twentieth century (Knudson 1913). The early approaches were also based on the systematic separation between early wood and late wood structures in conifers (Mork 1928). Growth was also periodically monitored using callipers (Buckhout 1907), a technique that would subsequently be replaced by manual and, later, automatic dendrometers. In the 1950s, new tools (dendrographs) were developed with the aim of continuously recording the radial changes of the stem, allowing the first accurate descriptions of diurnal radius variations (Fritts and Fritts 1955; Kozlowski and Winget 1964). Wolter (1968) proposed and implemented the fascinating idea of piercing the cambium with a thin needle. Although it remained latent for several years, this technique is now known as 'pinning' and is applied worldwide (Nobuchi et al. 1995; Schmitt et al. 2004; Seo et al. 2007, 2008). After the definition of field and lab procedures, several papers were published in the 1970s with more emphasis on data analysis and the relationships with environmental factors (Denne 1971, 1974; Wodzicki 1971). At the same time, Eckstein et al. (1977) began to explore the spatial structure of the tree-ring components, focusing on the size of the cells and their walls. This topic was examined more analytically by Vaganov (1990) with the tracheidogram method.

Investigations have directly attempted to link environmental factors with xylogenesis (Antonova and Stasova 1997; Deslauriers and Morin 2005; Seo et al. 2008) or with the radial variation of the stem (Downes et al. 1999; Deslauriers et al. 2003b, 2007a; Bouriaud et al. 2005; Zweifel et al. 2006). Several methods are now available to perform such relationships on a short time scale: they can be divided into direct observations or indirect measurements, according to the type of data collection. The direct observations are based on thin sections of wood samples prepared and stained to examine the xylem cells for assessing the phenology of xylem formation (Makinen et al. 2003; Schmitt et al. 2004; Seo et al. 2008; Deslauriers et al. 2008; Rossi et al. 2010a), with some works focusing on a precise event, such as the onset (Oribe et al. 2001; Gričar et al. 2006) or end of tracheid formation (Gindl et al. 2000; Gričar et al. 2005), or both (Lupi et al. 2010). Indirect measurements employ tools (manual or automatic dendrometers) installed on the stem to measure the variation in size over time. By providing time series composed of both tree growth and circadian rhythms of water storage and depletion, dendrometers can be used to indirectly quantify the growth, although care should be taken when interpreting their results (Deslauriers et al. 2007a, b; Giovannelli et al. 2007; Drew et al. 2008, 2009; Turcotte et al. 2009).

9.3 Dynamics of Wood Formation at Cell Level

Intra-annual analyses of wood formation decompose the growing season into short periods through sampling at different temporal intervals (from minutes to weeks), according to the technique used. This approach produces successive pictures of the condition of the tissues arranged in sequential order. Each picture reflects the state of growth at the moment of sampling. When reporting all these pictures along the temporal axis, xylogenesis, or stem growth, can be visualized and then analyzed. The intra-annual analyses of wood formation provide chronologies of the developing cell numbers or stem radial variation by taking into account the growth dynamics and allowing some key questions in ecophysiology to be answered: what is the phenology of cambial activity? How does xylogenesis vary between species? How does climate influence xylogenesis?

9.3.1 Dynamics of Cambial Activity

In spring, the initial cells divide periclinally, producing xylem and phloem mother cells that divide again through repeated tangential divisions occurring on both sides of the radial file leading the cambial zone to appear expanded. Derivative cells gradually develop, assuming typical characteristics of a phloem or xylem cell and achieving the level of specialization and physiological stability characterizing the component plant tissues.

In boreal and temperate regions, the number of cambial cells changes during the year (Fig. 9.1). In autumn and winter, when there is no cell production, dormant cambium is constituted by a minimum number and close-together group of cells, generally six to eight in alpine conifers. In spring, cambial cells swell and expand radially, radial cell walls grow thin, cytoplasm assumes shinier and less densely granulate features and the nuclei enlarge (Fahn and Werker 1990). Inside the protoplasts, the numerous small vacuoles gradually merge, forming one or two larger vacuoles (Oribe and Kubo 1997). The swelling is followed by periclinal divisions and production of phloem and xylem derivatives. The cambial zone in *Pinus cembra* can attain nine to ten cells by the end of April, while in *Larix decidua* and *P. abies*, the number of cambial cells increases in mid-May. The cell number increases to 10–14 during the period of maximum growth activity in May–June, as also reported by Bannan (1962) for slow growth rate trees. Once annual activity has ended, the cambium stops dividing and the number of cells in the cambial zone reduces to a minimum value corresponding to quiescence conditions around the middle of August (Fig. 9.1).

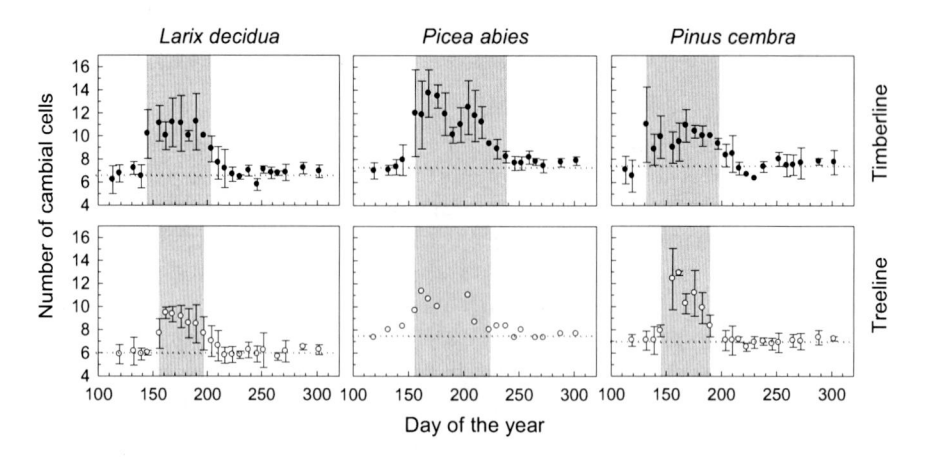

Fig. 9.1 Numbers of cells in the cambial zone of *Larix decidua*, *Picea abies* and *Pinus cembra* during 2004 at the alpine timberline (*black dots*) and treeline (*white dots*). *Error bars* and *horizontal dotted line* indicate one standard deviation among trees and the number of dormant cambial cells, respectively. Periods of cambial activity, when deviation bars do not cross the *horizontal dotted line*, are highlighted in *grey* (Adapted from Rossi et al. 2007)

9.3.2 **Cambial Cells Versus Differentiating Cells**

During the period of tree-ring formation, all species show clear trends of xylem production and differentiation, with the radial rows of differentiating cells having a similar pattern of variation during the year (Camarero et al. 1998; Deslauriers et al. 2003a, 2009; Rossi et al. 2006a). At the start of the growing season (i.e. from the end of April to the beginning of May for *L. decidua*, Fig. 9.1), when only cell division occurs, the cambial zone increases. Once cell enlargement begins (May 12th in *L. decidua*, Fig. 9.2), an equilibrium between cell division and differentiation determines a temporary arrangement of the width of the cambial region. When the division rate slows down (June 16th–June 23rd in *L. decidua*), cell differentiation occurs faster than cell division, the cambial zone begins to narrow and the radial number of enlarging cells for *L. decidua* increases to 11–12. Analogous linkages can be observed between the successive developmental phases.

These observed variations in cell numbers during the development phases are distinctive and characteristic. Along a row, cells in the same differentiation phase constitute a queue and are strictly interconnected with the other developing cells (Ford et al. 1978; Fahn and Werker 1990). During wood formation, the variations in rate of cell division and differentiation determine variations among the cell queues, which result in a typical annual pattern consisting of three delayed bell-shaped curves (cambial, cell enlargement and wall thickening cells) and a growing s-shaped curve (mature cells) (Figs. 9.1 and 9.2). The bell-shaped patterns are connected with the number of cells passing through each differentiation phase, whereas the s-shaped curves are associated with the gradual accumulation of mature cells in the tree ring.

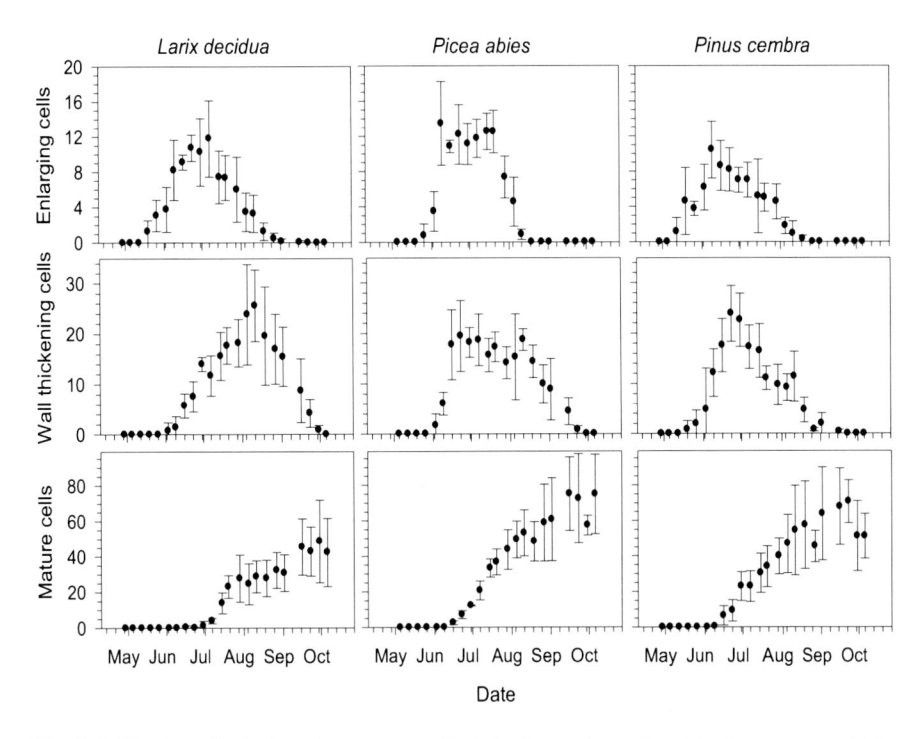

Fig. 9.2 Number of cells in enlargement, wall thickening, and number of mature xylem cells in *Larix decidua*, *Picea abies* and *Pinus cembra* during 2003. *Dots* and *bars* represent average number of cells and standard deviations among five trees, respectively (Adapted from Rossi et al. 2006a)

9.3.3 Dynamics of Cell Differentiation

During development, the cambial derivatives alter both morphologically and physiologically, progressively assuming definite features. In other words, the derivatives differentiate into the specific elements of the stem tissues. In the xylem cells, this process is associated with changes in both cell wall composition and organization. Each species shows specific timings of tree-ring formation, which are maintained over the years (Rossi et al. 2007c). In general, the start of radial enlargement at the alpine timberline is observed in mid-May, while the end of xylogenesis, which corresponds to the conclusion of lignification, occurs at the end of September. The duration of wood formation includes all the differentiation phases from the onset of cell enlargement to the end of maturation and, in these environments, lasts about 4 months.

During cell enlargement, the derivatives consist of a protoplast still enclosed in the thin and elastic primary wall. Following positive turgor increase by water movement into the vacuoles, the cell wall stretches, increasing the radial diameter of the tracheid and consequently, the lumen area. This process occurs despite strong compression, the cells being enclosed between xylem tissues and bark and deformed files of tracheids can generally be observed (Rossi et al. 2006a). In cross section, observations under polarized light discriminate between enlarging and cell wall thickening cells. Because of the arrangement of the cellulose microfibrils, the developing secondary walls shine when observed under polarized light (Rossi et al. 2006a). Instead, no shining is observed in enlargement zones where cells are still composed of just primary wall (Donaldson 1991; Abe et al. 1997).

Once their final size had been reached, the cells begin maturing through cell wall thickening and lignification. These processes are coordinated by the expression of several genes specifically involved in the synthesis of cell wall constituents, such as polysaccharides, structural lignins and proteins (Plomion et al. 2001). The deposition of microfibrils of cellulose and hemicelluloses starts the formation of the three layers constituting the secondary wall. The lignin is deposited in the intercellular layers and the intermicrofibril spaces of the wall, thereby cementing the microfibrils together. The progress of cell wall lignification can be directly detected by staining sections with cresyl violet acetate that reacts with lignin (Kutscha et al. 1975; Antonova and Shebeko 1981). The observations of lignin incorporation with light microscopy have been confirmed by results using electron microscopy and UV-microspectrophotometry (Gričar et al. 2005). Lignification is shown by a color change from violet (unlignified secondary cell walls) to blue (lignified cell walls). At first, lignin deposition is observed at the cell corners on the primary wall, then extending on intercellular layers and, finally, on the secondary wall. The lignification pattern is the same in all species and is related to the spread of the lignin precursors diffusing through the secondary wall to the external parts of the cell wall (Donaldson 1991). The lignin distribution within developing cells corresponds to the delay between the polysaccharide deposition and lignin incorporation (Gindl et al. 2000; Gričar et al. 2005). The final stages of maturation involve autolysis of the protoplast. A blue color over the whole cell wall indicates the end of lignification and the attainment of the mature stage for the tracheid.

9.4   Temperature, Photoperiod, and Their Interaction

Unlike other organisms, plants are unable to escape the unfavorable periods of the year so the meristems suspend their activity. As the photoperiod shortens during late summer and autumn, cells produce sugars and amino acids that protect tissues from cold damage, so supplying an excellent survival tool and adapting plants to the conditions prevailing in the cooler temperate zones of the world (Stern et al. 2003). Protective scales develop on buds, leaves become senescent on deciduous trees and cell metabolism slows down, initiating dormancy, which lasts from a few days to several months. During this state, growth inactivity of seeds, buds, bulbs, and other plant organs is maintained even if environmental requirements of temperature, water, or day length are met. In late winter, a change from dormancy to a new state occurs, the quiescence, where growth cannot occur unless the environmentally-favorable conditions required are present.

It is not surprising that temperature is one of the key ecological factors controlling growth in temperate and boreal ecosystems. However, although numerous hypotheses have been proposed and discussed (Stevens and Fox 1991; Körner 2003; Sveinbjörnsson 2000), neither the control mechanisms (e.g. gradual influences versus threshold effects) nor the physiological processes (e.g. carbon assimilation versus allocation) involved have yet been clearly and definitively demonstrated. One of the main questions regarding trees in cold environments is whether the observed abrupt reduction in growth at low temperatures is caused by an insufficiency of assimilated carbon or a reduction of carbon investment in structural growth (the sink limitation hypothesis, Körner 1998). A definitive explanation on the role of temperature at the cell level (i.e. at the resolution at which the formation of the new tissues occurs) is still missing, even if some recent findings indicate that cold temperatures limit sink (carbon allocation) rather than source (carbon assimilation) activity in plant tissues, repeatedly validating Körner's hypothesis (Cavieres et al. 2000; Hoch et al. 2002; Hoch and Körner 2003; Körner and Paulsen 2004; Piper et al. 2006; Alvarez-Uria and Körner 2007).

The intra-annual analyses are suitable tools for investigating the relationships between temperature and stem growth. In Finland, the beginning of xylogenesis in *Pinus sylvestris* and *Betula pendula* is strongly controlled by temperature (Schmitt et al. 2004). Deslauriers and Morin (2005) demonstrated that in earlywood of *A. balsamea* in Canada, the higher temperatures corresponded to higher rates of cell production. Rossi et al. (2007c) compared xylem phenology in three conifers at the alpine treeline with air, soil and stem temperatures to define at daily scale the thermal thresholds attained when growth processes were occurring. In *L. decidua*, *P. cembra* and *P. abies*, xylogenesis is active in spring, when the minimum air and stem temperatures reach 2–4 and 4 °C, respectively.

Analyses at a wide geographical scale comparing several conifer species confirm the convergence of thermal thresholds around specific critical values, demonstrating the existence of precise thermal limits in radial growth and tree-ring formation (Rossi et al. 2008b, 2010a; Deslauriers et al. 2008). An explanation of the results is provided by Fig. 9.3 that shows the temperature distributions for xylem differentiation in Canadian and European conifers by means of box plots (Rossi et al. 2008b). When the daily temperatures occurring during the year are distinguished based on active xylogenesis, a clear separation into groups is observed. The overlapping within each pair of distributions involves only low percentages of data points and is calculated as 10–15 % of all daily temperatures. In a few cases (i.e. for mean temperature in *A. balsamea*), no overlapping is observed between whiskers, meaning that less than 10 % of the values are in common between the two distributions. Despite the huge difference in temperature between the study sites analyzed by Rossi et al. (2010a), and although location and amplitude of temperature distributions are species-specific, overlapped areas tend to be on similar values of temperatures: these common values for xylem differentiation can be roughly and visually evaluated on the graphs of Fig. 9.3 as ranging between 2–6, 6–8 and 10–14 °C for minimum, mean and maximum air temperature, respectively.

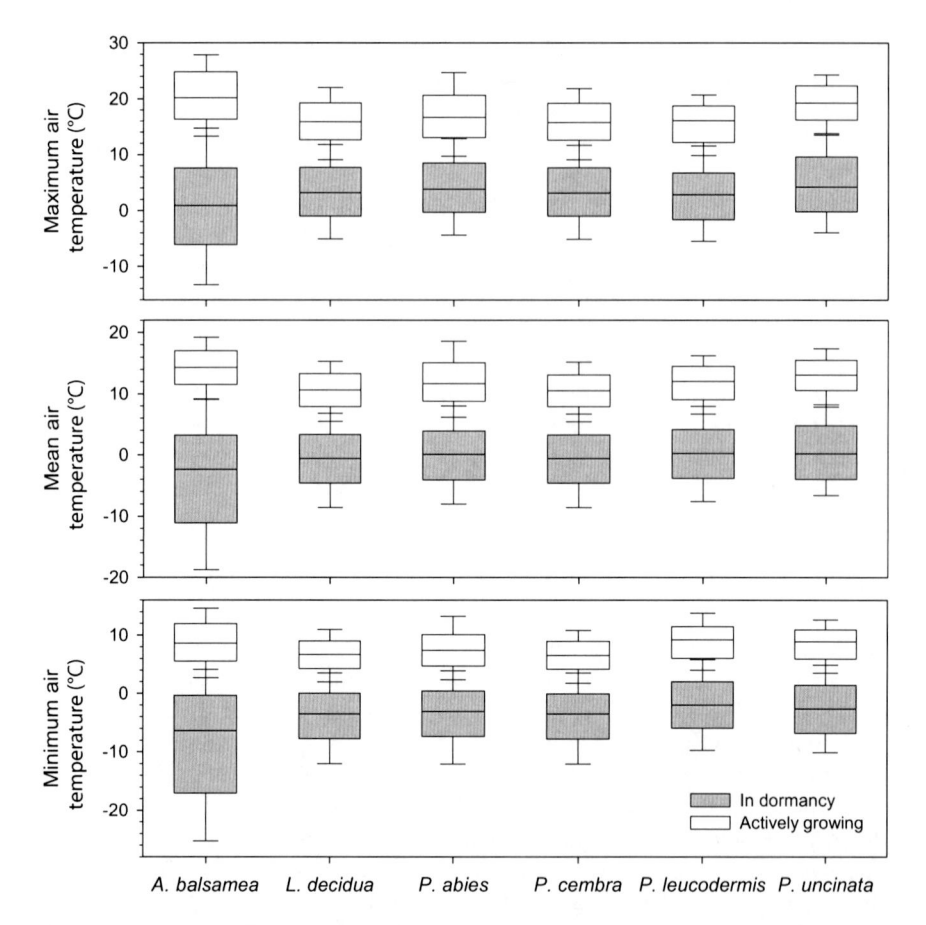

Fig. 9.3 Distribution of maximum, mean and minimum daily temperatures classified according to the status of the xylem, actively growing or in dormancy in *Abies balsamea*, *Larix decidua*, *Picea abies*, *Pinus cembra*, *Pinus leucodermis* and *Pinus uncinata*. *Boxes* represent upper and lower quartiles, whiskers show the 10th and 90th percentiles, median is drawn as *horizontal solid line* (From Rossi et al. 2008b)

Cambium is a sink for non-structural carbohydrates, and cambial activity requires a continuous supply of energy in the form of sucrose, extracted from the storage tissues for the first formed cells, or produced by photosynthesis (Hansen and Beck 1990, 1994; Oribe et al. 2003). During cell maturation, trees assign a large amount of carbon obtained from photosynthesis to the deposition of cellulose microfibrils in order to provide the developing cells with secondary walls (Hansen et al. 1997). According to Rossi et al. (2007c, 2008b), the demand for photo-assimilates by the metabolic processes involved in cell growth should be limited by daily mean temperatures lower than 6–8 °C. Surprisingly, similar temperatures have also been estimated for other plant organs: shoot extension in Scots pine was inhibited by air temperatures lower than 6–8 °C (James et al. 1994), and 5 % of maximum rate of root growth occurred at a soil temperature of 6 °C (Turner and

Streule 1983 in Schönenberger and Frey 1988). It seems that a critical mean temperature of between 6 and 8 °C exists that affects the growth processes in all parts of the tree (shoots, stem and roots). Temperatures below 6 °C modify the phase transition of lipids of cell membranes and dramatically affect water transport in trees (Running and Reid 1980). The convergence of the threshold temperatures estimated in different apical and lateral meristems indicates the existence of similar temperature-dependent mechanisms of growth at cell level in all plant tissues.

In cold climates, is the temperature sufficient to comprehensively explain stem growth? If so, xylem cell production should closely follow the seasonal thermal trend. We are familiar with the bell-shaped pattern of air temperature in most temperate and boreal regions of the northern hemisphere, with a gradual increase in spring, a maximum around the second part of July, and a successive reduction during summer and autumn (Fig. 9.4). It should be expected that all processes of cell production occurring in plants of temperature-limited environments take place in synchrony with the temperature pattern, hence with a culmination around the second part of July. However, a curious aspect of xylogenesis appears when comparing the annual pattern of stem growth at both microscopic (anatomical) and macroscopic (stem variation in size) resolution (Rossi et al. 2006b). The analysis of the curve of xylem cell accumulation and stem variation during the year shows that the production of the xylem tissues takes place during the first part of the growing season, and all studied species attain maximum growth rates in the same period of the year, during the summer solstice, when day length, and not temperature, culminates (Fig. 9.4). After 21st June, cell production gradually decreases until ceasing. These findings suggest that the influence of temperature on cambium interacts with photoperiod, which is a factor independent of temperature and constant in time.

Trees would have evolved by synchronizing their rates of growth with day length, which represents an authentic and reliable natural calendar. The advantages in connecting cambial activity with photoperiod are explained by the need to avoid the period thermally unfavorable to growth. So, increases in stem temperature can induce cell division only in certain periods of the year.

In partially-heated stems of *Cryptomeria japonica*, the extent of cambial activity within the treated regions of the stem varies with the season. The response to heating tends to increase as the period of cambial dormancy progress (Oribe and Kubo 1997). Thus, there is no cambial response to the treatment in December, which corresponds to the dormancy state. Instead, the treatment shows major effects in late winter and spring, when cambium is quiescent and a local increase in temperature can successfully reactivate its cell division and differentiation. The application of high temperatures in *P. abies* drastically affects the rate of cell division in cambium at the beginning of the growing season, although no effect is observed in the second part of the season, indicating that the influence of temperature on cambial activity varies during the year (Gričar et al. 2007). Also, heated trees show higher amounts of tracheids in the xylem and earlier beginning of cell differentiation than control trees. Similar results have been observed at the treeline in southern Italy, where onset and duration of cell production and differentiation of

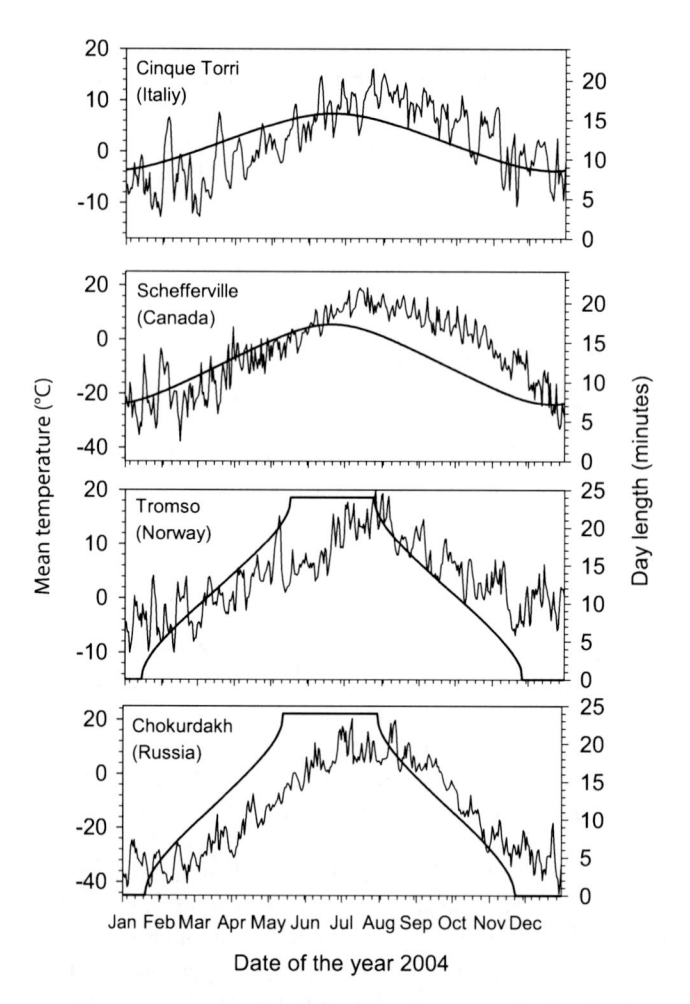

Fig. 9.4 Mean daily temperature and day length measured during 2004 at different latitudes and elevations: alpine treeline [Cinque Torri (46°27′N, 12°08′E, 2,080 m a.s.l.), Cortina d'Ampezzo (BL) Italy], North-American boreal forest [Schefferville (54°48′N, 66°48′W, 518 m a.s.l.), Quebec, Canada], North-European boreal forest [Tromso (69°39′N, 18°56′E, 114 m a.s.l.), Tromsø, Norway], and Russian timberline [Chokurdakh (70°37′N, 147°53′E, 61 m a.s.l.), Yakutia, Russia] (Adapted from Rossi and Deslauriers 2007b)

Pinus leucodermis were drastically affected by the hot spring in 2003 (Deslauriers et al. 2008).

Budburst occurrence in species of temperate and boreal regions is also a function of temperature, but an interesting interaction has been demonstrated with day length (Nizinski and Saugier 1988; Partanen et al. 1998). According to Nizinski and Saugier (1988), the cumulative temperature inducing budburst in *Quercus petraea* decreases during spring with maximum and minimum values observed in April and June, respectively (Fig. 9.5). The relationship between the day length

Fig. 9.5 Relationship between the cumulative mean daily air temperature (during the 10 days before budburst) and the day-length of the budburst date. The eight dates of budburst of *Quercus petraea* were recorded from 1976 to 1983. The fitted line is described by the equation $S = (2.42 \times 10^{-2}D)/[(1.41 \times 10^{-3})D - 1]$, where S and D correspond to heat sum and day length, respectively (From Nizinski and Saugier 1988)

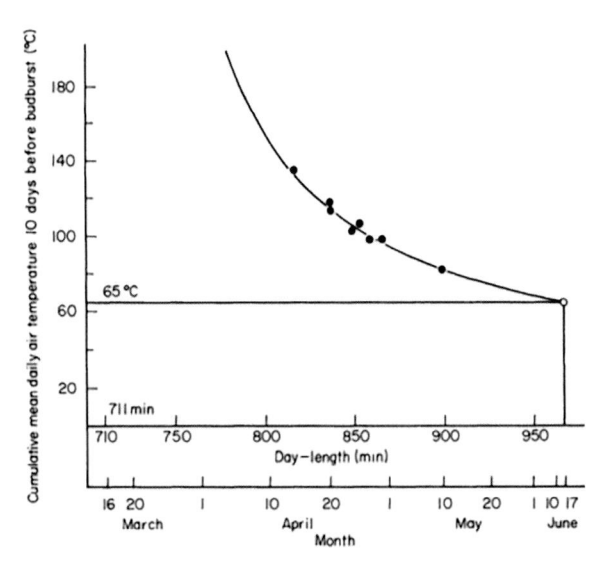

corresponding to the day of budburst and the mean daily temperature summed on the 10 days before budburst was examined. A fascinating hypothesis was supplied by Nizinski and Saugier (1988) that the heat sum (*S*) required for budburst is a decreasing function of day length (*D*) according to:

$$S = \frac{2.42 \times 10^{-2}D}{\left[(1.41 \times 10^{-3})D - 1\right]} \tag{9.1}$$

The asymptotes of the function have a specific biological significance. The boundary to an early budburst is analytically defined by the vertical asymptote: budburst cannot occur when the denominator is negative, i.e. with *D* lower than 711 min, which corresponds to the day length threshold of March 16th. The proportionally-higher temperatures required for budburst in early spring indicate a conservative mechanism leading plants to avoid growth resumptions during the brief late-winter episodes of warmer temperatures. At the latitudes where the study was carried out, the maximal day length is 967 min at the summer solstice. Thus, because of the horizontal asymptote, budburst could not occur if the mean daily air temperatures during the 10 days prior to budburst are below 6.4 °C. In such a condition, no budburst should be observed in *Q. petraea*.

Trees have to guarantee the complete formation of all xylem cells before the adverse season. The possibility of completion of cell division in early summer allows their resources to be relocated to the maturation of the newly-produced xylem tissues. As previously reported in this chapter, a latewood cell of conifer requires between 25 and 40 days to complete differentiation and to be functionally active in the transporting system (Deslauriers et al. 2003a; Rossi et al. 2003, 2006a). Cell divisions prolonged or delayed during late summer or maximum growth rates occurring in July, during the warmest period of the year, imply the maintenance of secondary wall synthesis and lignin deposition of latewood close to winter, when climatic conditions, especially temperature, might no longer be favorable and affect lignin content. Instead, cell production and enlargement occurring early in the growing season represent the better adaptation to allow an optimal functionality of the transport system and the minimum sensibility to biotic and abiotic stresses.

The recent findings from intra-annual analyses of wood formation in cold environments have shown the complex interaction between two climatic factors. Temperature allows the metabolic activities like cell production in cambium and cell maturation in the xylem to be maintained. Photoperiod is involved as signal for the regulation of the growth rate, for synchronizing xylogenesis during the most suitable period of the year. Therefore, in temperate and boreal ecosystems, temperature and photoperiod represent two key factors in the ecology of the radial growth of trees, although their influence has been demonstrated to have effects at different time scales.

9.5 Snow and Soil Nitrogen: Old Hypotheses, New Challenges

9.5.1 Dates of Snowmelt and Cambial Growth

In the boreal forest, the soil is covered by snow during the long winter, with temperatures close to 0 °C. With these thermal conditions, metabolism slows down as much as possible because only minimal physiological activity is allowed (Decker et al. 2003; Körner 2003; Rossi et al. 2008b). Several studies report that, where temperature limits tree growth, the upper soil layer must have started to thaw before radial growth can resume (Graumlich and Brubaker 1986; Cairns and Malanson 1998). Soils at temperatures close to or below 0 °C inhibit root activity and water uptake. Thus cambium, which has synchronous activity in stem and roots (Thibeault-Martel et al. 2008), begins to divide only after the snow has disappeared and soil begun to warm up (Rossi et al. 2007c; Turcotte et al. 2009; Lupi et al. 2010).

In spring, the insulating properties of snow prevent the soil from warming up and maintain roots at a temperature below the minimum threshold for growth (Decker et al. 2003; Körner and Hoch 2006). Only after the snow has completely disappeared and soil has dried out, soil temperature increases abruptly and cambium can activate its irreversible process of cell division, which will stop only in mid-summer (Rossi et al. 2008b). In fact, in cold environments xylem formation commonly occurs during one, uninterrupted, and short period of time (Rossi et al. 2008b). This behavior diverges from other metabolic activities in trees. Needles of evergreen conifers can be temporarily active, showing photosynthetic activity and producing photoassimilates during the warmest days of winter and early spring if liquid water is available for the roots (Bergh and Linder 1999; Goodine et al. 2008).

Cell production is interconnected with the phenological phases of xylem according to causal links. The date of onset of xylogenesis affects the number of cells produced by the cambium which, in turn, influences the ending of cell differentiation (Lupi et al. 2010). As a result, earlier cambial resumptions lengthen the period available for cell division in the secondary meristem, increasing the growth potential during the year (Gričar et al. 2005; Rossi et al. 2007, 2008a; Deslauriers et al. 2008). Also, the higher amount of cells produced by cambium leads to larger accumulations of cells in the developing xylem, increasing the time for differentiation and maturation of the tracheids, and delaying the end of wood formation (Lupi et al. 2010). Based on this causal relationship, it is logically deduced that a climatic factor affecting the date of resumption of growth could indirectly influence cell production.

High air temperatures promote earlier snowmelt, making water rapidly available to the roots for stem rehydration and photosynthesis, and allowing the temperature in the upper layer of soil to rise (Goodine et al. 2008; Turcotte et al. 2009). High spring temperatures could also affect cambial activity by increasing the rate of cell division and the amount of xylem produced (Deslauriers and Morin 2005). However, despite the marked increasing trend of temperatures detected at northern latitudes, a tree-ring and climate analysis in subarctic Eurasia has shown a tendency towards reduced growth in conifers. On the one hand, the observed thinner tree rings have been linked by Vaganov et al. (1999) to a significant increase in snowfall that would have delayed the date of complete snowmelt and, consequently, growth resumption in spring. On the other, the warming is mainly associated with a stronger increase in the minimum temperatures and fewer cold events (Vincent and Mekis 2006) that could had increased nighttime respiration and carbohydrate loss, thus reducing the resources available for growth.

It has been argued that the length of the snow-free period indirectly influences tree growth by acting on the carbon-dioxide uptake and nutrient cycling and availability (Jarvis and Linder 2000). However, the effect of spring temperatures on soil decomposition and mineralization is likely to be delayed until summer, when the production of new cells by cambium is concluded, and to cumulate in time, becoming apparent only in the long term (Bergh et al. 1999). At the heart of the question lies the unresolved problem of disentangling the direct and indirect effects of climatic factors on cambial phenology and cell production in xylem.

The gradual reduction in length of the growing season and wood production at increasing elevations and latitudes is well known (Moser et al. 2010; Rossi et al. 2010a, b). Similarly, gradients of decomposition and mineralization of soil organic matter exist, which modify the availability of nutrients for trees (Jarvis and Linder 2000). Therefore, in the long term, the main direct constraints to growth of boreal forests could be the capture of carbon and nutrients from the soil. Wintertime nutrient cycling and microbial processes are known to strongly affect the annual pattern of internal transfer of nutrients in plants of cold ecosystems (Grogan and Jonasson 2003; Groffman et al. 2009). Although the importance of the date of snowmelt on cambial growth seems unquestionable, the determinant role of nutrient cycling, and especially of nitrogen availability, in the biomass production of the boreal forest could hold some surprises and should be considered and investigated more deeply. So, the role of the most important nutrient for plant growth, soil nitrogen, is examined in detail in the following paragraphs.

9.5.2 Role of Soil Nitrogen on Growth

With its low rates of evapotranspiration and decomposition, the boreal forest exhibits a slow cycle of nutrients, which results in an accumulation of organic matter during stand development (Read et al. 2004; Kielland et al. 2007). Nitrogen (N) is an essential element for plant nutrition and, together with phosphorus, is copiously required for all essential metabolic processes of trees. Surprisingly, despite the huge accumulation of organic matter in its soils, the growth of the boreal forest is considered to be limited by N (Vitousek and Howarth 1991).

Soil N affects size, structure and distribution of the root system. In *Picea abies* and *Betula pendula*, when N is limiting, starch accumulates in leaves, and additional amounts of photosynthates are translocated to increase the size of the root system (Ericsson 1995). It is assumed that, in roots and shoots, the balance between N and C influences the processes associated with C fixation or the formation of new tissues, and determines the allocation of resources between belowground and aboveground components (Ericsson 1995). The proportion of resources allocated to stem and roots can be affected by high exposures to pollutants such as O_3 (Grulke and Balduman 1999). Factors like soil temperature, moisture and nutrient availability play a role in the timings and duration of root growth, while root longevity is controlled by microsite conditions, patterns of development and length of the growing season (Gill and Jackson 2000).

The current increases in temperature and N-depositions due to human activities are expected to strongly modify the mineralization and supply of N, thus affecting the acquisition of resources and cambial phenology (Binkley and Hogberg 1997; Galloway et al. 2004; Campbell et al. 2009) and consequently, tree growth (Nord and Lynch 2009; Lupi et al. 2010). Changes in snowfall can alter the seasonal pattern of N availability, so increased snowfalls are observed to shift the peak of N mineralization from mid-summer to winter in an arctic tundra system (Nord and Lynch 2009). The positive effect of soil temperature on mineralization is well documented (Campbell et al. 2009). With the increase in temperature and changes in distribution and type of precipitations (e.g. rain vs. snow), climate change will likely affect evapotranspiration and nutrient cycling, altering both N and water availability (Campbell et al. 2009). Since N and water regulate plant growth, changes in their availability could affect the acquisition of other resources (Nord and Lynch 2009). The acquisition of NO_3^-, a mobile nutrient form, should increase with the lengthening of leaf growing season, while the acquisition of less mobile nutrients, like NH_4^+ and organic N, which are especially important in the acidic soils of the boreal forest, are more related to the duration of root growth (Nord and

Lynch 2009). This probably explains why, in the first half of the growing season, growth in cold ecosystems (e.g. arctic) often relies more on the nutrients stored in the plant tissues than on those absorbed from the soil (Weintraub and Schimel 2005). In fact, in these ecosystems, even after the air temperature has risen above freezing and solar radiation has been rapidly increasing, soils remain cold (sometimes frozen), inhibiting root growth and nutrient uptake (Weintraub and Schimel 2005; Alvarez-Uria and Körner 2007). So, it has been observed that production and elongation of fine roots starts later and lasts longer than shoot growth (Steinaker and Wilson 2008; Steinaker et al. 2010), although secondary growth of superficial roots can start as early as in the stem. For example, in *Picea mariana* and *A. balsamea*, despite similar onset of xylem formation, xylogenesis lasts 22 days longer in shallow adventitious roots than in the stem, because of a longer duration of cell wall thickening and lignification (Thibeault-Martel et al. 2008).

In boreal species, nutrient contents and the abundance of amino acids exhibit a pronounced seasonality and vary according to tree growth (Millard and Proe 1992; Linder 1995). These patterns should be carefully considered when analyzing tree growth at short time scales. Within plants, the concentrations of amino acids and organic P gradually decrease from May, when the growing season starts, to July. Accordingly, decreases in the concentrations of amino acids and NH_4^+ are observed in the soil, suggesting an intense use of N (Weintraub and Schimel 2005). In spring, the proteins and amino acids stored in stem, roots and older leaves are rapidly remobilized to sustain the reactivation of the meristems (Millard and Proe 1992; Linder 1995). In mature conifers, the internal cycling can provide between 30 % and 60 % of the N contained in the new foliage (Millard and Proe 1992). A fascinating hypothesis suggests that, once plant reserves are depleted, plant growth slows or stops, and root uptake replenishes the internal stores with nutrients for the following year's growth (Weintraub and Schimel 2005). However, high N supplies can abundantly contribute to the complete replenishment of the pre-existing shoots, which are depleted when N is remobilized, and can sustain a potential second flush of growth (Millard and Proe 1992; Kaakinen et al. 2004; Weintraub and Schimel 2005). In consequence, N can play several roles in the growth processes by increasing the photosynthetic capacity as well as stimulating foliage production of trees (Anttonen et al. 2002).

Wood is influenced both directly and indirectly by N nutrition (Makinen et al. 2002; Meyer et al. 2008). As well as stimulating photosynthesis, N is indirectly implied in cell lignification (i.e. amino acids as C-skeleton for lignin synthesis) and affects growth rates (Reich et al. 1997; Canovas et al. 2007; Kaakinen et al. 2007). N availability affects biomass allocation by modifying the root-shoot ratio in seedlings in the second part of the growing season (Kaakinen et al. 2004). A high N availability increases the biomass of the stem in seedlings of *Picea abies*, suggesting an influence on cell division during cambial activity and/or cell size during cell enlargement (Kaakinen et al. 2004). Plants fertilized with N show wider tree rings, which correspond to increased rates of cell division or longer durations of cambial activity, or both (Makinen et al. 2002; Kostiainen et al. 2004). Fertilization also increases the diameter of tracheid lumen and reduces cell wall

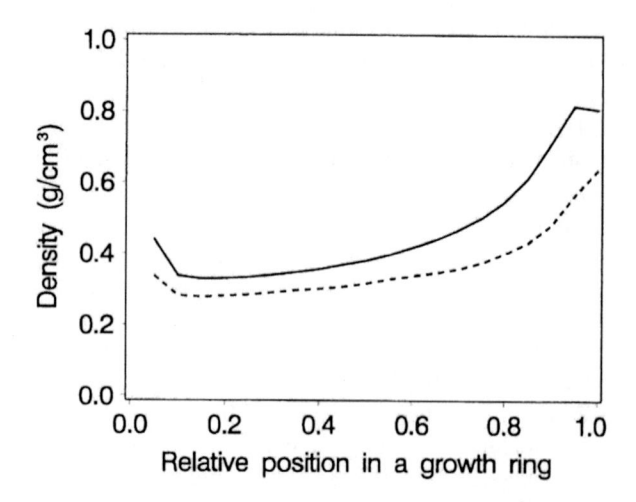

Fig. 9.6 Mean density of earlywood to latewood of control trees (*solid line*) and fertilized trees (*broken line*) in the rings formed during the fertilization experiment. Data at breast height and at a height of 4 m are combined (From Makinen et al. 2002)

thickness and latewood proportion, which affects the density of the resulting wood (Anttonen et al. 2002; Makinen et al. 2002; Kostiainen et al. 2004). The observed increases in ring-width and decreases in wood density are in part the result of the increased proportions of earlywood, which has a lower density than latewood (Kostiainen et al. 2004, Fig. 9.6). The decrease in density is mainly observed towards the outer edge of the rings in *Picea abies* subjected to a nutrient optimization experiment in northern Sweden (Fig. 9.6). Also, in a 70-year-old stand of lodgepole pine in western Canada, N-addition affected the length of the tracheids, with the shorter tracheids observed at the highest amounts of N supplied (Yang et al. 1988). Wood density decreases with fertilization, but only during the first 5 years following N-enrichment, and this decline coincides with greater growth in volume (Yang et al. 1988). Decreases in tracheid length and density produce softer tree rings, with important consequences for the mechanical properties (Meyer et al. 2008). For example, Meyer et al. (2008) observed that *Picea abies* broken by windstorm showed a higher concentration of N in the wood, although the responses can vary with site according to soil features and N-availability (Kaakinen et al. 2007; Meyer et al. 2008).

Even if N is not present in lignin, amino acids are involved in its biosynthesis: they supply C-skeletons for lignin formation and release NH_4^+, which is later reassimilated (Canovas et al. 2007). Fertilization can directly affect wood chemistry by increasing the concentration of lignin, as observed in fertilized *Picea abies* (Anttonen et al. 2002). Similarly, extractives, soluble sugars, sterols and dehydroabietic acid have also been observed to increase with fertilization (Anttonen et al. 2002). However, the effects are site-dependent: Kaakinen et al. (2007) found that in the more northern site, where fertilization resulted in increased height and diameter, the ratios lignin:N and C:N decreased, while in the southern site, extractives increased although growth was unaffected.

Changes in nutrient supply can also alter the availability of photosynthates and, consequently, the chemistry of the growing tissues, according to strengths of the sinks and other environmental conditions: fertilization of *Picea abies* seedlings affected soluble sugars and starch, which are associated with cold hardiness (Kaakinen et al. 2004). The increase in soluble sugars in fertilized trees can reflect an increased availability of carbohydrates for wood formation, e.g. lignification may continue later in autumn after the end of cell enlargement and cell wall thickening (Anttonen et al. 2002). For example, in poplar cultivars, the more productive clone was also that with higher content of soluble sugars, especially during the period of maximum growth (Deslauriers et al. 2009). Moreover, at the time of decreased xylem differentiation, the total non-structural soluble sugar was still higher in the cambium of the more productive clone, while similar contents were observed after the completion of xylogenesis, supporting the hypothesis advanced by Anttonen et al. (2002) that sugar availability and wood formation are positively related. Fertilization usually increases lignin and N concentration, and decreases the C:N ratio in wood (Anttonen et al. 2002; Kostiainen et al. 2004). In conifers the increase in lignin concentration can be partly explained by the increase in earlywood, which has greater lignin concentrations than latewood (Anttonen et al. 2002). Moreover, fertilization reduces the thickness of cell walls, thus increasing the proportion of middle lamella, the layer with the highest concentration of lignin (Anttonen et al. 2002; Makinen et al. 2002).

The increased N concentration in wood agrees with the increased amount of living cells observed in fertilized trees, since N is present at greater concentrations in the living parenchyma cells and xylem sap (Stockfors and Linder 1998; Anttonen et al. 2002). On the other hand, N deficiency increases earlier leaf senescence, while N fertilization can extend the growing season (Kaakinen et al. 2004; Nord and Lynch 2009). This might be a consequence of the stimulation of stem growth by N, in agreement with a recent study in which the authors observed a positive causal relation between xylem production and the end of xylem differentiation (Kaakinen et al. 2004; Lupi et al. 2010). According to Lupi et al. (2010), an earlier onset of wood formation in *Picea mariana* is associated with higher wood production. This may be due to a longer period with favorable conditions for growth, since earlier onsets occur at higher spring temperatures, which allows a higher or longer acquisition of resources (e.g. nutrients and water) and C-fixation. The higher number of cells produced requires more time to end differentiation, leading to a later completion of xylogenesis (Lupi et al. 2010).

The increase in anthropogenic depositions during the last century seems to have alleviated the N limitations of the boreal forest and to have stimulated its growth (Binkley and Hogberg 1997; Hyvonen et al. 2008). When coupled with climate change, the anthropogenic depositions may affect snowfall and snowmelt, and consequently nutrient cycling, by increasing the decomposition of soil organic matter and the availability of N, thus possibly altering the dynamics and timings of tree growth. The faster-growing earlier-successional species could be more responsive to these changes because of their closer relationship between N and photosynthesis and because of their higher rates of nutrient uptake, and could gain advantage over slower-growing species (Reich et al. 1998). Future studies should aim to define and verify the importance of and interaction between snow and N cycling and its availability, in order to understand the role of these factors on tree growth at short and long time scales, and their effects on the evolution of the boreal ecosystems.

9.6   Current and Future Research Directions

Since the beginning of this century, wood production and the seasonal dynamics of xylem formation in trees of cold climates have been analyzed thoroughly. However the factors influencing cambial activity and their mechanisms of action still remain to be clearly identified and precisely quantified. On the one hand, the influence of environment, well documented in dendroecology through climate-growth relationships (Schweingruber 1996), has been demonstrated to be complicated by the asynchronous and long-lasting production and maturation of xylem cells (Deslauriers et al. 2007a, 2008; Fonti et al. 2007; Seo et al. 2008). On the other, xylogenesis is strongly determined by internal factors, such as cambial age, carbohydrate availability and hormone distribution (Uggla et al. 1998, 2001; Rossi et al. 2008a; Deslauriers et al. 2009).

Although many questions on the ecology of tree growth still remain unresolved, the recent findings from monitoring wood formation have provided valuable results in several research fields, divided here into fundamental research, climate-growth relationships, and modeling. Nevertheless, these categories are not mutually exclusive and, in practice, authors navigate between the different aspects.

9.6.1   Fundamental Research

Fundamental research aims to assess and quantify timings and rates of each phenological component of the multifaceted process of xylem growth. In the last decade, a revived interest in Europe and North-America has set the basis of wood formation by determining periods of radial growth (Thibeault-Martel et al. 2008; Gruber et al. 2010; Moser et al. 2010), timings of earlywood and latewood formation (Deslauriers et al. 2003a; Rossi et al. 2006a), and rates of tree-ring formation (Rossi et al. 2006b; Deslauriers et al. 2008). In temperate and boreal environments, the wide degrees of plasticity occurring in the patterns of xylem formation and rates of cell development provide great advantages in the context of a fast changing climate. Variations of up to 1 month have been observed among years in xylem phenology in both European and Canadian conifers (Deslauriers et al. 2003a; Rossi et al. 2007, 2008b; Gruber et al. 2010). In turn, these variations affect forest productivity, in terms of quantity (wood production) and quality (cell features) (Gričar et al. 2005; Deslauriers et al. 2008; Lupi et al. 2010). There is evidence that the key period in wood formation is the onset in spring: this crucial event influences the number of cells produced by the meristems, which influence the ending of cell differentiation according to a linear causal relationship (Lupi et al. 2010). Further investigations of the links between cambial phenology and wood production may supply tools for new and more suitable strategies of plant breeding.

Several features of trees and their surrounding environment, such as vitality (Marion et al. 2007), age (Rossi et al. 2008a) or competition (Grotta et al. 2005) can strongly modify timings and rates of xylem growth, forcing research to better investigate plant responses at a more integrated level. It has been demonstrated that metabolic activities in the stem react in synchrony with wood formation: the cambial zone is a powerful carbon sink breaking down carbohydrate to build the new cell walls during division. The rates of cell production and differentiation have been linked with the concentration of soluble carbohydrates in meristems and developing tissues (Deslauriers et al. 2009). This explains the positive relationships between living xylem cells and CO_2 efflux from the stem (Lavigne et al. 2004; Gruber et al. 2009), which is an important parameter for the carbon budget of forests.

9.6.2 Climate-Growth Relationships

Habitually, the relationships between tree growth and climate have focused on two main factors in dendroclimatology: temperature and precipitation. However, with an intra-annual monitoring these relationships can be explored at shorter time scales, thus allowing different and higher levels of precision and more complete understandings of the roles of environmental factors on the mechanisms and components of wood formation. The impressive progress of the research in the ecology of growth registered in the last years has demonstrated some temperature-limited mechanisms such as temperature thresholds (Rossi et al. 2008b), quiescent and dormant stages of cambium in winter (Oribe et al. 2001; Gričar et al. 2006; Begum et al. 2010), and cambial reactions to extreme years (Deslauriers et al. 2008). The longer time window for xylogenesis observed in older trees is assumed to lead to a dilution of the climatic signal during growth and to reduce the climate-growth relationships (Szeicz and MacDonald 1994, 1995; Carrer and Urbinati 2004; Rossi et al. 2008a). The higher temperatures expected in spring and early summer will also increase the period of wood formation in temperature-limited environments, and, in turn, cell production, mainly earlywood, which is characterized by larger diameters and higher hydraulic efficiency than latewood (Domec and Gartner 2002; Deslauriers et al. 2008). The consequences of such modifications in xylem anatomy for the physiology of trees remain to be studied.

The effects on growth of photoperiod (Rossi et al. 2006b), water stress (Giovannelli et al. 2007; van der Werf et al. 2007; Rossi et al. 2009b; Gruber et al. 2010) and defoliation (Rossi et al. 2009a) are just the first challenges tackled by the analyses on cambial phenology. The discovery of other environmental signals acting alone or in synergy in the process of xylem formation will isolate the fine-scale ecological factors influencing tree growth and survival and will strengthen the understanding of forest dynamics and productivity experiencing climatic change.

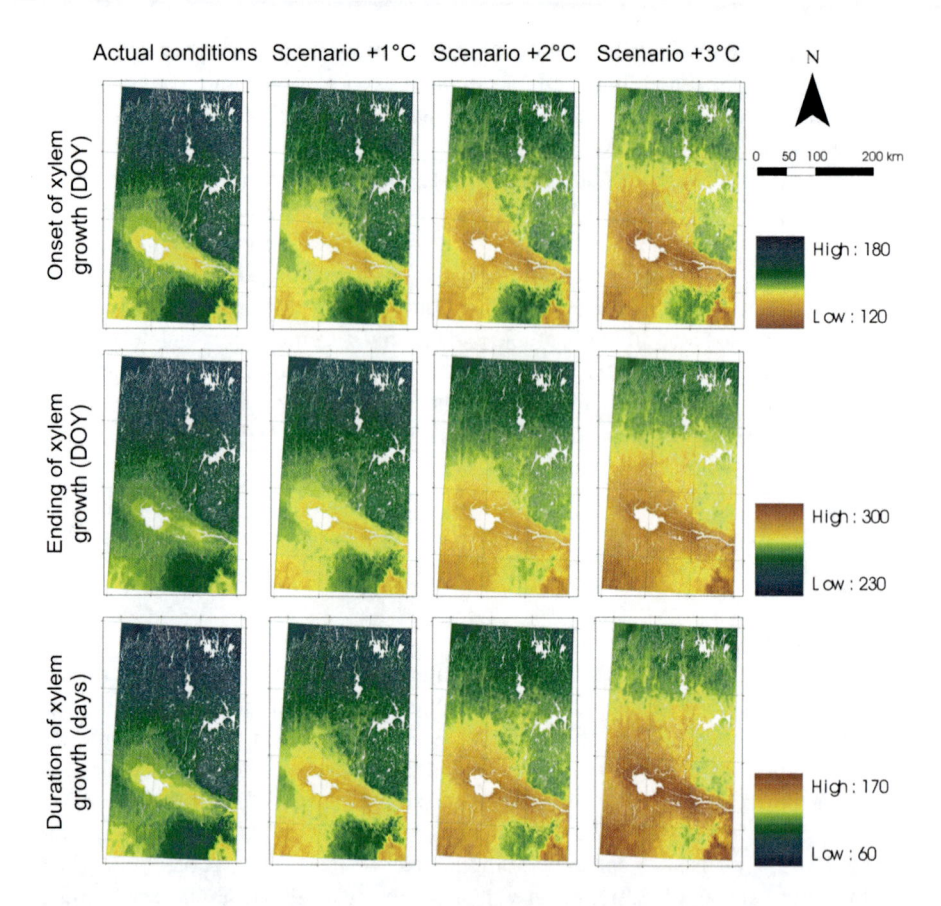

Fig. 9.7 Onset, ending and duration of xylem growth of *Picea mariana* in the Saguenay-Lac-Saint-Jean area (Quebec, Canada) estimated applying the logistic models obtained from the minimum temperatures of the existing temperature pattern and of three future climatic scenarios assuming uniformly warming increases in the daily temperatures of 1, 2 and 3 °C (From Rossi et al. 2010a)

9.6.3   Modeling

This level of investigation applies the previous findings to reconstruct or simulate past, current or future traits of wood formation and establishes links with other disciplines, such as ecology, forestry, climatology and geography. Punctual episodes affecting the typical sequence of events of the growth process can be suitably reconstructed with simple non-linear sigmoid functions. Kaczka et al. (2010) compared the proportion of the formed tree ring to estimate the date of the 1996 debris flow in Quebec, Canada. So, scars from unknown debris flow or forest fires can be dated with a weekly precision, which allows the weather conditions causing the event to be investigated more thoroughly. As the phases of xylem growth can be mathematically defined using thermal thresholds or sums (Seo et al. 2008; Rossi et al. 2007c, 2008b), simulations of cambial phenology have been produced to estimate future growth periods of *Picea mariana* under possible warming scenarios (Rossi et al. 2010a). A spatial simulation of the timings of xylem growth on a whole region shows durations of between 110 and 135 days in the valley and durations shorter than 95 days on the mountains (Fig. 9.7). In the scenarios with temperatures rising by 1–3 °C, the duration of xylem growth is predicted to reach 130–160 days of growth in the valley. At the higher elevations and with temperatures rising by up to 3 °C, xylem growth is predicted to last 120 days, ca. 30 days more than at the existing conditions (Fig. 9.7). Interestingly, reconstructions of cambium phenology can be carried out based on historical time series of environmental factors, such as temperature, precipitation, concentrations of CO_2 and other atmospheric pollutants, to assess the long-term pattern of variation. Large stems and old roots could represent an important buffer in storing carbohydrates, thus modifying the allocation of carbon to secondary growth. Influences of the preceding years on the current growth could also be investigated when adequately-long chronologies of cambial activity become available. For dendroclimatology, such reconstructions could allow a more precise relationship with climate, the period of wood formation being precisely defined.

References

Abe H, Funada R, Ohtani J, Fukazawa K (1997) Changes in the arrangement of cellulose microfibrils associated with the cessation of cell expansion in tracheids. Trees 11:328–332

Alvarez-Uria P, Körner C (2007) Low temperature limits of root growth in deciduous and evergreen temperate tree species. Funct Ecol 21:211–218

Antonova GF, Shebeko VV (1981) Applying cresyl violet in studying wood formation. Khim Drev 4:102–105

Antonova GF, Stasova VV (1997) Effects of environmental factors on wood formation in larch (*Larix sibirica* Ldb.) stems. Trees 11:462–468

Anttonen S, Manninen AM, Saranpaa P, Kainulainen P, Linder S, Vapaavuori E (2002) Effects of long-term nutrient optimisation on stem wood chemistry in *Picea abies*. Trees 16:386–394

Bannan MW (1962) The vascular cambium and tree ring development. In: Kozlowski TT (ed) Tree growth. Ronald Press, New York

Begum S, Nakaba S, Oribe Y, Kubo T, Funada R (2010) Cambial sensitivity to rising temperatures by natural condition and artificial heating from late winter to early spring in the evergreen conifer *Cryptomeria japonica*. Trees 24:43–52

Bergh J, Linder S (1999) Effects of soil warming during spring on photosynthetic recovery in boreal *Picea abies* stands. Glob Chang Biol 5:245–253

Bergh J, Linder S, Lundmark T, Elfving B (1999) The effect of water and nutrient availability on the productivity of *Picea abies* in northern and southern Sweden. For Ecol Manag 119:51–62

Bernoulli M, Körner C (1999) Dry matter allocation in treeline trees. Phyton Ann Rei Bot 39:7–11

Binkley D, Hogberg P (1997) Does atmospheric deposition of nitrogen threaten Swedish forests? For Ecol Manag 92:119–152

Bouriaud O, Leban J-M, Bert D, Deleuze C (2005) Intra-annual variations in climate influence growth and wood density of *Picea abies*. Tree Physiol 25:651–660

Buckhout WA (1907) The formation of the annual ring of wood in European larch and the pine. J For 5:259–267

Cairns DM, Malanson GP (1998) Environmental variables influencing the carbon balance at the alpine treeline: a modeling approach. J Veg Sci 9:679–692

Camarero JJ, Guerrero-Campo J, Gutiérrez E (1998) Tree-ring growth and structure of *Pinus uncinata* and *Pinus sylvestris* in the Central Spanish Pyrenees. Arct Alp Res 30:1–10

Campbell JL, Rustad LE, Boyer EW, Christopher SF, Driscoll CT, Fernandez IJ, Groffman PM, Houle D, Kiekbusch J, Magill AH, Mitchell MJ, Ollinger SV (2009) Consequences of climate change for biogeochemical cycling in forests of northeastern North America. Can J For Res 39:264–284

Canovas FM, Avila C, Canton FR, Canas RA, de la Torre F (2007) Ammonium assimilation and amino acid metabolism in conifers. J Exp Bot 58:2307–2318

Carrer M, Urbinati C (2004) Age-dependent tree-ring growth responses to climate in *Larix decidua* and *Pinus cembra*. Ecology 85:730–740

Cavieres LA, Rada F, Azócar A, García-Núñez C, Cabrera HM (2000) Gas exchange and low temperature resistance in two tropical high mountain tree species from the Venezuelan Andes. Acta Oecol 21:203–211

Decker KLM, Wang D, Waite C, Scherbatskoy T (2003) Snow removal and ambient air temperature effects on forest soil temperatures in northern Vermont. Soil Sci Soc Am J 67:1234–1243

Denne MP (1971) Temperature and tracheid development in *Pinus sylvestris* seedlings. J Exp Bot 22:362–370

Denne MP (1974) Effects of light intensity on tracheid dimensions in *Picea sitchensis*. Ann Bot 38:337–345

Deslauriers A, Morin H (2005) Intra-annual tracheid production in balsam fir stems and the effect of meteorological variables. Trees 19:402–408

Deslauriers A, Morin H, Begin Y (2003a) Cellular phenology of annual ring formation of *Abies balsamea* in the Quebec boreal forest (Canada). Can J For Res 33:190–200

Deslauriers A, Morin H, Urbinati C, Carrer M (2003b) Daily weather response of balsam fir (*Abies balsamea* (L.) Mill.) stem radius increment from dendrometer analysis in the boreal forests of Québec (Canada). Trees 17:477–484

Deslauriers A, Anfodillo T, Rossi S, Carraro V (2007a) Using simple causal modelling to understand how water and temperature affect daily stem radial variation in trees. Tree Physiol 27:1125–1136

Deslauriers A, Rossi S, Anfodillo T (2007b) Dendrometer and intra-annual tree growth: what kind of information can be inferred? Dendrochronologia 25:113–124

Deslauriers A, Rossi S, Anfodillo T, Saracino A (2008) Cambium phenology, wood formation and temperature thresholds in two contrasting years at high altitude in Southern Italy. Tree Physiol 28:863–871

Deslauriers A, Giovannelli A, Rossi S, Castro G, Fragnelli G, Traversi L (2009) Intra-annual cambial activity and carbon availability in stem of poplar. Tree Physiol 29:1223–1235

Domec JC, Gartner BL (2002) How do water transport and water storage differ in coniferous earlywood and latewood? J Exp Bot 53:2369–2379

Donaldson LA (1991) Seasonal changes in lignin distribution during tracheid development in *Pinus radiata* D. Don. Wood Sci Technol 25:15–24

Downes G, Beadle C, Worledge D (1999) Daily stem growth patterns in irrigated *Eucalyptus globulus* and *E. nitens* in relation to climate. Trees 14:102–111

Downes GM, Wimmer R, Evans R (2002) Understanding wood formation: gains to commercial forestry through tree-ring research. Dendrochronologia 20:37–51

Downes G, Wimmer R, Evans R (2004) Interpreting sub-annual wood and fibre property variation in terms of stem growth. In: Schmitt U, Ander P, Barnett JR, Emons AMC, Jeronimidis G, Saranpää P, Tschegg S (eds) Wood fibre cell walls: methods to study their formation, structure and properties. Swedish University of Agricultural Sciences, Uppsala

Drew DM, O'Grady AP, Downes G, Read J, Worledge D (2008) Daily patterns of stem size variation in irrigated and unirrigated *Eucalyptus globulus*. Tree Physiol 28:1573–1581

Drew DM, Downes G, Grzeskowiak V, Naidoo T (2009) Differences in daily stem size variation and growth in two hybrid eucalypt clones. Trees 23:585–595

Eckstein D (2004) Change in past environments – secrets of the tree hydrosystem. New Phytol 163:1–4

Eckstein D, Frisse E, Quiehl F (1977) Holzanatomische Untersuchungen zum Nachweis anthropogener Einflüsse auf die Umweltbedingungen einer Rotbuche. Angew Bot 51:47–56

Ericsson T (1995) Growth and shoot-root ratio of seedlings in relation to nutrient availability. Plant Soil 168:205–214

Fahn A, Werker E (1990) Seasonal cambial activity. In: Iqbal M (ed) The vascular cambium. Wiley, New York

Fonti P, Solomonoff N, García-González I (2007) Earlywood vessels size of *Castanea sativa* record temperature before their formation. New Phytol 173:562–570

Ford ED, Robards AW, Piney MD (1978) Influence of environmental factors on cell production and differentiation in the earlywood of *Picea sitchensis*. Ann Bot 42:683–692

Fritts HC (1976) Tree rings and climate. Academic, New York

Fritts HC, Fritts EC (1955) A new dendrograph for recording radial changes of a tree. For Sci 1:271–276

Galloway JN, Dentener FJ, Capone DG, Boyer EW, Howarth RW, Seitzinger SP, Asner GP, Cleveland CC, Green PA, Holland EA, Karl DM, Michaels AF, Porter JH, Townsend AR, Vorosmarty CJ (2004) Nitrogen cycles: past, present, and future. Biogeochemistry 70:153–226

Gill RA, Jackson RB (2000) Global patterns of root turnover for terrestrial ecosystems. New Phytol 147:13–31

Gindl W, Grabner M, Wimmer R (2000) The influence of temperature on latewood lignin content in treeline *Picea abies* compared with maximum density and ring width. Trees 14:409–414

Giovannelli A, Deslauriers A, Fragnelli G, Scaletti L, Castro G, Rossi S, Crivellaro A (2007) Evaluation of drought response of two poplar clones (*Populus x canadensis* Mönch 'I-214' and *P. deltoides* Marsh. 'Dvina') through high resolution analysis of stem growth. J Exp Bot 58:2673–2683

Goodine GK, Lavigne MB, Krasowski MJ (2008) Springtime resumption of photosynthesis in balsam fir (*Abies balsamea*). Tree Physiol 28:1069–1076

Graumlich LJ, Brubaker LB (1986) Reconstruction of annual temperature (1590–1979) for Longmire, Washington, derived from tree-rings. Quat Res 25:223–234

Gričar J, Čufar K, Oven P, Schmitt U (2005) Differentiation of terminal latewood tracheids in silver fir trees during autumn. Ann Bot 95:959–965

Gričar J, Zupancic M, Čufar K, Koch G, Schmitt U, Oven P (2006) Effect of local heating and cooling on cambial activity and cell differentiation in the stem of *Picea abies* (*Picea abies*). Ann Bot 97:943–951

Gričar J, Zupančič M, Čufar K, Oven P (2007) Regular cambial activity and xylem and phloem formation in locally heated and cooled stem portions of *Picea abies*. Wood Sci Technol 41:463–475

Groffman PM, Hardy JP, Fisk MC, Fahey TJ, Driscoll CT (2009) Climate variation and soil carbon and nitrogen cycling processes in a northern hardwood forest. Ecosystems 12:927–943

Grogan P, Jonasson S (2003) Controls on annual nitrogen cycling in the understory of a subarctic birch forest. Ecology 84:202–218

Grotta AT, Gartner BL, Radosevich SR, Huso M (2005) Influence of red alder competition on cambial phenology and latewood formation in Douglas-fir. IAWA J 26:309–324

Gruber A, Wieser G, Oberhuber W (2009) Intra-annual dynamics of stem CO_2 efflux in relation to cambial activity and xylem development in *Pinus cembra*. Tree Physiol 29:641–649

Gruber A, Strobl S, Veit B, Oberhuber W (2010) Impact of drought on the temporal dynamics of wood formation in *Pinus sylvestris*. Tree Physiol 30:490–501

Grulke NE, Balduman L (1999) Deciduous conifers: high N deposition and O_3 exposure effects on growth and biomass allocation in ponderosa pine. Water Air Soil Pollut 116:235–248

Hansen J, Beck E (1990) The fate and path of assimilation products in the stem of 8-year-old Scots pine (*Pinus sylvestris* L.) trees. Trees 4:16–21

Hansen J, Beck E (1994) Seasonal changes in the utilization and turnover of assimilation products in 8-year-old Scots pine (*Pinus sylvestris* L.) trees. Trees 8:172–182

Hansen J, Türk R, Vogg G, Heim R, Beck E (1997) Conifer carbohydrate physiology: updating classical views. In: Rennenberg H, Eschrich W, Ziegler H (eds) Trees: contributions to modern tree physiology. Backhuys Publishers, Leiden

Hoch G, Körner C (2003) The carbon charging of pines at the climatic treeline: a global comparison. Oecologia 135:10–21

Hoch G, Popp M, Körner C (2002) Altitudinal increase of mobile carbon pools in *Pinus cembra* suggests sink limitation of growth at the Swiss treeline. Oikos 98:361–374

Hyvonen R, Persson T, Andersson S, Olsson B, Agren GI, Linder S (2008) Impact of long-term nitrogen addition on carbon stocks in trees and soils in northern Europe. Biogeochemistry 89: 121–137

James JC, Grace J, Hoad SP (1994) Growth and photosynthesis of *Pinus sylvestris* at its altitudinal limit in Scotland. J Ecol 82:297–306

Jarvis P, Linder S (2000) Constraints to growth of boreal forests. Nature 405:904–905

Kaakinen S, Jolkkonen A, Iivonen S, Vapaavuori E (2004) Growth, allocation and tissue chemistry of *Picea abies* seedlings affected by nutrient supply during the second growing season. Tree Physiol 24:707–719

Kaakinen S, Saranpaa P, Vapaavuori E (2007) Effects of growth differences due to geographic location and N-fertilisation on wood chemistry of *Picea abies*. Trees 21:131–139

Kaczka RJ, Deslauriers A, Morin H (2010) High-precision dating of debris-flow events within the growing season. In: Stoffel M, Bollschweiler M, Butler DR, Luckman BH (eds) Tree rings and natural hazards: a state of the art. Springer, Dordrecht/Heidelberg

Kielland K, McFarland JW, Ruess RW, Olson K (2007) Rapid cycling of organic nitrogen in taiga forest ecosystems. Ecosystems 10:360–368

Knudson L (1913) Observations on the inception, season, and duration of cambium development in the American larch [*Larix laricina* (Du Roi) Koch.]. Bull Torrey Bot Club 40:271–293

Körner C (1998) A re-assessment of high elevation treeline positions and their explanation. Oecologia 115:445–459

Körner C (2003) Carbon limitation in trees. J Ecol 91:4–17

Körner C, Hoch G (2006) A test of treeline theory on a montane permafrost island. Arct Antarct Alp Res 38:113–119

Körner C, Paulsen J (2004) A world-wide study of high altitude treeline temperatures. J Biogeogr 31:713–732

Kostiainen K, Kaakinen S, Saranpaa P, Sigurdsson BD, Linder S, Vapaavuori E (2004) Effect of elevated [CO_2] on stem wood properties of mature *Picea abies* grown at different soil nutrient availability. Glob Chang Biol 10:1526–1538

Kozlowski TT, Winget CH (1964) Diurnal and seasonal variation in radii of tree stems. Ecology 45:149–155

Kutscha NP, Hyland F, Schwarzmann JM (1975) Certain seasonal changes in Balsam fir cambium and its derivatives. Wood Sci Technol 9:175–188

Lavigne MB, Little CHA, Riding RT (2004) Changes in stem respiration rate during cambial reactivation can be used to refine estimates of growth and maintenance respiration. New Phytol 162:81–93

Linder S (1995) Foliar analysis for detecting and correcting nutrient imbalances in *Picea abies*. Ecol Bull 44:178–190

Lupi C, Morin H, Deslauriers A, Rossi S (2010) Xylem phenology and wood production: resolving the chicken-or-egg dilemma. Plant Cell Environ 33:1721–1730

Makinen H, Saranpaa P, Linder S (2002) Wood-density variation of *Picea abies* in relation to nutrient optimization and fibre dimensions. Can J For Res 32:185–194

Mäkinen H, Nöjd P, Saranpää P (2003) Seasonal changes in stem radius and production of new tracheids in *Picea abies*. Tree Physiol 23:959–968

Marion L, Gričar J, Oven P (2007) Wood formation in urban Norway maple trees studied by the micro-coring method. Dendrochronologia 25:97–102

Meyer FD, Paulsen J, Körner C (2008) Windthrow damage in *Picea abies* is associated with physical and chemical stem wood properties. Trees 22:463–473

Millard P, Proe MF (1992) Storage and internal cycling of nitrogen in relation to seasonal growth of Sitka spruce. Tree Physiol 10:33–43

Mork E (1928) Die Qualität des Fichtenholzes unter besonderer Rücksichtnahme auf Schleif- und Papierholz. Pap-Fabrikant 26:741–747

Moser L, Fonti P, Buentgen U, Franzen J, Esper J, Luterbacher J, Frank D (2010) Timing and duration of European larch growing season along altitudinal gradients in the Swiss Alps. Tree Physiol 30:225–233

Nizinski JJ, Saugier B (1988) A model of leaf budding and development for a mature *Quercus* forest. J Appl Ecol 25:643–652

Nobuchi T, Ogata Y, Siripatanadilok S (1995) Seasonal characteristics of wood formation in *Hopea odorata* and *Shorea henryana*. IAWA J 16:361–369

Nord EA, Lynch JP (2009) Plant phenology: a critical controller of soil resource acquisition. J Exp Bot 60:1927–1937

Oribe Y, Kubo T (1997) Effect of heat on cambial reactivation during winter dormancy in evergreen and deciduous conifers. Tree Physiol 17:81–87

Oribe Y, Funada R, Shibagaki M, Kubo T (2001) Cambial reactivation in locally heated stems of the evergreen conifer *Abies sachalinensis* (Schmidt) masters. Planta 212:684–691

Oribe Y, Funada R, Kubo T (2003) Relationships between cambial activity, cell differentiation and the localisation of starch in storage tissues around the cambium in locally heated stems of *Abies sachalinensis* (Schmidt) masters. Trees 17:185–192

Partanen J, Koski V, Hänninen H (1998) Effects of photoperiod and temperature on the timing of bud burst in *Picea abies* (*Picea abies*). Tree Physiol 18:811–816

Piper FI, Cavieres LA, Reyes-Díaz M, Corcuera LJ (2006) Carbon sink limitation and frost tolerance control performance on the tree *Kageneckia angustifolia* D. Don (Rosaceae) at the treeline in central Chile. Plant Ecol 185:29–39

Plomion C, Leprovost G, Stokes A (2001) Wood formation in trees. Plant Physiol 127:1513–1523

Purnell B (2003) To every thing there is a season. Science 301:325

Read DJ, Leake JR, Perez-Moreno J (2004) Mycorrhizal fungi as drivers of ecosystem processes in heathland and boreal forest biomes. Can J Bot 82:1243–1263

Reich PB, Grigal DF, Aber JD, Gower ST (1997) Nitrogen mineralization and productivity in 50 hardwood and conifer stands on diverse soils. Ecology 78:335–347

Reich PB, Walters MB, Tjoelker MG, Vanderklein D, Buschena C (1998) Photosynthesis and respiration rates depend on leaf and root morphology and nitrogen concentration in nine boreal tree species differing in relative growth rate. Funct Ecol 12:395–405

Rossi S, Deslauriers A (2007a) Intra-annual time scales in tree rings. Dendrochronologia 25:75–77

Rossi S, Deslauriers A (2007b) Scale temporali d'azione di temperatura e fotoperiodo sulla xilogenesi al limite superiore del bosco. Forest 4:6–10

Rossi S, Deslauriers A, Anfodillo T, Carraro V (2007c) Evidence of threshold temperatures for xylogenesis in conifers at high altitude. Oecologia 152:1–12

Rossi S, Deslauriers A, Morin H (2003) Application of the Gompertz equation for the study of xylem cell development. Dendrochronologia 21:1–7

Rossi S, Deslauriers A, Anfodillo T (2006a) Assessment of cambial activity and xylogenesis by microsampling tree species: an example at the Alpine timberline. IAWA J 27:383–394

Rossi S, Deslauriers A, Anfodillo T, Morin H, Saracino A, Motta R, Borghetti M (2006b) Conifers in cold environments synchronize maximum growth rate of tree-ring formation with day length. New Phytol 169:279–290

Rossi S, Deslauriers A, Anfodillo T, Carrer M (2008a) Age-dependent xylogenesis in timberline conifers. New Phytol 177:199–208

Rossi S, Deslauriers A, Gričar J, Seo J-W, Rathgeber CBK, Anfodillo T, Morin H, Levanic T, Oven P, Jalkanen R (2008b) Critical temperatures for xylogenesis in conifers of cold climates. Glob Ecol Biogeogr 17:696–707

Rossi S, Simard S, Deslauriers A, Morin H (2009a) Wood formation in *Abies balsamea* seedlings subjected to artificial defoliation. Tree Physiol 29:551–558

Rossi S, Simard S, Rathgeber CBK, Deslauriers A, De Zan C (2009b) Effects of a 20-day-long dry period on cambial and apical meristem growth in *Abies balsamea* seedlings. Trees 23:85–93

Rossi S, Morin H, Deslauriers A, Plourde P-Y (2010a) Predicting xylem phenology in black spruce under climate warming. Glob Chang Biol 17:614–625

Rossi S, Morin H, Tremblay M-J (2010b) Growth and productivity of black spruce (*Picea mariana*) belonging to the first cohort in stands within and north of the commercial forest in Quebec, Canada. Ann For Sci 67:807–816

Running SW, Reid CP (1980) Soil temperature influences on root resistance of *Pinus contorta* seedlings. Plant Physiol 65:635–640

Schmitt U, Jalkanen R, Eckstein D (2004) Cambium dynamics of *Pinus sylvestris* and *Betula* spp. in the northern boreal forest in Finland. Silva Fenn 38:167–178

Schweingruber FH (1996) Tree rings and environment. Dendroecology. Swiss Federal Institute for Forest, Snow and Landscape, Haupt, Berne/Stuttgart/Vienna

Seo J-W, Eckstein D, Schmitt U (2007) The pinning method: from pinning to data preparation. Dendrochronologia 25:79–86

Seo J-W, Eckstein D, Jalkanen R, Rickbusch S, Schmitt U (2008) Estimating the onset of cambial activity of Scots pine in northern Finland by means of the heat-sum approach. Tree Physiol 28:105–112

Shönenberger W, Frey W (1988) Untersuchungen zur Ökologie und Technik der Hochlagenauf-forstung. Forschungsergebnisse aus dem Lawinenanrissgebiet Stillberg. Schweiz Z Forstwes 139:735–820

Steinaker DF, Wilson SD (2008) Phenology of the roots and leaves in forest and grassland. J Ecol 29:1222–1229

Steinaker DF, Wilson SD, Peltzer DA (2010) Asynchronicity in root and shoot phenology in grasses and woody plants. Glob Change Biol 16:2241–2251

Stern KR, Bidlack J, Jansky S (2003) Introductory plant biology. McGraw Hill, New York

Stevens GC, Fox JF (1991) The causes of treeline. Ann Rev Ecol Syst 22:177–191

Stockfors J, Linder S (1998) Effect of nitrogen on the seasonal course of growth and maintenance respiration in stems of *Picea abies* trees. Tree Physiol 18:155–166

Sveinbjörnsson B (2000) North American and European treelines: external forces and internal processes controlling position. Ambio 29:388–395

Szeicz JM, MacDonald GM (1994) Age-dependent tree-ring growth responses of subarctic white spruce to climate. Can J For Res 24:120–132

Szeicz JM, MacDonald GM (1995) Dendroclimatic reconstruction of summer temperatures in northwestern Canada since A.D. 1638 based on age-dependent modeling. Quat Res 44:257–266

Thibeault-Martel M, Krause C, Morin H, Rossi S (2008) Cambial activity and intra-annual xylem formation in roots and stems of *Abies balsamea* and *Picea mariana*. Ann Bot 102:667–674

Turcotte A, Morin H, Krause C, Deslauriers A, Thibeault-Martel M (2009) The timing of spring rehydration and its relation with the onset of wood formation in black spruce. Agric For Meteorol 149:1403–1409

Turner H, Streule A (1983) Wurzelwachstum und Sprossentwicklung junger Koniferen im Klimastress der alpinen Waldgrenze, mit Berücksichtigung von Mikroklima, Photosynthese und Stoffproduktion. In: Böhm W, Kutschera L, Lichtenegger E (eds) Wurzelökologie und Ihre Nutzanwendung. Irding, Gumpenstein

Uggla C, Mellerowicz EJ, Sundberg B (1998) Indole-3-acetic acid controls cambial growth in Scots pine by positional signaling. Plant Physiol 117:113–121

Uggla C, Magel E, Moritz T, Sundberg B (2001) Function and dynamics of auxin and carbohydrates during earlywood/latewood transition in Scots pine. Plant Physiol 125:2029–2039

Vaganov EA (1990) The tracheidogram method in tree-ring analysis and its application. In: Cook R, Kairiukstis L (eds) Methods of dendrochronology. Kluwer, Dordrecht

Vaganov EA, Hughes MK, Kirdyanov AV, Schweingruber FH, Silkin PP (1999) Influence of snowfall and melt timing on tree growth in subarctic Eurasia. Nature 400:149–151

van der Werf GW, Sass-Klaassen UGW, Mohren GMJ (2007) The impact of the 2003 summer drought on the intra-annual growth pattern of beech (*Fagus sylvatica* L.) and oak (*Quercus robur* L.) on a dry site in the Netherlands. Dendrochronologia 25:103–112

Vincent LA, Mekis E (2006) Changes in daily and extreme temperature and precipitation indices for Canada over the twentieth century. Atmos Ocean 44:177–193

Vitousek PM, Howarth RW (1991) Nitrogen limitation on land and in the sea – how can it occur. Biogeochemistry 13:87–115

Weintraub MN, Schimel JP (2005) The seasonal dynamics of amino acids and other nutrients in Alaskan Arctic tundra soils. Biogeochemistry 73:359–380

Wimmer R, Downes GM, Evans R (2002) High-resolution analysis of radial growth and wood density in *Eucalyptus nitens*, grown under different irrigation regimes. Ann For Sci 59:519–524

Wodzicki TJ (1971) Mechanism of xylem differentiation in *Pinus silvestris* L. J Exp Bot 22:670–687

Wolter EK (1968) A new method for marking xylem growth. For Sci 14:102–104

Yang RC, Wang EIC, Micko MM (1988) Effects of fertilisation on wood density and tracheid length of 70-year-old lodgepole pine in west-central Alberta. Can J For Res 18:954–956

Zweifel R, Zimmermann L, Zeugin F, Newbery DM (2006) Intra-annual radial growth and water relations of trees: implication towards a growth mechanism. J Exp Bot 57:1445–1459

Chapter 10
Treelines in a Changing Global Environment

Gerhard Wieser, Friedrich-Karl Holtmeier, and William K. Smith

Abstract Over the last century the global mean surface temperature has increased by about 0.6 °C and was most pronounced at high elevation and high latitude. Because the elevations and latitudes of treelines are strongly correlated with the occurrence of heat deficiency, climate warming is expected to generate denser forests below the treeline, as well as treeline movement to greater elevations and poleward. Herein, conclusions are presented about the future of treeline movement following a review of mechanisms and limiting factors for tree growth, differences between tall trees and low stature vegetation (including seedlings), and seedling establishment and growth to forest tree stature.

10.1 Introduction

Alpine (elevational) and polar (latitudinal) timberlines or treelines are some of the most obvious vegetation boundaries. Rather than being abrupt boundaries, treelines usually form an ecotone between uppermost closed montane and subarctic forests and the treeless alpine zone and tundra above (Däniker 1923; Holtmeier 1974, 2009; Wardle 1974; Tranquillini 1979; Slatyer and Noble 1992; Wieser and Tausz 2007; McDonald et al. 2008). Within this transition zone above the closed forest, trees become flagged and stunted, forming the *scrub*-like trees traditionally referred

G. Wieser (✉)
Department of Alpine Timberline Ecophysiology, BFW, Innsbruck, Austria
e-mail: gerhard.wieser@uibk.ac.at

F.-K. Holtmeier
Institute of Landscape Ecology, University of Münster, Münster, Germany

W.K. Smith
Department of Biology, Wake Forest University, Winston-Salem, NC, USA

M. Tausz and N. Grulke (eds.), *Trees in a Changing Environment*, 221
Plant Ecophysiology 9, DOI 10.1007/978-94-017-9100-7_10,
© Springer Science+Business Media Dordrecht 2014

to as *krummholz* mats. This severe alteration in growth form is presumed now to be due to abiotic climatic severity. Thus, researchers have attempted to define more precisely treelines as the upper elevational or latitudinal limit of trees greater than 2 m in height (Kullman 2001; Wieser 2012) where tree crowns are much more coupled to atmospheric conditions such as air temperature measured at a standard height (2 m) for weather station instruments. This height requirement is despite recognition that the clustering of sun-exposed stems on the same branch, as well as needles on individual shoots, can result in needle temperatures well above air temperature (Hadley and Smith 1987; Smith and Carter 1988; Smith and Brewer 1994; Grace et al. 2002).

Above the occurrence of 2 m tall trees within the treeline ecotone individual trees are often deformed into the classic krummholz growth form that is character-ized by severely stunted, bush-like habits, that experiencing microclimates more similar to low-growing vegetation (e.g. dwarf-shrubs, grassland, and meadows) that dominate the next higher elevational or latitudinal vegetation belt (Grace et al. 2002). It is also known that krummholz tree islands can be composed of both mats and flagged trees often greater than 2 m in height and usually found on the leeward edges of larger mats and islands lower in the ecotone. These severely flagged trees gradually disappear at the highest edge of the treeline ecotone where low mats usually prevail (Holtmeier 1996, 2009; Smith et al. 2003, 2009; Holtmeier and Broll 2010a). It is also important to recognize that even the smallest of krummholz mats are still taller than the surrounding tundra plant cover. Moreover, the developmental capability for treeline tree species for forming the compact krummholz habit is crucial for efficient snow collection and, thus, needle protection from the severe winter climate above the snowpack (Hadley and Smith 1987; Smith et al. 2003, 2004, 2009). Not all conifer tree species, even those found in the subalpine forests, have this capability for shifting growth to lateral buds as apical buds are killed.

Although the elevational/latitudinal position of a treeline may vary with respect to site conditions, treeline formation and maintenance appears correlated with air or soil temperatures on both a continental and global scale. In the last century, it was considered that treelines roughly coincide with the 10 °C air temperature isotherm of the warmest month (Daubenmire 1954; LaMarche 1973; Grace 1977, 1989; Ives and Hansen-Bristow 1983; Yanagimachi and Ohmori 1991), which holds for continental mountain ranges in the northern hemisphere while it is up to 13 °C in maritime regions (Holtmeier 2009). Tropical and subtropical treelines coincide with summer isotherms as low as 3–6 °C (Körner 1998a). Other authors found coincidences of the tritherm or tetratherm and the elevational treeline position (Odland 1996; Wardle 1968). Most recently, a re-examination globally revealed that alpine treelines occur at an elevation matching a growing season mean air temperature ranging from 5.5 to 7.0 °C (Körner 1998a, 2003; Hoch et al. 2002) and a growing season mean soil temperature at 10 cm soil depth of 5–8 °C, with a global mean of 6.7 ± 0.8 °C (Körner and Paulsen 2004; Körner 2007a). Such mean temperatures however, do not exist in nature and hence should be considered rather

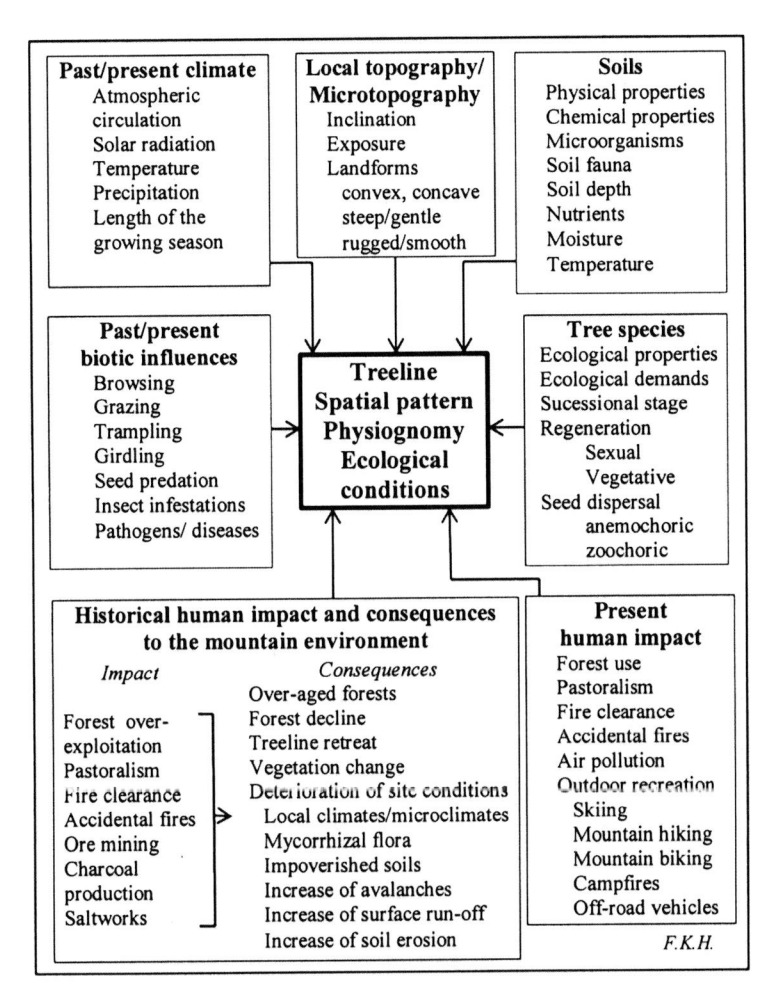

Fig. 10.1 Factors influencing treeline spatial pattern, physiognomy, and ecological conditions (Modified from Holtmeier 2009)

as an indicator of thermal deficiency than a causal factor (Holtmeier 1974, 2009; Tuhkanen 1980; Ohsawa 1990).

On a global scale, the upper limit of tree life appears ultimately dependent on heat balance, which everywhere becomes increasingly unfavourable with rising altitude and latitude. The possible role of heat-deficiency is clearly reflected in a gradual decline in treeline elevation from its maximum in the subtropics towards the poles, and in tree stature with increasing elevation, e.g. *krummholz* mats (Holtmeier and Broll 2007). Nevertheless, there are exceptions such as in arid and semi-arid mountains areas where alpine treeline elevation may be controlled by low precipitation and limited soil water availability (Troll 1973; Wardle 1974; Gieger and Leuschner 2004). Beside temperature there are also an abundance of regional and more local factors influencing treeline altitude, spatial pattern, and dynamics (Holtmeier 1994, 2009; Holtmeier and Broll 2007; Fig. 10.1).

Especially in dry continental mountain environments, low soil water availability and the occurrence of ecological facilitation has been shown to limit seedling establishment and growth in treeline ecotones of the Rocky Mountains, USA (Daniels and Velben 2004; Bekker 2005; League and Veblen 2006, review by Smith et al. 2009). In addition, human impacts have depressed treelines in most continents worldwide for millennia (Holtmeier 1974, 1986, 2009; Burga 1988; Burga and Perret 1998; Holtmeier and Broll 2005, 2010b; Kaltenrieder et al. 2005; Wesche et al. 2008; Fig. 10.1).

Currently, there is much interest in treelines because they are expected to undergo significant alterations due to climate change, especially climate warming (Holtmeier 1994; Grace et al. 2002; Holtmeier and Broll 2005, 2007; Walther et al. 2005; Smith et al. 2009; Wieser et al. 2009). During the last century the global mean surface temperature increased by about $0.6 \pm 0.2\,°C$ (Jones et al. 1988) and global change models predict even further temperature increase ($1.4–5.8\,°C$) during the approaching decades (IPCC 2007). Because observed changes seem to be most pronounced at high elevation (Beniston et al. 1997; Diaz and Bradley 1997) and high latitude (Lugina et al. 2006), and considering that the altitudes and latitudes of treelines are strongly correlated with the occurrence of heat deficiency, it might be hypothesized that global warming could generate denser, higher-growth forests below the treeline, as well as treeline movement to higher elevations and poleward (Grace et al. 2002; Dullinger et al. 2004).

More mechanistically, any upward or northerly movement of treelines will be dependent on successful seedling establishment and growth to maturity in the treeline ecotone, closest to the timberline first and then upward to treeline and beyond (Smith et al. 2003, 2009). Yet, studies of the mechanisms of seedling establishment and growth to maturity at treeline have been far fewer than studies on already established, mature trees (Wieser and Tausz 2007). In fact, no mention of this basic requirement was made in a recent review of treeline theories by Richardson and Friedland (2009), wherein it was concluded that a recently proposed idea based on the more exposed height of the tree habit (and self-shading), plus a greater cold-temperature limitation to carbon processing rather than gain, were the most reasonable, universal explanation for the elevations and latitudes observed for treelines globally. As discussed below, these conclusions have come from studies on adult trees and not the critical seedling establishment and growth phase.

It seems apparent that a fundamental requirement for understanding the causation of elevational and latitudinal treeline limits, and possible changes in response to climate change, is the mechanisms driving new tree establishment and growth to forest-tree stature across the treeline ecotone. Because much new information about the causes of treeline elevation is now available, this chapter summarizes the most current knowledge about the potential complex effects of climate change on alpine and polar treelines. Tropical treelines, however, are considered only in passing, primarily because knowledge is still deficient.

10.2 Mechanisms and Limiting Factors for Mature Tree Growth

Growth is a traditional measure of how well a species is adapted to a particular environment such as found at the treeline. To date, two of the latest views of what drives treeline elevations, and possibly latitudes as well, have originated from studies on mature trees. These newly proposed causes are (1) limitations to growth due to low temperature effects on carbon processing, not photosynthesis, and (2) effects of being arborescent and taller in height and, thus, exposure to the abiotic environment (especially during winter) is severe. Yet, the survival of a species is ultimately dependent on its reproductive success, including seed production, dispersal, germination, and successful establishment of reproductively viable individuals. For treeline trees, these early life stages must solve the severe environmental challenges of the more exposed treeline ecotone, a formidable barrier to higher elevation migration. Yet, much less is known about the early life stages of treeline trees that, as for most tree species, have much higher mortality and observed damage compared to mature trees. Episodic events such as drought, disease and pest outbreaks, fire, and anthropogenic disturbance may be also be critically involved in this upward migration of which relatively little unified theory exists on a world scale (but see review by Richardson and Friedland 2009).

10.2.1 Gas Exchange at High Elevation

Leaf gas exchange is the fundamental process supporting growth and survival that is dependent on a complex array of environmental variables. Because high elevation and latitude locations experience lower air temperatures, and increased elevation also results in lower ambient pressure (thus lower CO_2 partial pressure), treeline trees could be expected to be significantly impacted by global change, i.e. warmer temperatures and elevated atmospheric CO_2. Just as air temperatures decrease with greater latitude, at any latitude an increase in elevation is associated with an additional decline due to the air temperature lapse rate (a dry adiabatic maximum of approximately 0.98 °C per 100 m increase in altitude, but more typically 0.6 °C per 100 m = subadiabatic or saturated adiabatic lapse rate), and along with a decline in the atmospheric air pressure of approximately 13.3 kPa per 100 m of elevation (Baumgartner 1980; Smith and Geller 1979). In parallel, the partial pressures of its component gases, including N_2, O_2, and CO_2, decreases in the same proportion. In contrast, saturation water vapour pressure decreases with temperature because of the inability of colder air to hold water vapour, a colligative property of temperature that is independent of pressure changes. All these changes affect the driving forces for gas diffusion via the stomata, as recently discussed in Smith and Johnson (2009). Although temperature and pressure effects are in opposite directions, the pressure effect is dominant and hence diffusion rates increase significantly with elevation (Smith and Knapp 1990; Jones 1991; Smith and Donahue 1991; Smith et al. 2009). However, other energy balance considerations that include effects on leaf temperatures show that a rather complex interaction of several variables will dictate ultimate elevation effects on plant transpiration and CO_2 uptake (Smith and Johnson 2009). Except for ambient pressure effects, microclimate variability at any elevation can alter any biophysical parameters typically associated with elevation, e.g. air temperature lapse rate (Smith and Geller 1979).

Although the diffusion of water vapor from the leaf at any degree of stomatal opening tends to increase with elevation, only limited data are available on the effect of elevation on plant (Smith and Geller 1979) or tree transpiration (Smith and Johnson 2009). There is evidence that tree transpiration will increase with increasing elevation in more humid mountain regions (Matyssek et al. 2009), and a similar finding was reported by Smith and Johnson (2009) using a biophysical modeling approach that predicted substantial decreases (over 50 %) in water use efficiency at higher elevation (Fig. 10.2). This substantial decline in estimated water use efficiency was due to predicted increases in transpiration (constant stomatal conductance) with little change in photosynthetic performance. However, the role of differences in plant structure and microsite conditions were also mentioned as possible, but currently unknown, sources of impact on the interaction of elevation and plant gas exchange.

With respect to CO_2 assimilation, trees at treeline have been found to have a similar in situ net photosynthetic capacity compared to that of their relatives at lower elevation sites (Mooney et al. 1964; Benecke et al. 1981; Rada et al. 1998; Wieser and Tausz 2007; Wieser et al. 2010). Yet, elevation related differences in leaf structure have also been reported, e.g. timberline trees often produce greater specific leaf mass per unit area (Benecke et al. 1981; Hurtin and Marshall 2000; Richardson et al. 2001; Wieser 2012), and also have similar or even higher nitrogen concentrations per unit dry mass when compared to lower elevation sites (Sparks and Ehleringer 1997; Richardson et al. 2001; Birmann and Körner 2009; Wieser 2012). Thus, higher elevations may be driving adaptations in leaf and plant structure that would enhance CO_2 uptake, beyond the more rapid diffusion at lower pressure that nearly compensates for the smaller leaf-to-air CO_2 gradient (Smith and Hughes 2009).

The rise in atmospheric CO_2 concentration is another global phenomenon (Keeling and Whorf 2005) that could alter gas exchange physiology, especially at higher elevations. Results of a 3-year free air CO_2 enrichment at the alpine timberline provided light-saturated net photosynthesis and non-structural carbohydrate concentrations of 30-year-old *Larix decidua* and *Pinus uncinata* trees to be significantly higher under elevated CO_2 (Hättenschwiler et al. 2002; Handa et al. 2005). Nevertheless, CO_2 enhancement caused a significant stimulation in lateral shoot growth, and in radial stem increment of deciduous *Larix decidua*, whereas no response was observed in evergreen *Pinus uncinata* (Handa et al. 2005, 2006). Because differences in branch growth perhaps relates to foliage type (Cornelissen et al. 1999), it is likely that other factors than elevated CO_2, such as

Fig. 10.2 Simulated changes in transpiration (E) and water use efficiency (WUE) with elevation at a computed as the product of leaf-to-air vapor difference and a constant stomatal conductance of 400 mmol m^{-2} s^{-1} (*circles*) and where stomatal conductance is proportional to the diffusion coefficient for water vapor in air (*squares*) at a wet laps rate of 0.3 °C per 100 m increase in elevation (Modified from Smith and Johnson 2009)

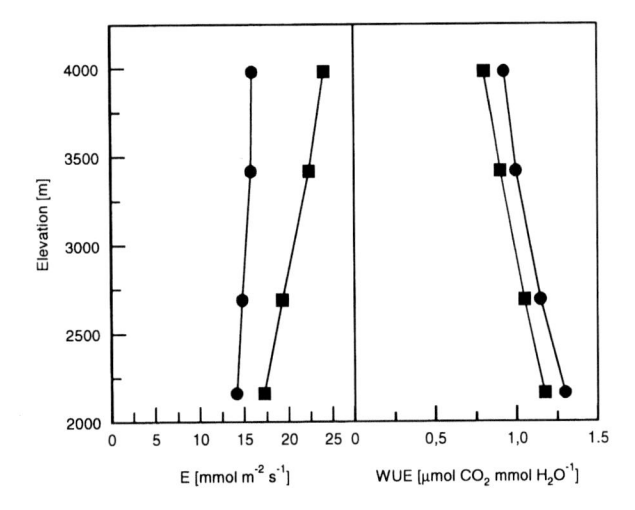

temperature and hence the length of the growing season might also be involved in the observed growth stimulation of deciduous larch as compared to evergreen pine (c.f. also Crawford 2008). Evidence exists that CO_2 effects will decline over time due to metabolic acclimation (Körner 2006) and age-related stand dynamics in competition for above and belowground resources (Hättenschwiler et al. 1997).

10.2.2   Carbon Gain Saturation or Limitation at Treeline?

Some studies have suggested that that annual tree growth at treeline is not carbon limited, but carbon saturated under the present environmental conditions (Körner 2003). This idea is based on reports of the constancy of stored soluble carbo-hydrates regardless of elevation in tree species along elevational gradients (Hoch et al. 2002, 2003; Hoch and Körner 2003; Shi et al. 2006, 2008, see also review by Smith et al. 2009). However, the idea of carbon saturation has been questioned recently in several studies that provide opposing data (Millard et al. 2007; Susiluoto et al. 2007; Ow et al. 2008; Li et al. 2008; Gruber et al. 2011) and in particular when applied to young seedling growth in the alpine treeline ecotone (Bansal and Germino 2008).

High non-soluble carbohydrate contents of treeline trees suggests that it is not photosynthetic carbon supply that limits tree growth at treeline, but the rate at which glucose can be processed for growth (Grace et al. 2002; Körner 2003). This indicates that these particular trees will likely not respond to elevated CO_2 because they are not limited by photosynthetic CO_2 supply. As the latter is more temperature driven, low temperatures rather than CO_2 supply has been proposed as limiting tree growth at high elevation and latitude (Körner 2003). For example, at the treeline in the Central Austrian Alps Wieser et al. (2009) showed that warm springs increased the carbon gain of *Pinus cembra*, while Kronfuss and Havranek (1999) showed growth of *Pinus cembra* to be positively related to the length of the growing season. Year-to-year variations in net ecosystem exchange of a boreal forest were also strongly dominated by variations in spring temperature (Suni et al. 2003). Artificial air and soil warming also significantly enhanced tree growth at the alpine and the subarctic treeline (Strömgren and Linder 2002; Danby and Hik 2007; Gruber et al. 2010). This is because low air or soil temperatures at treeline are known to limit tissue formation and growth when temperature drops below 5–7.5 °C (Loris 1981; Turner and Streule 1983; Tranquillini 1973; Juntilla and Nilsen 1993; James et al. 1994; Häsler et al. 1999; Alvarez-Uria and Körner 2007; Oberhuber 2007). Low root zone temperatures also affect shoot function such as photosynthesis, transpiration (Havranek 1972), and leaf conductance (Day et al. 1990; Körner 1994; Wieser 2000), as well as mycorrhizal activity (Hasselquist et al. 2005).

10.2.3   Soil Water and Nutrients

At alpine and polar treelines, cold air and soil temperatures and their effects on biochemical processes and growth of adult trees have been proposed to be the key factor in determining treeline at a particular elevation or latitude. The influence of elevation on soil water limitations, however, has not been systematically studied in treeline species. Nevertheless, there is evidence of soil moisture deficits causing drought stress and significantly decreased net photosynthesis and transpiration of potted high altitude provenances of *Pinus cembra*, *Picea abies* and *Larix decidua* seedlings (Havranek and Benecke 1978; Wieser and Kronfuss 1997). This is because the primary physiological response to water limitation is stomatal closure, thus diminishing gas-exchange. In all three species gas exchange began to decline at a soil water potential of −0.4 MPa and approached zero at soil water potentials of −1.5 (*Pinus cembra*, *Picea abies*) to −1.9 (*Larix decidua*) MPa due to complete stomatal closure as well as to direct effects of water shortage (Kaiser 1987; Larcher 2001) and possible effects on electron transport and photophosphorylation. Yet, greater demands on transpiration at higher elevation, along with cold temperature effects on soil water uptake (i.e. roots and stems), implicates a strong potential for drought limitations. On the other hand, severe waterlogging may hamper decomposition and nutrient uptake in shallow snow-rich depressions or in the active layer of permafrost affected soils (Larsen 1989). Also, mycorrhizae do not tolerate lasting water-logging. In dry regions by contrast, water supply in the active layer may favour rather than prevent tree establishment (Kryuchkov 1973).

In the treeline ecotone low soil temperatures result in a low metabolic activity, including low decomposition rates. Generally, at temperatures <5 °C mineralization and nutrient uptake rapidly decline (Retzer 1974). Tree growth at its northern distribution limit has also been shown to be affected by soil nutrient availability in particular nitrogen supply (Steinbjörnsson et al. 1992; Schulze et al. 1994; Karlsson and Weih 2001). At the upper tree limit of the mountain birch forest in northern Sweden (Abisko), at soil temperatures <5 °C nutrient uptake and seedling growth almost ceased (Karlsson and Nordell 1996; Karlsson and Weih 2001). Similar results were found at the upper *Nothofagus solandri* treeline in New Zealand (Benecke and Havranek 1980; Wardle 1985). Thus, enhanced nitrogen deposition with adequate moisture availability may be considered growth stimulating. Nevertheless, a time series analysis did not reveal any positive correlation between annual ring width in the timberline ecotone of the Austrian and Swiss Alps and nitrogen deposition (Nicolussi et al. 1995; Paulsen et al. 2000). In general, additional nitrogen will not enhance growth as long as other nutrients such as phosphorus, for example, are not available at sufficient quantities. Holtmeier and Broll (1992) found plant available phosphorus to be a probable limiting factor at the treeline on the Colorado Front Range. Similar results have been reported from the mountain birch treeline in northernmost Finland (Schreiber 1991). In this case an increase of other 'fertilizers' (e.g., nitrogen) would exacerbate the situation. However, plant available phosphorus might have been low because mycorrhiza had taken it already up (see also Haimi et al. 1992). Soil nutrient limitations during new seedling establishment in the treeline ecotone may be even more critical for such shallow root systems, and may play a role in the limitation of carbon processing hypothesized for treeline tree species (Chapman et al. 2006).

Understanding the dynamics of treeline altitude must consider seedling establishment and limitations to growth (Cui and Smith 1991; Germino et al. 2002), as well as the temporal dynamics that are possibly episodic in nature (Holtmeier et al. 2003; Smith et al. 2003, 2009; Holtmeier and Broll 2007; Holtmeier 2009). Very little is known at treeline concerning the seed population demographics of germination and establishment, either in the ecotone or just below the closed forest (timberline) boundary. With respect to nutrient and water requirement seedlings differ considerably from mature trees. In contrast to relatively deep rooting of mature trees, seedlings and saplings mainly depend on soil moisture and nutrient supply from the litter layer and the topsoil. On permeable and rapidly draining substrates (e.g., sandy skeletal basal till) drought and insufficient nutrient supply may impede seedlings and saplings, in particular if the organic (humus) layer, which is the main nutrient reservoir, is eroded (Holtmeier et al. 2003; Holtmeier and Broll 2006; Broll et al. 2007; Anschlag et al. 2008). Nutrient and soil water availability strongly depend on soil type. Although pedogenetic processes and also the effects of soil temperature, soil moisture, decomposition, and nutrient supply on vegetation partly depend on the elevational change of climate, there is no single soil type which can be considered typical of the treeline ecotone. Generally, mosaics of different soils exhibiting different physical (e.g., texture, bulk density, moisture content) and chemical properties (e.g. pH, cation exchange capacity, nutrients) characterize the treeline ecotone (Blaser 1980; Broll et al. 2007; Stöhr 2007; Holtmeier 2009).

The influence of the winter snowpack as a nutrient source in the treeline ecotone should not be neglected because snow pack accumulates nutrients by deposition of aeolian dust ('alpine loess'; Thorn 1978; Litaor 1987; Holtmeier and Broll 1992; Broll and Holtmeier 1994). Needles, bud scales, and bark particles also accumulate in the deepening snowpack. Aeolian dust and organic debris act as a nutrient source mainly during snow melt (for a review see Tranter and Jones 2001). Additionally, canopy litter contributes to relative high concentrations of readily plant available nutrients such as NO_3, NH_4^+, $H_2PO_4^-$, and HPO_4^{2-} in melt waters (Tranter 1991).

10.3   What Makes Trees Different from Low Stature Vegetation?

Given their arborescent life form, trees at the timberline are generally perceived as experiencing a more severe climate compared to the low stature vegetation of the nearby alpine and arctic tundra. Because their canopy is aerodynamically rougher and experiences higher wind speeds, trees generally operate closer to air temperature during both the day and night throughout the year (Grace et al. 1989). In conifers within the timberline ecotone daily mean leaf-to-air temperature differences usually did not exceed approximately 4 °C under normally windy conditions (Turner 1958; Tranquillini and Turner 1961; Tranquillini 1964; Loris 1981; Hadley and Smith 1987; Gross 1989; Wieser 1997, 2002; Loris et al. 1999; Wieser and Bahn 2004). However, depending on canopy position and season, periods may occur when daytime needle temperatures during the growing season in the sunlit upper parts of tree crowns are commonly well above air temperature (>10 °C) and, due to radiative cooling under clear skies, needle temperature can be <10 °C below air temperature (Tranquillini and Turner 1961; Tranquillini 1964; Baig and Tranquillini 1980; Hadley and Smith 1987; Wilson et al. 1987; Gross 1989; Jordan and Smith 1994). These over- and under-temperatures measured in the field are the result of tight needle packing characteristic of sun shoots, as well as increased secondary branch packing (Smith and Carter 1988). Compared to needle temperatures maximum overheating (9 °C) and undercooling (5 °C) were significantly smaller in aboveground woody tissues (Loris 1981; Gross 1989; Wieser 1997, 2002; Loris et al. 1999; Wieser and Bahn 2004) and nil in coarse roots (Wieser and Bahn 2004).

A decrease in height growth with increasing elevation in mature trees has to be attributed to a shorter growth season and/or corresponding increases in wind damage (Kronfuss and Havranek 1999). Seedlings, however, are less influenced by wind because of their initially short stature and greater coupling to canopy heating (Hadley and Smith 1987; Grace et al. 2002), at least as long as not being exposed to strong winds (Wilson et al. 1987; Holtmeier and Broll 2010a). Under dense krummholz canopies however, soil temperatures are usually relatively low (Holtmeier and Broll 1992). Although low temperatures may affect carbon use and nutrient uptake, they do not appear to limit tree growth as suggested by Körner (1998a, b). Even the uppermost krummholz mats usually produce annual growth rings. Dwarfed trees, often expand gradually by branch layering, and have survived for centuries as suppressed growth forms (mats, table-trees) 'waiting' for more favourable conditions e.g. improving climate, facilitation by increasing tree population, reduced wind effects) that will enable release of upright stems (cf. Holtmeier 2009).

Windthrow, mechanical damage due to breaking of twigs and branches by wind and snow (Däniker 1923), as well as mechanical damage due to ice-blasting (Müller-Stohl 1954; Lindsay 1971; Holtmeier 1974, 1980, 2009; Hadley and Smith 1983, 1986, 1990; Van Gradingen et al. 1991; Dahms 1992; Streule and Häsler 2006; Holtmeier and Broll 2010a) is a remarkable occurrence in the timberline ecotone.

Ice particle abrasion may remove even fully developed cuticles (Hadley and Smith 1983, 1986; Dahms 1992), which is in contrast to earlier observations and experimental studies at treeline in the European Alps (Holzer 1959; Turner 1968). There is evidence from the Rocky Mountains, Mt., Washington (New Hampshire), Canada, southern Greenland and from the European Alps that ice-particle abrasion and abrasion by mineral particles is often a more important factor than thinner or less developed cuticle and may cause needle loss even prior to winter desiccation (Marchand and Chabot 1978; Holtmeier 1980; Hadley and Smith 1983, 1986; Perkins et al. 1991, for additional references see Holtmeier 2009; Holtmeier and Broll 2010a). Winter desiccation of incompletely developed needles projecting beyond the winter snowpack provides a possible precondition for subsequent needle loss by wind damage.

Abrasion by windborne mineral particles (sand blast) may locally occur downwind of exposed rapidly drying mineral soil, mainly during the summer and snow-free dry periods in winter (Holtmeier et al. 2004; Holtmeier and Broll 2010a). Needle loss due to strong winds during summer reduces photosynthetically active tissue, and thus, may also cause nutrient loss (Sveinbjörsson et al. 1996).

At temperate treelines, snow distribution by wind and snow burial is often of paramount importance to site conditions (Holtmeier et al. 2003; Holtmeier 2005b, 2009; Holtmeier and Broll 2010a). Wind redistributes fallen snow from exposed convex topography to snow-collecting downwind sites such as leeward slopes of small ridges, depressions, gullies and similar concave topography (Holtmeier 2005b). However, notwithstanding destruction by avalanches, a long lasting snow cover in temperate mountains also abbreviates the growing season length. In addition, seedlings and small saplings buried below snow are protected from mechanical abrasion, animal herbivory, photo-oxidative stress, and evaporative water loss during the winter. On the other hand, trees under snow lose their frost resistance and can easily be damaged by low temperatures after being released (Tranquillini 1959). Moreover, the high humidity of the air together with a constant temperature of about 0 °C below the snow surface, favours the attack of weakened trees by parasitic snow fungi (Donaubauer 1963; Bazzigher 1978; Nierhaus-Wunderwald 1996) which can cause widespread damage in winters with an exceptionally long snow cover (Aulitzky and Turner 1982). Wind also increases the leaf boundary layer conductance, but also decreases leaf temperature, and thus affecting transpiration oppositely (Larcher 1985). Cold winds promote the penetration of low temperatures and soil frost in snow-free habitats and rocks (Aulitzky 1961;

Bonan 1992) and winter desiccation may affect younger trees, even in these sheltered locations. In snow-rich sites (e.g., lee slopes of convex topography, depressions, open tree stands) seedlings and saplings may be distorted by mechanical stress or even destroyed by snow creep and settling snow (Holtmeier 2005b).

While frost tolerance is generally high in treeline forming adult trees (Larcher 1985; Holtmeier 2009 and literature therein), seedlings and saplings may be severely affected if not sufficiently protected by snow burial (Sakai and Weiser 1973; Slatyer 1976; Sutinen et al. 2001). Thus, removal of snow by strong winds may be fatal to young trees (Frey 1983). However, young trees buried beneath the winter snow pack are usually less frost tolerant than those unburied (Neuner 2007). A particular risk for tree growth and survival within the treeline ecotone may also arise from early frost dehardening in spring in response to warmer winter temperatures (Neuner 2007) as well as frost drought (Tranquillini 1976, 1982). The latter is assumed to be one of the main causes for the upper limit of tree growth in high mountains outside the tropics (Turner 1968; Tranquillini 1976, 1979; Baig and Tranquillini 1980; Sowell et al. 1982; Schwarz 1983; Christersson et al. 1988) because of insufficient maturation of cuticles and buds during the preceding summer (Holtmeier 1974; Hadley and Smith 1989). Frost drought or winter desiccation can occur throughout late winter and early spring when water losses by cuticular transpiration cannot be compensated due to frozen soil and above-ground tissues (Michealis 1934a, b; Larcher 1957, 1963, 1985; Tranquillini 1976, 1979, 1982; Sakai and Larcher 1987; Hadley and Smith 1990; Havranek and Tranquillini 1995). Winter desiccation, however, is observed primarily in young trees unprotected from snow burial and, thus, has been considered with a determinant of the krummholz altitude limit (Tranquillini 1979, 1980; Sowell et al. 1982; Barclay and Crawford 1982; Delucia and Berlyn 1984; Cairns 2001). Because winter desiccation probably diminishes with increasing tree age and height above the snow pack, it also appears to distort tree growth at treeline (Holtmeier 2009; Holtmeier and Broll 2010a). More work is needed to fully elucidate the role of winter desiccation on treeline tree growth and existing spatial patterns.

10.4 Specific Evidence for Growth Limitations in Mature Trees

There is extensive, correlative evidence that temperature has a strong effect on new tissue formation and tree growth at treeline (Loris 1981; Kronfuss 1994; Körner 1998a; Jobbagy and Jackson 2000; Grace et al. 2002; Oberhuber 2007; Gruber et al. 2010) as well as for a tight coupling between temperature and meristem activity (Hellmers et al. 1970; Juntilla 1986; Scott et al. 1987a; James et al. 1994). Kronfuss (1994) documented that height growth in *Pinus cembra* in the Austrian Alps had a strong correlation with growing season mean air temperature. Also shoot and stem growth ceased when temperatures fell below 5–7.5 °C (Loris 1981;

Kanninen 1985; Juntilla and Nilsen 1993; Häsler et al. 1999). Also at treeline, soil temperature has been regarded as the most important limiting factor for root growth, including the initiation and cessation of seasonal growth (c.f. Pregitzer et al. 2000). Soil temperatures below 5–7 °C substantially reduced root elongation rates and root growth of treeline-associated conifer and deciduous tree species (Lyr and Hoffmann 1967; Tryon and Chapin 1983; Turner and Streule 1983; Andersen et al. 1986; Lyr 1996; Häsler et al. 1999; Anschlag et al. 2008).

Within the treeline ecotone annual height growth of *Pinus cembra* individuals exceeding a height of >0.5 m has been shown to decrease with elevation (Li et al. 2003; Li and Yang 2004) corresponding to a decrease in the length of the growing season and an increase in wind velocity (Kronfuss and Havranek 1999). This effect is more marked with increasing elevation and latitude when forest stands open up and trees are isolated from each other (Tranquillini 1979; James et al. 1994; Paulsen et al. 2000; Li et al. 2003). Thus, in extremely windy treeline climates (e.g. Rocky Mountains) a decrease in height growth may be dependent more on wind exposure than on elevation (Kronfuss and Havranek 1999; Holtmeier 2009; Holtmeier and Broll 2010a). In the early seedling stage, however, neither elevation nor topography strongly affected height growth (Li et al. 2003; Li and Yang 2004); simply because they experience a microclimate comparable to short-stature plants. However, as pointed out in Smith et al. (2003, 2009), both microsite and crown architecture appear to have major impacts on absorbed sunlight, photosynthetic carbon gain, root growth, mycorrhizal infection, and corresponding survival in newly emerged tree seedlings in the subalpine forest understory and across the treeline ecotone.

10.5 Differences Between Adult Trees and Establishing Seedlings

Smith et al. (2003, 2009) proposed that seedling establishment and growth in the treeline ecotone must be prolific enough to generate mutual facilitation of microclimate that will then facilitate seedling/sapling growth to forest-tree stature and, ultimately, the formation of new of subalpine forest at higher altitude. Although studies on adult trees could provide insight into understanding treeline position, e.g. cone/seed production, it is critical that environmental controls of seedling establishment and growth be understood before the impact of future global change can be predicted. For example, it is relatively well known that the life stage of most trees with the greatest mortality is seedling establishment. Studies on subalpine trees in the understory showed a high mortality of first-year seedlings that appeared related to microsite sun exposure, either too much or too little (Knapp and Smith 1982; Cui and Smith 1991; Germino et al. 2002). Moreover, in both cases, lack of carbon gain was associated with poor root growth, poor mycorrhizal infection, and death by apparent high temperature and desiccation at sunny microsites (Cui and

Smith 1991). These same findings were similar to a series of studies on treeline seedlings of the same age-class, but now implicating photoinhibition of photosynthesis generated by high sunlight exposure, especially following cold clear nights (Jordan and Smith 1995), a common occurrence in the treeline ecotones of high elevations (Germino and Smith 1999, 2001; Germino et al. 2002, 2006).

As another example of fundamental differences between adult trees and establishing seedlings, the question of carbon source (photosynthetic) versus carbon sink (processing) limitation has recently been evaluated in treeline seedlings during their first years of establishment (Bansal and Germino 2008). These results showed a greater carbon sink limitation at higher elevation that generated a greater photosynthesis/respiration ratio. However, and importantly, photosynthesis and dry mass gain were positively correlated. Thus, carbon gain limitation also occurred with greater elevation, even though photosynthetic carbon gain was somewhat less influenced than respiratory carbon processing. To our knowledge, no other studies at treeline have involved establishing seedlings, especially the first few years of growth when mortality can be >90 %. In fact, in the species studied (*Abies lasiocarpa*; *Picea engelmannii*) mortality in seedlings greater than 3 years-old was nil, and remained so for the life of the tree (see also Mellmann-Brown 2005). Only fire and disease losses occurred episodically in the older trees of these species.

While height and diameter growth of adult trees may well correlate with increasing thermal deficiency at higher elevation (see above) and regional and local climatic fluctuations (temperature, periods of drought), seedling establishment and survival may be more closely related to topographically controlled microclimates and soil conditions (temperature, moisture, nutrient availability). Also extreme climatic events such as severe storms, strong frosts, snow break, and insect infestations may reduce growth for many years. Furthermore, competition with, but also facilitation by the surrounding dwarf-shrub and grassland vegetation play an important role by providing protection from wind, excessive solar radiation, and night-time radiation frost.

Facilitation by adult trees may encourage seedling establishment and development (Germino and Smith 1999, 2001; Hättenschwiler and Smith 1999). The feedback of the trees on their environment, in concert with competition, dissemination, and availability of amenable seed beds, cause a locally varying mosaic of tree groves and scattered trees surrounded by dwarf shrub, grassland, non-vegetated patches and/or bog vegetation. Shelter from wind, radiation (Germino and Smith 1999; Germino et al. 2002; Johnson et al. 2004), and grazing or browsing by mammal herbivores favour seedlings and saplings and thus promotes infilling of treeless patches in the treeline ecotone. Increased snow pack however, may also have negative impacts. For example, as the growth season is shortened a delayed rise of soil temperature may affect germination and growth of seedlings. Moreover, the risk of infection by parasitic snow fungi may increase with increasing seedling density (Holtmeier 2005b; Holtmeier and Broll 2010a). In dry climates by contrast, meltwater supply from late-laying snow associated with drifting around tree islands appeared to improve growth conditions for tree seedlings (e.g. Hättenschwiler and Smith 1999; Holtmeier 2009). Finally, shading by increased tree cover reduces soil temperature (Patten 1963; Holtmeier 1982) and, thus, may also affect germination and seedling establishment. Körner (1998a, b) argued that taller trees would therefore eliminate themselves by shading the ground, although this would not apply to seedlings located away from the forest edge into the treeline ecotone (Holtmeier 2009; Malanson et al. 2007, 2009; Kullman 2010). In fact, treeline tree seedlings of the Rocky Mountains, USA, were found more often in the shade cast by krummholz mats and tree islands, providing both protection from excessive sunlight and longwave radiation loss to the cold night sky (Germino et al. 2002).

At tropical treelines the situation is completely different due to a diurnal climate. No seasonal snowpack prevents seedlings and young growth from adverse climatic influences. Thus, young trees that became established outside closed forest stands are exposed all year round to excessive solar radiation, drought, and night frosts (Larcher 1980b; Smith 1980; Crawford 1989, 2008; Goldstein et al. 1994; Bader et al. 2007; Richter et al. 2008). Climatic stress is higher in seedlings and saplings than in tall trees coupled to the free atmosphere. Moderate shade seems to have promoted seedling establishment in open tree stands (Cui and Smith 1991; Bader et al. 2007), while under closed forest canopies, young seedling growth was slow, probably due to sub-optimal light intensities and lower soil temperatures (Lange et al. 1997; Wesche et al. 2008).

10.6 Seedling Establishment and Growth to Forest Tree Stature

Seed production, dispersal, and germination success are critical life stages that have received only limited study at treeline. In general, the upper limit of production of viable seeds is located below the physiological tree limit (Holtmeier 2009). As one approaches the treeline the amount and quality of seeds decrease (Holtmeier 1974, 2005a, 2009). For example, Dahms (1984) found germination capacity of *Picea engelmanni* seeds to be <1 %, in the treeline ecotone (3,350–3,500 m) on the Colorado Front Range as compared to 24 % in the subalpine belt (3,150–3,350 m). In the long-term the upper limit of production of viable seeds is located below the physiological tree limit and the elevational limit for viable seeds fluxuates. Uppermost trees in the ecotone may occasionally produce fertile seeds following several subsequent, sufficiently warm growing seasons without extreme events (e.g. strong frost) and new trees may become established and survive at suitable microsites. Thus, regeneration at the alpine treeline ecotone can vary temporally and spatially at both the local and regional scale (Holtmeier 2009).

Young growth in the treeline ecotone was represented by only a few age classes and was attributed to an insufficient supply of viable seeds rather than to a lack of favourable sites for seedling establishment (Holtmeier 1974, 2005a; Juntunen and Neuvonen 2006). If a lack of suitable sites is the only factor impeding seed-based regeneration, complete age classes could be expected. Nevertheless, both poor seed quality and a lack of favourable sites or years for seedling establishment have to be considered as primary causes of the episodic regeneration observed at treeline (League and Veblen 2006). This however, may change in a future warmer environment.

In general, seed-based regeneration inside and beyond the treeline ecotone may be considered as a series of hurdles (Holtmeier 2009) because the regeneration process from bud formation to seed germination and seedling establishment takes at least several years and is controlled by many physical and biotic factors (Holtmeier 1993, 2009; Holtmeier et al. 2003; Smith et al. 2009; Fig. 10.3). Moreover, disturbances that inhibit regeneration are possible at any time and may override positive effects of climate warming (Dullinger et al. 2004). Interestingly, regeneration does not appear to require exceptionally warm summers. In Scots pines (*Pinus sylvestris*) in northernmost Finnish Lapland regeneration above the present pine forest occurred during the relatively cold periods from the mid-1970s through the 1980s (Holtmeier 1974, 2005a; Holtmeier et al. 2003; Juntunen et al. 2002; Juntunen and Neuvonen 2006). Furthermore, cold spells with mean day temperature $<5\,^\circ\text{C}$ occurred with relatively high frequency during the growth seasons of the mid-1980s to the early 1990s (Holtmeier et al. 2003), yet did not seriously impair regeneration (Holtmeier and Broll, unpublished data). However, the relatively warm summers in 1972–1974 and in 2004 could have favoured the production of both viable seeds and seedling establishment.

As to successful regeneration from seeds the way of seed dispersal (by wind or animals) could play an important role. Wind-mediated dispersal depends on weather conditions, topography, and plant cover and, thus, is very irregular. Also strong winds may carry conifer seeds over long distances (Aario 1940; Payette and Delwaide 1994). Most windborne seeds however, reach the ground within a distance two to three times the height of the seed trees. Moreover, wind carries light seeds over greater distances than heavy seeds. As energy content and germination capacity of wind-dispersed seeds are positively correlated with seed size (Sveinbjörsson et al. 1996), the amount of viable seeds decreases with distance from the seed sources. Thus, a long distance from a seed source could explain the scarcity of tree seedlings and saplings in a given location in the treeline ecotone, rather than a paucity of suitable sites. More research on these contrasting limitations to seedling establishment however is necessary (Marr 1977; Batllori et al. 2009).

10.6.1 Seed Dispersal

Some treeline forming species such as *Pinus pinea* (e.g., *Pinus cembra, Pinus koraiensis, Pinus albicaulis*), juniper (e.g. *Juniperus indica, Juniperus turkestanica*), and mountain ash (*Sorbus aucuparia, Sorbus microphylla*) rely on seed dispersal by birds either endozoochoric (e.g. crows, jackdaws, alpine choughs and thrushes) or synzoochoric birds (nutcrackers). Endozoochoric dispersal is more incidental because it depends on defecation. Moreover, seeds accumulated on the

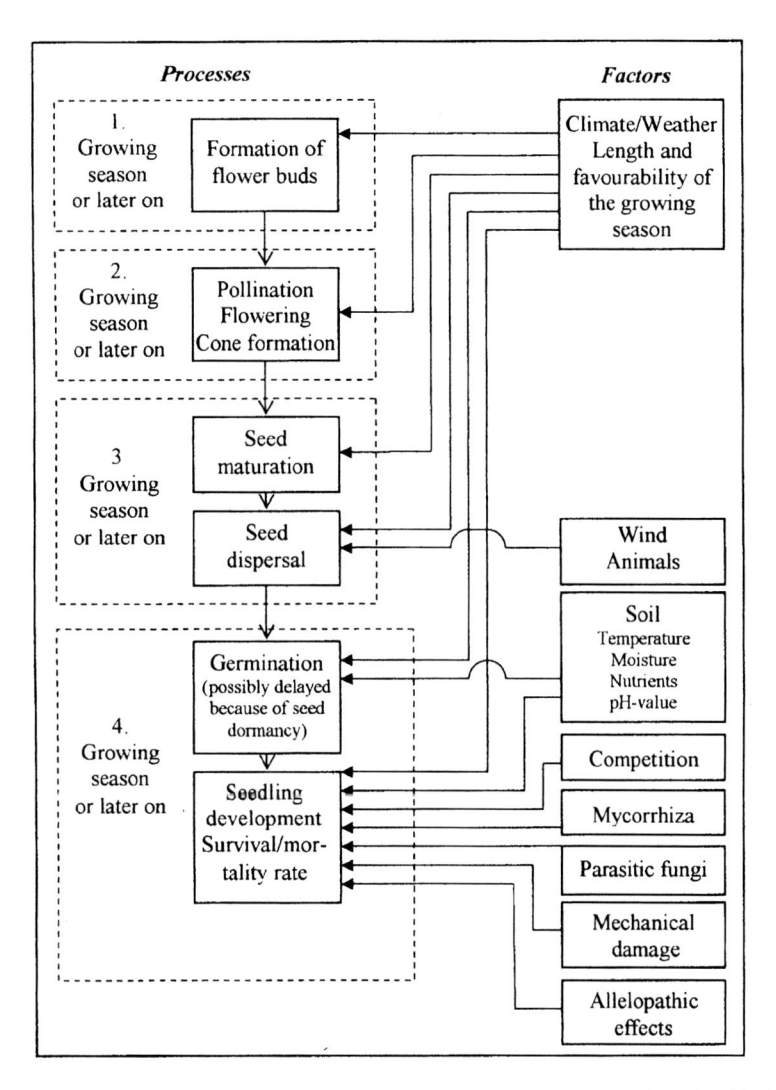

Fig. 10.3 Factors and processes controlling seed-based regeneration at treeline (Modified from Holtmeier 2009)

ground are easily accessible to seed predators. Thus, more or less great losses are likely. Nutcrackers (*Nucifraga caryocatactes* with several subspecies and races in Eurasia, and *N. columbiana* in North America) by contrast, select specific sites to cache the heavy wingless or almost wingless *Pinus* seeds on which they and their nestlings rely for food in winter and spring (Holtmeier 1996, 2009; Tomback 1977; Mattes 1978, 1982). The total treeline ecotone and the alpine zone above are within the nutcrackers' home range and area of activity (Sutter and Amann 1953; Holtmeier 1974, 2009; Tomback 1977; Mattes 1978). Nutcrackers hoard seeds mainly in underground seed caches (in the organic layer) where seeds are invisible to predators. Unused seed caches my give origin to seedling clusters. As the seed caches are established mainly on convex topography, not covered too long with snow, they are relatively favourable for seedling establishment. Moreover, the hoarded seeds are usually rich in nutrients which allow rapid seedling development. Altogether, seedling establishment at and beyond treeline seed dispersal by nut-crackers seems to be more effective than endozoochoric and wind-mediated distri-bution of seeds.

Inside *Stone pine* clusters, the individual tree is relatively protected from cli-matic injuries, even thought these clusters usually thin out in the long term due to root competition. Occasional root grafting which is most frequent between tree individuals genetically closely related to each other (Tomback 1977; Tomback et al. 1994) can reduce competition. On the other hand, increasing snow accumu-lation inside the taller growing groups favours infection by snow fungi which may destroy the smaller individuals (Holtmeier 1974, 2009; Holtmeier and Broll 2010b).

10.6.2 Germination

Viable seeds of some tree species may persist for several years before germination. For example, seeds of *Pinus albicaulis* (Tomback et al. 1993; McCaughey 1994), mountain ash (*Sorbus aucuparia*) and mountain birch (*Betula pubescens* ssp. *czerepanovii*) remained viable for several years, while Scots pine seeds germinated shortly after released from cones (Granström 1987). Prolongation of diapause makes tree regeneration a little more independent of climatically-controlled fluc-tuations in the production of viable seeds.

Abbreviated growth seasons due to late-lying winter snowpack or missing snowpack, waterlogging, or excessively high (Turner 1958; Lauer and Klaus 1975; Barry 1978; Scott et al. 1987; Rien et al. 1998) or low soil temperatures (Holtmeier et al. 2003; Anschlag 2008) may impair or even prevent germination, as well as seedling growth. Also, prolonged strong winds may impair growth and establishment (Holtmeier 2005b; Kullman 2005a, 2010; Anschlag et al. 2008; Holtmeier and Broll 2010a; Paus 2010).

10.6.3 Early Seedling Growth and Survival

Conditions favorable to germination do not necessarily enhance seedling growth. Thus, in snow-rich sites germination may be supported by melt-water supply while seedling establishment may fail due to low soil temperature and curtailed growing season (Turner 1958; Franklin and Dyrness 1973; Noble and Alexander 1977). In contrast, in dry sun-exposed places, germination may be impeded whereas seed-lings, once established, show improved growth.

In general, seedling mortality is particularly high during the first years (Cui and Smith 1991; Mellmann-Brown 2005; Juntunen and Neuvonen 2006). The main adverse factor is competition with dwarf shrubs and grasses (Wardle 1985). In addition, allelopathic effects of the associated vegetation (e.g., *Empetrum hermaphroditum*, some lichen species) may impede germination and seedling growth (Farmer 1997; Nilsson et al. 1993; Weih and Karlsson 1999).

10.6.4 Escape from the Boundary Layer

After seedlings have survived the initial years of establishment they enter a particularly critical life stage associated with growth through the sheltereding boundary layer provided by adjacent low stature vegetation (dwarf-shrubs and grasslands). When no longer growing in a 'chamaephyte environment', survival depends on the individual hardiness of the young trees fully exposed to severe climatic agents (wind, radiation, frost, winter desiccation, etc.) During the winter, blowing snow can cause ice abrasion of exposed (vertical leaders from prior summer growth) surface needles, and result in desiccation, and death. This same boundary layer however, may also be protective and growth enhancing (winter snow capture and summer shading) as well as providing thermal advantages that extend the summer growth period (see Smith et al. 2003, 2009 for detail).

For vertical stems to survive this 'escape' from the protective boundary layer apical buds, at least, must survive growth above the snowpack, e.g. above the surface of krummholz mats. This apical bud survival, despite needle death on the lower stem, can be seen commonly in the field on vertical leaders that extend beyond the krummholz mat surface, usually first on the leeward edge of mats where windflow and ice abrasion is lessened due to boundary layer effects (Smith et al. 2003). This successful growth through the boundary layer above the winter snowpack collected by krummholz mats or by low stature vegetation is enhanced by the facilitation of surrounding vegetation and the accompanying amelioration of microclimate. Thus, the survival of apical buds above stem sections of deceased and discarded needles, plus the facilitation of surrounding vegetation, once again, plays a critical role in seedling growth to forest tree stature. This growth, both vertically and laterally, is a key prerequisite for the movement of timberline edge into the ecotone and the upward migration of subalpine forest to establish new timberlines and treelines.

10.7 Treeline Advancement

Driven by changing climates elevational and polar treelines have been fluctuating throughout postglacial times (Burga and Perret 1998). After the thermal optimum (for example, in northern Scandinavia, 4,000–5,000 years ago) treelines gradually receded because of general climatic cooling. The warming after the end of the 'Little Ice Age' (about 1900) brought about treeline rise in many regions of the northern hemisphere, in particular during the twentieth century (Holtmeier 2009 and further references therein). Treelines have been advancing in pulses by infilling treeless gaps within the treeline ecotone and tree establishment beyond the existing tree limit. These processes peaked during the favourable 1920s to the 1940s and, after a break, resumed since the 1970s. However, warming and treeline advance were not always synchronous and occurred at different regional and local intensities (Kullman 2001, 2004, 2010; Juntunen et al. 2002; Holtmeier et al. 2003; Dalen and Hofgaard 2005; Danby and Hik 2007; Dial et al. 2007; Ninot et al. 2008; Batllori et al. 2009; Holtmeier 2009; Kullman and Öberg 2009).

Information on treeline response to climate change in the tropics as well as in the southern hemisphere so far is still scarce. Bader et al. (2008a) suppose rapid response of treeline to climate warming in the Ecuadorian Andes to be unlikely though the present elevational position of the treeline apparently is located below its natural climatic limit (Lauer et al. 2001). The same holds true for East Africa. Regular burning to prevent invasion of trees into high-elevation pastures or for creating new pastures, prevented seedling establishment above the existing forest (Bader et al. 2008a, b). Other factors such as excessive solar radiation loads and daily temperature amplitudes also appeared involved.

In New Zealand, silver beech (*Nothofagus menziesii*) and mountain beech (*Nothofagus solandri*) established above closed forest stands, usually within 9 m of the outer canopy edge. Obviously, upslope expansion of the beech forest has been far less than could be expected at the rise of temperature (0.6–1.0 °C) during the last 100 years (Wardle and Coleman 1992; Wardle 2007). Cullen et al. (2001) report no recent increase of seedlings or an upward treeline shift. At the treeline in Patagonia and Tierra del Fuego regeneration of *Nothofagus pumilio* also largely failed despite temperature at treeline has been increasing since 1970 (Cuevas 2002; Daniels and Veblen 2004).

In many areas, tree density has increased within the treeline ecotone while no or only a few new trees have become established beyond the current tree limit (c.f. Holtmeier and Broll 2007, 2010b). On the other hand, in the Swedish Scandes (Kullman 2004; Kullman and Öberg 2009) and locally in the Rocky Mountains (Butler et al. 2004; Holtmeier 2009) seedlings and saplings now occur far above the tree limit, even if many have moderately to severely disturbed growth forms. In northern Norway, treeline seems to be stable or advancing in the southernmost and probably in the middle regions while it is retreating in the northern part (Dalen and Hofgaard 2005). In the Alps, trees are invading former alpine pastures and alpine zones (Wieser et al. 2009).

In general, subalpine and northern forests will not advance beyond their current limits as a closed front because of the effects of varying site conditions on tree establishment. Moreover, in case of a rapid warming elevational and polar treelines would not respond immediately but with a time lag (c.f. Holtmeier 2009 for further references) as may be concluded from many treelines that have advanced less than could be expected based on the degree of climatic warming (Wardle and Coleman

1992; Holmgren and Tjus 1996; Lloyd and Graumlich 1997; Tasanen et al. 1998; Bader et al. 2008a). However, there are alternative hypotheses (Tinner and Kaltenrieder 2005). Treeline shift may not be synchronous, as may be concluded from climatically-driven treeline fluctuations of the past (c.f. Holtmeier 2009 for further references). In case of rapid temperature rise trees may not have had adequate time to respond to the new thermal conditions (Skre 1993).

Tree species respond differently to climate change. For example, on Beartooth Plateau (Montana/Wyoming), seedlings of whitebark pine (*Pinus albicaulis*) became established in the treeline ecotone, whereas seedlings of Engemann spruce (*Picea engelmannii*) and subalpine fir (*Abies lasiocarpa*) are almost missing (Mellmann-Brown 2005). At treeline on Niwot Ridge (Colorado Front Range) spruce seedlings prevail by far (73 %), followed by limber pine (*Pinus flexilis*; 10 %) and subalpine fir (9 %) (Holtmeier 1995, 2009). Along five transects crossing the treeline ecotone on Pallastunturi (western Finnish Lapland) seedlings of Norway spruce (Picea abies) occur at great numbers (up to almost 80 % of all sampled seedlings), followed by seedlings of Scots Pine (*Pinus sylvestris*; about 20 %) while mountain birch seedlings rarely occur (ca. 6 %) (Holtmeier et al. 2003). In the Swedish Scandes Norway spruce and Scots pine have migrated to greater elevation more rapidly than *Betula pubescens* ssp *czerepanovii* (Kullman and Öberg 2009). The proportions are likely to change in the long-term due to natural succession. In the southern Urals, Siberian spruce (*Picea obovata*) is invading the mountain tundra before birch (*Betula tortuosa*) and mountain ash (*Sorbus aucuparia*). Later, evergreen spruce may outcompete the deciduous more shade-intolerant tree species if no disturbances create open patches.

It is important to understand that treeline history (human use, extreme climatic events, climatic fluctuations, fire, etc.) is influencing current regional and local treeline dynamics and thus must be considered when predicting future change (Kullman 1997; Holtmeier 1993, 1994, 1995, 2009; Holtmeier and Broll 2005, 2010b; Fig. 10.4).

10.7.1 Driving Forces

Treeline advance begins with seedling establishment and survival, assuming seed sources are not limiting. Even during relatively cool periods in the past (e.g., Little Ice Age) seedlings have occasionally become established in and above the treeline ecotone (Caccianiga and Payette 2006). However, most of them became severely distorted or even died when they grew above the shelter provided by associated non-arborescent vegetation, microtopographical structures and the winter snow pack (Holtmeier 1974, 2005a; Holtmeier et al. 1996, 2003). Some survived as suppressed prostrate growth forms. Locally, spruce (e.g., *Picea engelmanii, Picea abies*) and fir (*Abies lasiocarpa*) have persisted for hundreds of years or even for millennia (Ives and Hansen-Bristow 1983; Holtmeicr 1985; Kullman 2005b) by layering (formation of adventitious roots). Clonal tree islands are very common at the upper treeline in

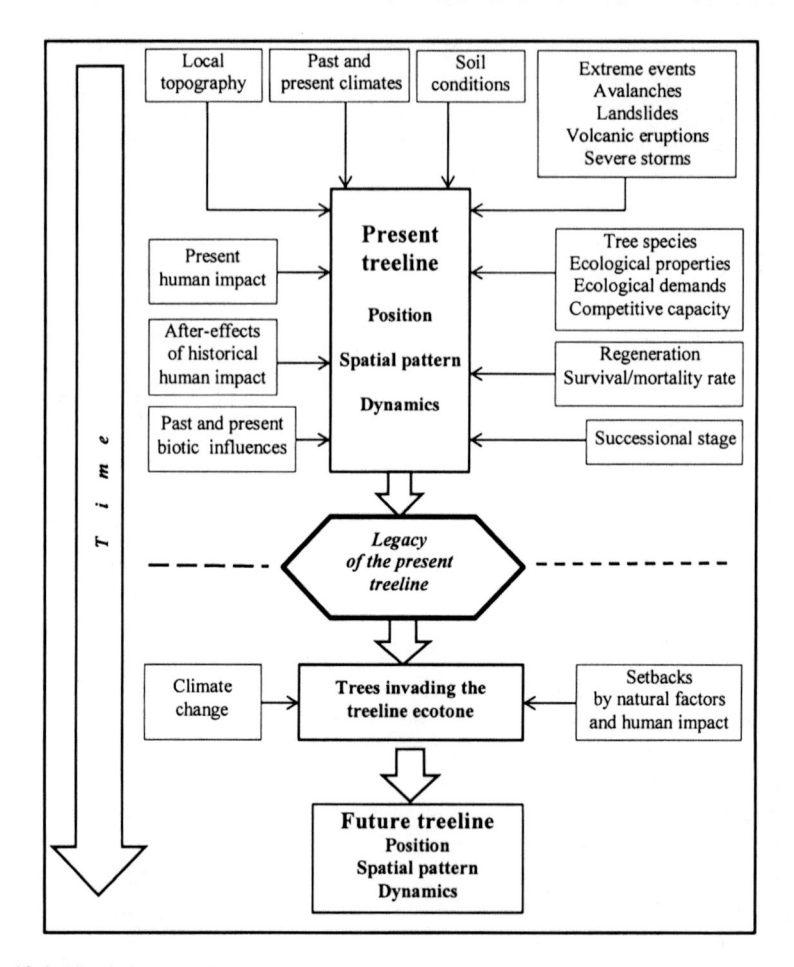

Fig. 10.4 The influence of historical legacy on the treeline ecotone (Modified from Holtmeier 2009)

the Rocky Mountains and other high mountain ranges in North America, in Fennoscandia, and in the Urals. They are also typical of the northern treeline in Eurasia and America and thus contrasting with the European Alps, where high-elevation forests became widely removed by human use. Layering is possible even at temperatures lower than that preventing seed-based regeneration (Larcher 1980a; Holtmeier 1999, 2009). Under the influence of climate warming these suppressed solitary trees as well as clonal groups may influence microclimate, as well as become effective seed sources enhancing seedling establishment above the alpine treeline and northern tundra (Caccianiga and Payette 2006).

Higher temperatures and longer growing seasons should provide more favourable conditions for regeneration, tree growth, and winter hardening, which are primarily driving factors in current treeline migration to higher elevation and northerly position. However, greater limitations due to drought effects occurring during warmer and longer growth seasons may curtail this advantage (Hu et al. 2010). In many regions of the European Alps and Norway, where pastoral use, mining, charcoal production and other human forest use has lowered the treeline to about 150–300 m below its maximum position during the Holocene (Holtmeier 1974, 2009; Burga and Perret 1998), the alpine zone expanded to lower elevation, and topographically-controlled contrasting microclimates ('pseudoalpine surrogate landscape' sensu Friedel 1967). The decline of human impact mainly during the last century has triggered forest invasion into former alpine pastures (Aas and Faarlund 1996; Bryn 2008; Hofgaard 1997; Stützer 2005; Wieser et al. 2009; Holtmeier 2009 and further references therein). Though climate warming is very likely supporting forest expansion, strongly contrasting microclimates and also degraded soils, impoverished mycorrhizal flora and other biotic factors such as snow fungi impede tree establishment more than might be expected at the present (anthropogenic) treeline. Tree density however, can be expected to increase more rapidly on former alpine pastures than near the current tree limit (Müterthies 2002, 2003; Holtmeier and Broll 2005; Vittoz et al. 2008).

At treeline in the northern Cairngorm Mountains (Scotland) where *Pinus sylvestris* has been expanding since the 1950s when regular burning was stopped and grazing ceased no response to climatic change is evident, however (French et al. 1997). This may change if climatic warming continues.

Solitary trees and compact clonal groups (e.g. *Picea, Abies*) that have survived as suppressed growth forms often far above the forest limit, and also new trees that became established at and beyond the existing treeline during the favourable periods of the previous century occasionally produce viable seeds now (c.f. also Caccianiga and Payette 2006) and thus may become effective drivers of treeline advance. However, as long as no cone bearing stems considerably project above the krummholz canopies the range of (anemochoric) seed dispersal remains small (e.g. Holtmeier 2009; Holtmeier and Broll unpublished data).

Körner (1998a, b, 1999) has hypothesized that it is limitation of carbon investment by low temperature rather than carbon gain that controls tree growth at the treeline. However, the effects may be different in mature trees compared to seedlings and saplings. In some cases seedlings may experience an occasionally negative carbon balance due to high respiratory losses (Germino and Smith 1999). However, a negative carbon balance was not measured in treeline trees and, thus, assuming that seedlings will only survive if being able to maintain a positive carbon balance in the long-term (Malanson et al. 2007, 2009). Moreover, carbon and nutrient cycles are closely linked and have to be considered as one functional system (Körner 2007b), similar to soil water availability and nutrients which are needed to match the carbon demand of the meristems and photosynthetic tissues. Nevertheless, any further assessment of treeline responses to future climate change must consider the importance of local topography (Aulitzky 1961; Aulitzky et al. 1982; Wieser and Tausz 2007) and potential feedbacks on microclimate-vegetation interactions (Holtmeier and Broll 2007), that are interdependent.

10.8   Conclusions

In spite of more than a hundred years of research, what specific factor determines the elevational or latitudinal position of treelines is still an open question. Although recent reviews on the causes of the upper elevational and latitudinal limits of treelife list several hypotheses, the role of temperature limitations on resource availability, and growth are probably the most obvious ones (Körner 1998a, 2003; Wieser and Tausz 2007; Holtmeier 2009). The fact that alpine and arctic tundra species grow and survive above treeline points to a need for experimental evaluation of the specific factors beyond low temperature limiting tree growth at higher elevations and latitudes (Grace et al. 2002; Smith et al. 2009).

With regard to climate change, elevated CO_2 may not influence mature tree growth significantly (Körner 2003). As the responsiveness of treeline trees also depends on foliage type (Handa et al. 2005, 2006), treelines are unlikely to respond dramatically to increasing CO_2. If treeline trees are not limited by photosynthetic CO_2 supply (Grace et al. 2002; Smith and Johnson 2009), but rather assimilated carbon processing, low temperature limitations on assimilated carbon processing, rather than CO_2 supply, is suggested to be the strongest limiting factor of growth in mature trees at high elevation and latitude (Wieser et al. 2009). Yet, the growth of established trees at treeline may not be as important to understanding treeline limits as is an understanding of limitations to seedling establishment and growth within the treeline ecotone (Holtmeier 2009; Smith et al. 2003, 2009).

Although a response of altitudinal and latitudinal treelines to climate warming since the end of the Little Ice Age appears omnipresent (Holtmeier 2009) the effects of elevated CO_2 and climate warming on seedling establishment, a prerequisite first step in treeline advancement, must be better understood. In this regard, limitations to growth which Körner (1998a, 2003) has referred to as "the growth limitation hypothesis" also needs more substantiation, especially for the processes of seedling establishment and growth (Millard et al. 2007; Li et al. 2008; Bansal and Germino 2008; Gruber et al. 2011).

Except for sites with uncovered mineral soils, seedlings become established and live for many years within a 'chamaephyte' microenvironment when soil temperatures during the growing season are high enough not to impair carbon investment. Low soil temperatures limiting seedling establishment and growth, the development of dwarf shrubs would also be inhibited. Other factors such as competition with the associated vegetation for light, nutrients and moisture, photoinhibition at high light intensity and low temperatures, may impair or even prevent seedling establishment and growth. Facilitation from adjacent vegetation and microtopography may also result in critical benefits (Smith et al. 2003, 2009). Young trees first projecting above the relatively favourable microclimate near the ground, or above the protective winter snowpack, become fully exposed to recurrent environmental injury (e.g. severe late and early frost, excessive sunlight, winter cuticle abrasion and desiccation). Escape and survival from the snow surface is an essential precondition for reaching 'tree size' (Smith et al. 2003). During subsequent favourable years or as a result of facilitation (mainly wind shelter) from neighbouring vegetation, suppressed 'krummholz' canopies may release vertical stems. Increased deposition of snow at the lee side of existing 'krummholz' and trees and within scattered tree stands may enforce this facilitation process. If not yet sufficiently protected from wind, new vertical stems will usually be first devoid of lateral stems and needles, become more or less flagged due to needle loss and shoot dieback on windward sides e.g., flagged table trees, skirted flagged trees; (c.f. Smith et al. 2003; Holtmeier 2009; Holtmeier and Broll 2010a, b). In the long-term and if no lasting setbacks occur almost undisturbed symmetric growth will be possible as a result the positive feedbacks of increasing tree population on the microclimate reduced wind velocity and sky exposure (cf. graphs in Holtmeier 2009; Holtmeier and Broll 2010b). Possible effects of increased CO_2 and carbon assimilation with continued global change cannot be excluded.

At large scales (global, zonal; cf. Holtmeier and Broll 2005) higher temperatures can be considered to be a primary driver in seedling establishment and thus treeline advancement. Current climate change models have greater difficulty in predicting alterations in moisture patterns than temperature predictions. At smaller scales (landscape, local scale, microscale), however, a host of factors must be considered (Holtmeier and Broll 2010b; Holtmeier 2009). Under certain conditions soil moisture may become a limiting factor as indicated already today at the boundary between treeline and the krummholz zone (Aulitzky 1984). Soil moisture deficiencies may impede seedling establishment particularly in dry climates, on sun-facing valley sides and mountain slopes as well as on sun-exposed microtopography (Leonelli et al. 2009; Camarero and Gutiérrez 2004; Brandes 2007; Batllori 2008). Drought stress is strongest during hot summers and is more likely on sun-exposed leeward slopes than on sun-facing windward slopes. Permeable soils (coarse texture, low humus content) may exacerbate moisture deficiency (e.g. Holtmeier et al. 2003; Broll et al. 2007). Moisture stress may be more critical at the seedlings stage, whereas tall trees may access soil moisture at greater depths. Thus, local factors may override the positive effects of the warming climate on seedling establishment.

10.9 Climate Change Perspectives

Future climate change might be of paramount importance for tree and neighbouring low stature vegetation within treeline ecotones at high elevation and latitude, where low temperatures generally limit carbon metabolism, growth, soil microbial activity and even ecosystem fluxes (Wieser and Tausz 2007; Holtmeier 2009; Smith et al. 2009). For example in the timberline ecotone of the Austrian Alps annual net ecosystem production (NEP) and evapotranspiration (ET) decreased from forests (NEP 260 g C m^{-2} year^{-1}, Wieser and Stöhr 2005; ET 480 mm year^{-1}, Matyssek et al. 2009) towards dwarf shrub (NEP 210–250 g C m^{-2} year^{-1}, Larcher 1977; ET 350 mm year^{-1}, Kronfuss 1997) to grassland systems (NEP 60 −1 40 g C

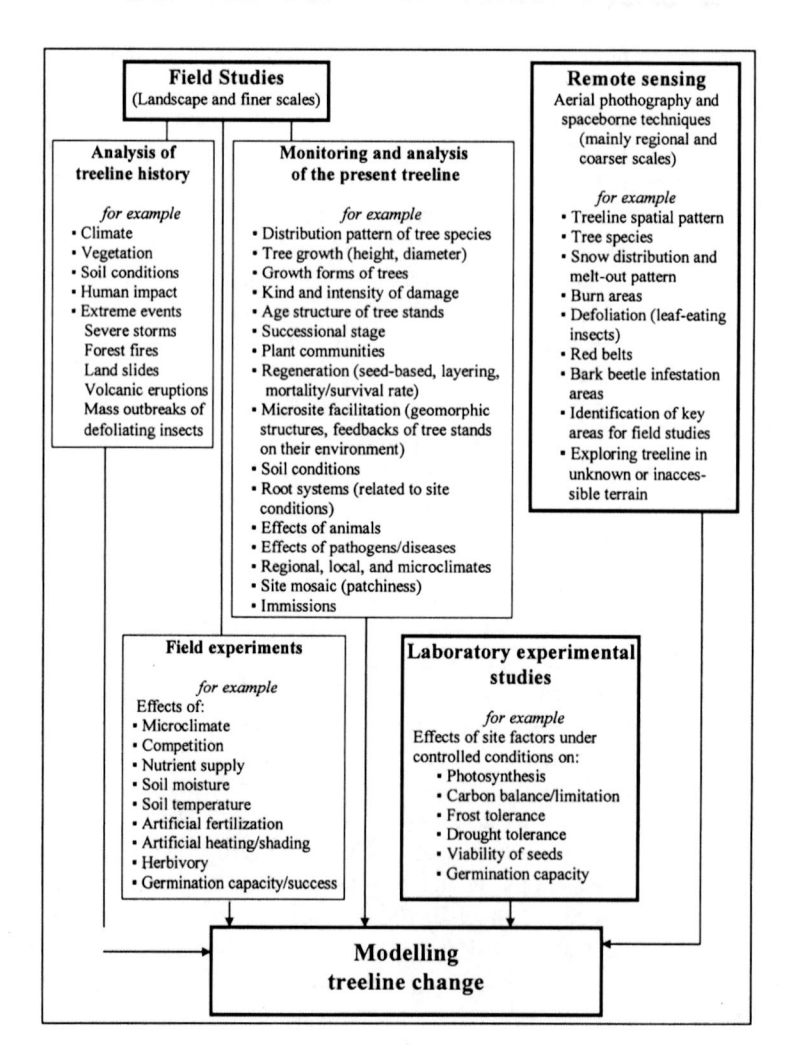

Fig. 10.5 Basic research needed for modelling possible treeline response to changing climate (Modified from Holtmeier 2009)

m^{-2} year^{-1}, Koch et al. 2008; ET 280 mm year^{-1}, Guggenberger 1980; Wieser et al. 2008).

However, there will be local variations due to historical legacies, different climatic zones, and land forms (c.f. Holtmeier 2009 and further references therein; Fig. 10.4), and thus treelines are not expected to advance in a closed front (Holtmeier and Broll 2010b). More studies are needed (Fig. 10.5) on both local and regional scales in order to predict future treeline dynamics. Natural thermal gradients within a common environment could provide excellent opportunities for ecological field research examining the response of various plant life forms to predicted global warming. Thus, natural abiotic gradients due to effects of elevation (alpine treelines) and latitude (polar treelines) provide powerful tools for global change (Wieser and Tausz 2007). Field studies on seedlings and young trees, rather than on mature trees need to be accompanied by the latest advances in laboratory, field monitoring and remote sensing techniques (Holtmeier 2009; Fig. 10.5).

Finally, artificial CO_2 enrichment and warming experiments, along with associated changes in the water and nutrient regimes may help in resolving the role of ecological facilitation to treeline advancement (Smith et al. 2009). Additionally, competition between seedlings and adjacent low stature vegetation also needs further investigation. As currently observed treeline advance rates do not agree well with rates from simple assumptions (Grace et al. 2002). Such studies and experiments will help predict the expected treeline advances to higher elevation and more northern and southern latitudes (Holtmeier and Broll 2007, Holtmeier 2009; Fig. 10.5) (Plates 10.1 and 10.2).

One has to be aware that treelines might be more susceptible to human impacts such as land use changes and management practices than will likely be the case under expected climate change. In addition, episodic climate extremes (summer drought, severe frost during the growing season) or biotic stresses (pathogens, herbivores) rather than a gradual increase in temperature will control tree population dynamics within treeline ecotones in a future warmer environment.

Plate 10.1 *Left* *top* to *bottom*: Krummholz- tree and forest limit at timberline on Mt. Patscherkofel Austrian Alps at 2,175, 2,100 and 1,950 m, respectively (Photos G. Wieser). *Right top*: Abraded stem of an Engelmann spruce in the treeline ecotone on Trail Ridge (Rocky Mountain National Park, Colorado at about 3,400 m (Photo Holtmeier July 22, 1987); *right bottom*: Winter injury in a Scots pine sapling at treeline in northernmost Finland at about 310 m (Photo Holtmeier August 15, 2003)

Plate 10.2 *Top left*: Faciliation by krummholz at treeline on Lee Ridge (Glacier National Park, Montana) Photo Holtmeier July 25, 2009; *top right* Vertical stem release in subalpine fir krummholz at treeline (about 2,026 m) on Mount Rainer (Photo Holtmeier August 4, 1985). *Middle left*: Survival of apical buds on vertical leaders of krummholz mats, despite needle death lower on the stem, leads to stem 'escape' and the ultimate growth of flagged stems on the leeward sides of larger mats (*middle right*) (Photos Smith). *Bottom left*: First-year *Abies lasiocarpa* seedling at end of summer growth period in a treeline ecotone (3,604 m elevation, Snowy Range Mountains, Wyoming). First primary bud has bolted, indicating a good chance of survival to the next summer growth period. Over 90 % mortality can occur during the first summer growth period following germination. Right Wind-exposed treeline showing flagged trees and krummholz mat formation (3,500 m, Snowy Range, southeastern Wyoming) (Photos Smith)

References

Aario L (1940) Waldgrenzen und subrezente Pollenspektren in Petsamo Lappland. Ann Acad Soc Fennoscandis Ser A 54:1–120

Aas B, Faarlund T (1996) The present and the Holocene birch belt in Norway. Paläoklimaforschung 20:18–24

Alvarez-Uria P, Körner C (2007) Low temperature limits of root growth in deciduous and evergreen temperate tree species. Funct Ecol 221:211–218

Andersen CP, Sucoff EI, Dixon RK (1986) Effects of root zone temperature on root initiation and elongation in red pine seedlings. Can J For Res 16:696–700

Anschlag K (2008) Regeneration der Fjellbirke (Betula pubescens ssp czerepanovii) und Wurzelsysteme ihres Jungwuchses im Waldgrenzökoton, Finnische Subarktis. PhD thesis, Münster University

Anschlag K, Broll G, Holtmeier F-K (2008) Mountain birch seedlings in the treeline ecotone, subarctic Finland: variation in above- and below-ground growth depending on microtopography. Arct Antarct Alp Res 40:609–616

Aulitzky H (1961) Die Bodentemperaturverhältnisse in der Kampfzone oberhalb der Waldgrenze und im subalpinen Zirben-Lärchenwald. Mitt der Forstlichen Bundesversuchsanstalt Mariabrunn 59:153–208

Aulitzky H (1984) The microclimatic conditions in a subalpine forest as basis for the management. GeoJournal 8:277–281

Aulitzky H, Turner H, Mayer H (1982) Bioklimatische Grundlagen einer standortsgemäßen Bewirtschaftung des subalpinen Lärchen-Arvenwaldes. Mitt der Eidgenössichen Anstalt für das Forstliche Versuchswesen 58:327–580

Bader MY, Rietkerk M, Bregt A (2007) Vegetation structures and temperature regimes of tropical alpine timberlines. Arct Antarct Alp Res 39:353–364

Bader MY, van Geloof I, Rietkerk M (2008a) High solar radiation hinders tree regeneration above the alpine treelines in northern Ecuador. Plant Ecol 191:33–45

Bader MY, Rietkerk M, Bregt AK (2008b) Simulated climate change in a simple spatial model of treeline dynamics influenced by excess solar radiation and fire. Arct Antarct Alp Res 40:269–278

Baig MN, Tranquillini W (1980) The effects of wind and temperature on cuticular transpiration of Picea abies and Pinus cembra and their significance in desiccation damage at the alpine treeline. Oecologia 47:252–256

Bansal S, Germino MJ (2008) Temporal variation of nonstructural carbohydrates in montane conifers: similarities and differences among developmental stages, species and environmental conditions. Tree Physiol 29:559–568

Barclay AM, Crawford RMM (1982) Winter desiccation stress and resting bud viability in relation to high altitude survival. Flora 72:21–34

Barry RG (1978) Diurnal effects on topoclimate on an equatorial mountain. Arb aus der Zentralanstalt für Meteorol und Geodynamik 32:1–8

Batllori E (2008) Regional assessment of recent Pinus uncinata alpine treeline dynamics in the Pyrenees. PhD dissertation, University of Barcelona

Batllori E, Camarero JJ, Ninot JM, Gutiérrez E (2009) Seedling recruitment and facilitation in alpine Pinus uncinata tree line ecotones. Implications and potential response to climate warming. Glob Ecol Biogeogr 18:460–472

Baumgartner A (1980) Mountain climates from a perspective of forest growth. In: Benecke U, Davis MR (eds) Mountain environments and subalpine forest growth, Proceedings IUFRO workshop November 1979, Christchurch, New Zealand Forest Service, Wellington, Technical paper 70

Bazzigher G (1978) Die Bekämpfung des Arven-Schneepilzes Phacidium infestans Karst. Eidgenössiche Anstalt für das Forstliche Versuchswesen Bericht 182

Bekker MF (2005) Positive feedback between tree establishment and patterns of subalpine forest advancemenmt, Glacier National Park, Montana, USA. Arct Antarct Alp Res 37:97–107

Benecke U, Havranek WM (1980) Phenological growth characteristic of trees with increasing altitude, Cragieburn Range, New Zealand. In: Benecke U, Davis MR (eds) Mountain environments and subalpine forest growth, Proceedings IUFRO workshop November 1979, Christchurch, New Zealand Forest Service, Wellington, Technical paper 70

Benecke U, Schulze E-D, Matyssek R, Havranek WM (1981) Environmental control of CO_2-assimilation and leaf conductance in *Larix decidua* Mill. I. A comparison of contrasting natural environments. Oecologia 50:54–61

Beniston M, Diaz HF, Bradley RS (1997) Climate change at high elevation sites: an overview. Clim Chang 36:233–251

Birmann K, Körner C (2009) Nitrogen status of conifer needles at the alpine treeline. Plant Ecol Divers 2:233–241

Blaser B (1980) Der Boden als Standortfaktor bei Aufforstungen in der subalpinen Stufe (Stillberg, Davos). Schweizerische Anstalt für das forstliche Versuchswesen Bericht 56

Bonan GB (1992) Soil temperature as an ecological factor in boreal forests. In: Shugart HH, Leemans R, Bonan GB (eds) A system analysis of the global boreal forest. Cambridge University Press, Cambridge

Brandes R (2007) Waldgrenzen griechischer Hochgebirge. Unter besonderer Berücksichtigung des Taygetos (Waldgrenzdynamik, dendrochronologische Untersuchungen). Erlanger Geographische Arbeiten, Sonderband 36

Broll G, Holtmeier F-K (1994) Die Entwicklung von Kleinreliefstrukturen im Waldgrenzökoton der Front Range (Colorado, USA) unter dem Einfluß leewärts wandernder Ablegergruppen (*Pice engelmanni* and *Abies lasiocarpa*). Erdkunde 48:48–59

Broll G, Holtmeier F-K, Anschlag K, Brauckmann H-J, Wald S (2007) Landscape mosaic in the treeline ecotone on Mt Rodjanoaivi, subarctic Finland. Fennia 185:89–105

Bryn A (2008) Recent forest limit changes in south-east Norway: effects of climate change or regrowth after abandoned utilization? Nor Geogr Tidsskr 62:251–270

Burga CA (1988) Swiss vegetation history during the last 18000 years New Phytol 110:581–602

Burga CA, Perret R (1998) Vegetation und Klima der Schweiz seit dem jüngeren Eiszeitalter. Ott Verlag, Thun

Butler DR, Malanson GP, Resler LM (2004) Turf-banked terrace treads and risers, turf exfoliation and possible relationship with advancing treeline. Catena 58:259–274

Caccianiga M, Payette S (2006) Recent advance of white spruce (*Picea glauca*) in the coastal tundra of the eastern shore of Hudson Bay (Québec, Canada). J Biogeogr 33:2120–2132

Cairns DM (2001) Patterns of winter desiccation in krummholz forms of *Abies lasiocarpa* at treeline sites in Glacier National Park, Montana, USA. Geogr Ann A 83:57–168

Camarero JJ, Gutiérrez M (2004) Pace and pattern of recent treeline dynamics: response of ecotones to climatic variability in the Spanish Pyrenees. Clim Chang 63:181–200

Chapman SK, Langley JA, Hart SC, Koch GW (2006) Plants actively control nitrogen cycling: uncorking the microbial bottleneck. New Phytol 169:27–34

Christersson L, von Fricks H, Sihe Y (1988) Damage to conifer seedlings by summer frost and winter drought. In: Sakai A, Larcher W (eds) Plant cold hardiness. Alan R Liss, Inc., New York

Cornelissen JHC, Carnelli AL, Callaghan TV (1999) Generalities in the growth, allocation and leaf quality responses to elevated CO_2 in eight woody species. New Phytol 141:401–409

Crawford RMM (1989) Studies in plant survival. Ecological case histories of plant adaptation to adversity. Blackwell, Oxford

Crawford RMM (2008) Plants at the margin. Ecological limits and climate change. Cambridge University Press, Cambridge

Cuevas JG (2002) Episodic regeneration at the *Nothofagus pumilio* alpine timberline in Tierra del Fuego, Chile. J Ecol 90:52–60

Cui M, Smith WK (1991) Seedling microenvironment, gas exchange and survival during first-year establishment in subalpine conifers. Tree Physiol 10:44–53

Cullen LE, Stewart GH, Duncan RP, Palmer JG (2001) Disturbance and climate warming influences on New Zealand Nothofagus tree-line population dynamics. J Ecol 89:1061–1071

Dahms A (1984) Die natürliche Vermehrung mehrerer Baumarten im oberen Waldgrenzbereich der Colorado Front Range in ökologischer Sicht. Master thesis, Institute for Geography, Westfälische Wilhelms-Universität, Münster

Dahms A (1992) Wachstumsbedingungen bei Picea engelmannii (Parry) Engelm. und Abies lasiocarpa (Hook) Nutt. an unterschiedlich windexponierten Standorten im Waldgrenzbereich der Colorado Front Range, U.S.A. dissertation Mathematisch-Naturwissenschaftliche Fakultät, Westfälische Wilhelms-Universität, Münster

Dalen L, Hofgaard A (2005) Differential regional treeline dynamics in the Scandes Mountains. Arct Antarct Alp Res 37:284–296

Danby RK, Hik DS (2007) Response of white spruce (Pinus glauca) to experimental warming at a subarctic alpine treeline. Glob Chang Biol 13:437–451

Daniels LD, Veblen TT (2004) Spatiotemporal influence of climate on altitudinal treeline in northern Patagonia. Ecology 85:1284–1296

Däniker A (1923) Biologische Studien über Baum- und Waldgrenzen, insbesondere über die klimatischen Ursachen und deren Zusammenhänge. Vierteljahresschrift Naturforsch Ges Zürich 68:1–102

Daubenmire R (1954) Alpine treelines in the Americas and their interpretation. Butler Univ Bot Stud 2:119–136

Day TA, DeLucia EH, Smith WK (1990) Effect of soil temperature on stem sap flow, shoot gas exchange and water potential in Picea engelmannii during snowmelt. Oecologia 84:474–481

Delucia EV, Berlyn GP (1984) The effect of increasing elevation on leaf cuticle thickness and cuticular transpiration in balsam fir. Can J Bot 62:2423–2431

Dial RJ, Berg EE, Timm K, McMahon A, Gecke J (2007) Changes in the alpine forest-tundra ecotone commensurate with recent warming in southcentral Alaska: evidence from orthophotos and field plots. J Geophys Res 112:G04015

Diaz HF, Bradley RS (1997) Temperature variations during the last century at high elevation sites. Clim Chang 36:253–279

Donaubauer E (1963) Über die Schneeschütte-Krankheit (Phacidium infestans Karst.) der Zirbe (Pinus cembra L.) und einiger Begleitpilze. Mitt der Forstlichen Versuchsanstalt Mariabrunn 60:147–166

Dullinger S, Dirnböck T, Grabherr G (2004) Modelling climate change-driven treeline shifts: relative effects of temperature increase, dispersal and invasibility. J Ecol 92:241–252

Farmer RE Jr (1997) Seed ecophysiology of temperate and boreal forest zone trees. St Lucie Press, Delray Beach

Franklin JF, Dyrness CT (1973) Natural vegetation of Oregon and Washington. USDA Forest Service, General technical report PNW-8

French DD, Miller GR, Cummins RP (1997) Recent development of high-altitude Pinus sylvestris scrub in the northern Cairngorm Mountains, Scotland. Biol Conserv 79:133–144

Frey W (1983) The influence of snow on growth and survival of planted trees. Arctic Alpine Res 15:241–251

Friedel H (1967) Verlauf der alpine Waldgrenze im Rahmen anliegender Gebirgsgelände. Mitt der Forstlichen Bundesversuchsanstalt Mariabrunn 75:81–172

Germino MJ, Smith WK (1999) Sky exposure, crown architecture, and low temperature photoinhibition inconifer seedlings at alpine tree line. Plant Cell Environ 22:407–415

Germino MJ, Smith WK (2001) Interactions of microsite, plant form and low-temperature photoinhibition in alpine plants. Arct Antarct Alp Res 32:388–396

Germino MJ, Smith WK, Resor AC (2002) Conifer seedling distribution and survival in an alpine treeline ecotone. Plant Ecol 162:57–168

Germino MJ, Hasselquist NJ, McGoilgle TM, Smith WK, Sheridan P (2006) Colonization of conifer seedling roots by fungal mycelium in an alpine treeline ecotone: relationships to microsite, developmental stage, and ecophysiology of seedlings. Can J For Res 36:901–909

Gieger T, Leuschner C (2004) Altitudinal change in needle water relations of *Pinus canariensis* and possible evidence of a drought-induced alpine timberline on Mt. Teide, Tenerife. Flora 199:100–109

Goldstein G, Meinzer FC, Rada F (1994) Environmental biology of a tropical treeline species, *Polylepis sericea*. In: Rundel PW, Smith AP, Meinzer FP (eds) Tropical alpine environments: plant form and function. Cambridge University Press, Cambridge

Grace J (1977) Plant response to wind. Academic, London

Grace J (1989) Tree lines. Philos Trans R Soc B 234:233–245

Grace J, Allen SJ, Wilson C (1989) Climate and the meristem temperatures of plant communities near the tree-line. Oecologia 79(1):98–204

Grace J, Berninger F, Nagy L (2002) Impact of climate change on the tree line. Ann Bot 90: 537–544

Granström A (1987) Seed viability of fourteen species during five years of storage in a forest soil. J Ecol 75:321–331

Gross M (1989) Untersuchungen an Fichten der alpinen Waldgrenze: Dissertationes Botanicae 139. Kramer, Berlin/Stuttgart

Gruber A, Wieser G, Oberhuber W (2010) Opinion paper: effects of simulated soil temperature on stem diameter increment of *Pinus cembra* at the alpine timberline: a new approach based on root zone roofing. Eur J For Res 129:141–144

Gruber A, Pirkebner D, Oberhuber W, Wieser G (2011) Spatial and seasonal variation in mobile carbohydrates in *Pinus cembra* in the timberline ecotone of the Central Austrian Alps. Eur J For Res 130:173–179

Guggenberger H (1980) Untersuchungen zum Wasserhaushat der alpine Zwergstrauchheide Patscherkofel. PhD thesis, Botany, University Innsbruck

Hadley JL, Smith WK (1983) Influence of wind exposure on needle desiccation and mortality for timberline conifers in Wyoming, U.S.A. Arctic Alpine Res 15:127–135

Hadley JL, Smith WK (1986) Wind effects on needles of conifers, seasonal influences on mortality. Ecology 67:12–19

Hadley JL, Smith WK (1987) Influence of krummholz mat microclimate on needle physiology and survival. Oecologia 73:82–90

Hadley JL, Smith WK (1989) Wind erosion of leaf surface wax in timberline conifers. Arctic Alpine Res 21:392–398

Hadley JL, Smith WK (1990) Influence of leaf surface wax and leaf area to water content ratio on cuticular transpiration in western conifers, U.S.A. Can J For Res 20:306–1311

Haimi J, Huhta V, Boucleham M (1992) Growth increase of birch seedlings under the influence of earth worms – a laboratory study. Soil Biol Biochem 24:1525–1528

Handa T, Körner C, Hättenschwiler S (2005) A test of the treeline carbon limitation hypothesis by in situ CO_2 enrichment and defoliation. Ecology 86:1288–1300

Handa T, Körner C, Hättenschwiler S (2006) Conifer stem growth at the altitudinal treeline in response to four years of CO_2 enrichment. Glob Chang Biol 12:417–2430

Häsler R, Streule A, Turner H (1999) Shoot and root growth of young *Larix decidua* in contrasting microenvironments near the alpine timberline. Phyton Ann Rei Bot 39:47–52

Hasselquist NJ, Germino MJ, McGoilgle TM, Smith WK (2005) Variability of *Cenococcum* colonization and its ecological significance for young conifers at alpine treeline. New Phytol 165:867–873

Hättenschwiler S, Smith WK (1999) Natural seedling occurrence in treeline conifers: a case study from the central Rocky Mountains, USA. Acta Oecol 20:219–224

Hättenschwiler S, Miglietta F, Raschi A, Körner C (1997) Thirty years of in situ growth under elevated CO_2: a model for future forest responses? Glob Chang Biol 3:463–471

Hättenschwiler S, Handa T, Egli L, Asshof R, Ammann W, Körner C (2002) Atmospheric CO_2 enrichment of alpine treeline conifers. New Phytol 156:363–375

Havranek WM (1972) Über die Bedeutung der Bodentemperatur für die Photosynthese und Transpiration junger Forstpflanzen und die Stoffproduktion an der Waldgrenze. Angew Bot 46:101–116

Havranek WM, Benecke U (1978) The influence of soil moisture on water potential, transpiration and photoynthesis of conifer seedlings. Plant Soil 49:91–103

Havranek WM, Tranquillini W (1995) Physiological processes during winter dormancy and their ecological significance. In: Smith WK, Hinckley TM (eds) Ecophysiology of coniferous forests. Academic, San Diego

Hellmers H, Genthe MK, Ronco F (1970) Temperature affects growth and development of Engelmann spruce. For Sci 16:447–452

Hoch G, Körner C (2003) The carbon charging of pines at the climatic treeline: a global comparison. Oecologia 135:10–21

Hoch G, Popp M, Körner C (2002) Altitudinal increase of mobile carbon pools in *Pinus cembra* suggests sink limitation of growth at the Swiss treeline. Oikos 98:361–374

Hoch G, Richter A, Körner C (2003) Non-structural carbon compounds in temperate forest trees. Plant Cell Environ 26:1067–1081

Hofgaard A (1997) Inter-relationships between treeline position, species diversity, land use and climate change in the central Scandes Mountains of Norway. Glob Ecol Biogeogr 6:419–429

Holmgren B, Tjus M (1996) Summer temperatures and tree line dynamics at Abisko. Ecol Bull 45:159–169

Holtmeier F-K (1974) Geoökologische Beobachtungen und Studien an der subarktischen und alpinen Waldgrenze in vergleichender Sicht. Steiner, Wiesbaden

Holtmeier F-K (1980) Influence of wind on tree physiognomy at the upper timberline in the Colorado Front Range. New Zealand Forest Service technical paper 70

Holtmeier F-K (1982) "Ribbon-forest" und "Hecken". Streifenartige Verbreitungsmuster des Baumwuchses an der oberen Waldgrenze in den Rocky Mountains. Erdkunde 36:142–153

Holtmeier F-K (1985) Die klimatische Waldgrenze – Linie oder Übergangssaum (Ökoton)? Ein Diskussionsbeitrag unter besonderer Berücksichtigung der Waldgrenzen in den mittleren und hohen Breiten der Nordhalbkugel. Erkunde 39:271–285

Holtmeier F-K (1986) Die obere Waldgrenze unter dem Einfluss von Klima und Mensch. Abh des Museums für Naturkunde 48:395–412

Holtmeier F-K (1993) Timberlines as indicators of climatic changes: problems and research needs. Paläoklimaforschung 9:211–222

Holtmeier F-K (1994) Ecological aspects of climatically-caused treeline fluctuations: review and outlook. In: Beniston M (ed) Mountain environments in changing climates. Routledge, London

Holtmeier F-K (1995) Waldgrenze und Klimaschwankungen – Ökologische Aspekte eines vieldiskutierten Phänomens. Geoökodynamik 16:1–24

Holtmeier F-K (1996) Der Wind als landschaftsökologischer Faktor in der subalpinen und alpinen Stufe der Front Range, Colorado. Arb Inst Landschaftsökol Westfälische Wilhelms-Univ Münster 1:19–45

Holtmeier F-K (1999) Ablegerbildung im Hochlagenwald und an der oberen Waldgrenze in der Front Range, Colorado. Mitt Deut Dendrolog G 84:39–61

Holtmeier F-K (2005a) Change in the timberline ecotone in northern Finnish Lapland during the last thirty years. Rep Kevo Subarctic Res Stn 23:97–113

Holtmeier F-K (2005b) Relocation of snow and its effects in the treeline ecotone – with special regard to the Rocky Mountains, the Alps, and Northern Europe. Erde 136:343–373

Holtmeier F-K (2009) Mountain timberlines. Ecology, patchiness, and dynamics, vol 36, Springer series: Advances in global change research. Springer, Dordrecht

Holtmeier F-K, Broll G (1992) The influence of tree islands and microtopography and pedoecological conditions in the forest-alpine tundra ecotone on Niwot Ridge, Colorado Front Range, U.S. Arctic Alpine Res 24:216–228

Holtmeier F-K, Broll G (2005) Sensitivity and response of northern hemisphere altitudinal and polar treelines to environmental change at landscape and local scales. Glob Ecol Biogeogr 14:395–410

Holtmeier F-K, Broll G (2006) Radiocarbon-dated peat and wood remains from the Finnish Subarctic: evidence of treeline and landscape history. The Holocene 16:743–751

Holtmeier F-K, Broll G (2007) Treeline advance – driving processes and adverse factors. Landscape 1:1–33

Holtmeier F-K, Broll G (2010a) Wind as an ecological agent at treelines in North America, the Alps, and in the European Subarctic. Phys Geogr 33:203–233

Holtmeier F-K, Broll G (2010b) Altitudinal and polar treelines in the northern hemisphere – causes and response to climate change. Polarforschung 79:139–153

Holtmeier F-K, Müterthies A, Stevens GE (1996) Effektive Verjüngung und Zuwachs der Kiefer (*Pinus sylvestris*) und der Fichte (*Picea abies*) an ihrer Höhengrenze in Finnisch-Lappland während der letzten 100 Jahre. Arb Inst Landschaftsökol Westfälische Wilhelms-Univ Münster 1:85–99

Holtmeier F-K, Broll G, Müterthies A, Anschlag K (2003) Regeneration of trees in the treeline ecotone: Northern Finnish Lapland. Fennia 181:103–128

Holtmeier FK, Broll G, Anschlag K (2004) Winderosion und ihre Folgen im Waldgrenzbereich und in der alpinen Stufe einiger nordfinnischer Fjelle. Geoökologie 25:203–224

Holzer K (1959) Winterliche Schäden an Zirben nahe der alpinen Baumgrenze. Centralblatt für das Ges Forstwes 76:232–244

Hu J, Moore DJP, Burns SP, Monson RK (2010) Longer growing seasons lead to less carbon sequestration by a subalpine forest. Glob Chang Biol 16:771–783

Hurtin KR, Marshall JD (2000) Altitude trends in conifer leaf morphology and stable carbon isotope composition. Oecologia 123:32–40

IPCC (2007) Climate change 2007. Cambridge University Press, Cambridge

Ives JD, Hansen-Bristow KJ (1983) Stability and instability of natural and modified upper timberline landscapes in the Colorado Rocky Mountains, U.S.A. Mt Res Dev 3:149–155

James JC, Grace J, Hoad SP (1994) Growth and photosynthesis of *Pinus sylvestris* at its altitudinal limit in Scotland. J Ecol 82:297–306

Jobbagy EG, Jackson RB (2000) Global controls of forest line elevations in the northern and southern hemispheres. Glob Ecol Biogeogr 9:253–268

Johnson DM, Germino MJ, Smith WK (2004) Abiotic factors limiting photosynthesis in *Abies lasiocarpa* and *Picea engelmannii* seedlings below and above the alpine timberline. Tree Physiol 24:377–386

Jones HG (1991) Plants and microclimate. A quantitative approach to environmental plant physiology. University Press, Cambridge

Jones PD, Wigley TML, Folland CK, Parker DE, Angelli JK, Jebedeff S, Hansen JE (1988) Evidence of global warming in the last decade. Nature 332:790

Jordan DN, Smith WK (1994) Energy balance analysis of nighttime leaf temperature and frost formation in a subalpine environment. Agr For Meteorol 71:359–372

Jordan DN, Smith WK (1995) Microclimate factors influencing the frequency and duration of growth season frost in subalpine plants. Agric For Meteorol 77:17–30

Juntilla O (1986) Effects of temperature on shoot growth in northern provenances of *Pinus sylvestris* L. Tree Physiol 1:185–192

Juntilla O, Nilsen J (1993) Growth and development of northern forest trees as affected by temperature and light. In: Alden J, Mastrantonio JL, Odum S (eds) Forest development in cold climates. Plenum Press, New York

Juntunen V, Neuvonen S (2006) Natural regeneration of Scots pine and Norway spruce close to the timberline in northern Finland. Silva Fenn 40:443–458

Juntunen V, Neuvonen S, Norokorpi Y, Tasanen T (2002) Potential of timberline advance in northern Finland, as revealed by monitoring during 1983–99. Arctic 55:348–361

Kaiser WM (1987) Effects of water deficit on photosynthetic capacity. Physiol Plant 77:142–149

Kaltenrieder P, Tinner W, Ammann B (2005) Zur Langzeitökologie des Lärchen-Arvengürtels in den südlichen Walliser Alpen. Bot Helv 115:137–154

Kanninen M (1985) Shoot elongation in Scots Pine: diurnal variations and response to temperature. J Exp Bot 36:760–1770

Karlsson PS, Nordell KO (1996) Effects of soil temperature on nitrogen economy and growth of mountain birch seedlings near its presumed low temperature distribution limit. Ecoscience 3:183–189

Karlsson PS, Weih M (2001) Soil temperature near the distribution limit of mountain birch (*Betula pubescens* ssp. *czerepanovii*): implications for seedling nitrogen economy and survival. Arct Antarct Alp Res 33:88–92

Keeling CD, Whorf TP (2005) Atmospheric CO_2 records from sites in the SIO air sampling network. http://cdiac.ornl.gov/trends/co2/sio-keel.htm. Accessed 30 Oct 2010

Knapp AK, Smith WK (1982) Factors influencing understory seedling establishment of Engelmann spruce (*Picea engelmannii*) and subalpine fir (*Abies lasiocarpa*) in southeast Wyoming. Can J Bot 60:2753–2761

Koch O, Tscherko D, Küppers M, Kandeler E (2008) Interannual ecosystem CO_2 dynamics in the alpine zone of the Eastern Alps, Austria. Arct Antarct Alp Res 40:487–496

Körner C (1994) Leaf conductance in the major vegetation types of the globe. In: Schulze E-D, Caldwell MM (eds) Ecophysiology of photosynthesis, vol 100, Ecological studies. Springer, Berlin

Körner C (1998a) A re-assessment of high elevation treeline positions and their explanation. Oecologia 115:445–459

Körner C (1998b) Worldwide position of alpine treelines and their causes. In: Beniston M, Innes JL (eds) The impacts of climatic variability on forest. Springer, Heidelberg

Körner C (1999) Alpine plant life. Functional plant ecology of high mountain ecosystems. Springer, Berlin/Heidelberg

Körner C (2003) Carbon limitation in trees. J Ecol 94:4–17

Körner C (2006) Plant CO_2 responses: an issue of definition, time and resource supply. New Phytol 172:393–411

Körner C (2007a) Climatic treelines: conventions, global patterns, causes. Erdkunde 61:316–324

Körner C (2007b) Drivers and the driven in a warming, high CO_2 world. BES annual meeting and AGM report, British Ecological Society, The Bulletin December 2007:19–20

Körner C, Paulsen J (2004) A world-wide study of high altitude treeline temperatures. J Biogeogr 31:713–732

Kronfuss H (1994) Der Einfluß der Lufttemperatur auf das Höhenwachstum der Zirbe. Centralblatt für das Ges Forstwes 111:165–181

Kronfuss H (1997) Das Klima einer Hochlagen aufforstung in der subalpinen Höhenstufe. FBVA Ber 100

Kronfuss H, Havranek WM (1999) Effects of elevation and wind on the growth of *Pinus cembra* L. in a subalpine afforestation. Phyton Ann Rei Bot 39:99–106

Kryuchkov VV (1973) The effect of permafrost on northern treeline. Proceedings of the international conference on Permafrost, USSR contribution, Washington, DC. Proc Natl Acad Sci U S A 1973:136–138

Kullman L (1997) Tree-limit stress and disturbance. A 25-year survey of geoecological change in the Scandes Mountains of Sweden. Geogr Ann A 79:139–165

Kullman L (2001) 20th century climate warming and treelimit rise in the southern Scandes in Sweden. Ambio 30:2–80

Kullman L (2004) Tree-limit landscape evolution at the southern fringe of the Swedish Scandes (Dalarna Province) – Holocene and 20th century perspectives. Fennia 182:73–94

Kullman L (2005a) Wind-conditioned 20th century decline of birch treeline vegetation in the Swedish Scandes. Arctic 58:286–294

Kullman L (2005b) Trädgränsen I Dalarfjällen. Del 1. Gamla och nya träd på Fulufjället vegetationshistoria på höh nivå. Miljövärdsenheten Rapport 2005:10

Kullman L (2010) One century of treeline change and stability-experiences from the Swedish Scandes, 1973–2005. Landscape 17:1–31

Kullman L, Öberg L (2009) Post-Little Ice Age treeline rise and climate warming in the Swedish Scandes: a landscape ecological perspective. J Ecol 97:415–429

LaMarche VC (1973) Holocene climatic variations, inferred from tree line fluctuations in the White Mountains, California. Quat Res 3:632–660

Lange S, Bussmann RW, Beck E (1997) Stand structure and regeneration of the subalpine Hagenia abyssinica forests of Mt. Kenya. Bot Acta 110:473–480

Larcher W (1957) Frosttrocknis an der Waldgrenze und in der alpinen Zwergstrauchheide. Veröff des Mus Ferdinandeum Innsbruck 37:49–81

Larcher W (1963) Zur spätwinterlichen Erschwerung der Wasserbilanz von Holzpflanzen an der Waldgrenze. Ber des Naturwiss Med Ver Innsbruck 53:25–137

Larcher W (1977) Ergebnisse des IBP-Projekts "Zwergstrauchheide Patscherkofel". Sitzungsbericht der Österr Akad der Wiss Wien Math Natwiss Kl Abt I 186:301–371

Larcher W (1980a) Ökologie der Pflanzen auf physiologischer Grundlage, 3rd edn. Ulmer, Stuttgart

Larcher W (1980b) Klimastreß im Gebirge – Adaptationstraining und Selektionsfilter für Pflanzen. Rhein Akad Wiss Vorträge 291:49–88

Larcher W (1985) Winter stress in high mountains. In: Turner H, Tranquillini W (eds) Establishment and tending of subalpine forests: research and management, Berichte der Eidgenössischen Anstalt für das Forstliche Versuchswesen 270

Larcher W (2001) Ökophysiologie der Pflanzen: Leben, Leistung und Stressbewältigung der Pflanzen in ihrer Umwelt. Ulmer, Stuttgart

Larsen JA (1989) The northern forest border in Canada and Alaska – biotic communities and ecological relationships, vol 70, Ecological studies. Springer, New York

Lauer W, Klaus D (1975) Geoecological investigations on the timberline of Pico de Orizaba, Mexico. Arctic Alpine Res 7:315–330

Lauer W, Rafigpoor MD, Theisen I (2001) Physiogeographie, Vegetation und Syntaxonomie der Flora des Páramo de Papallacta. Erdwissenschaftliche Forschung 29, Stuttgart

League K, Veblen T (2006) Climatic variability and episodic Pinus ponderosa establishment along the forest-grassland ecotones of Colorado. For Ecol Manag 228:98–107

Leonelli G, Pelfini M, Battapaglia G, Cherubini P (2009) Site-aspect influence on climate sensitivity over time of a high-altitude *Pinus cembra* tree-ring network. Climate Change 96:185–201

Li MH, Yang J (2004) Effects of microsite on growth of *Pinus cembra* in the subalpine zone of the Austrian Alps. Ann For Sci 61:319–325

Li MH, Yang J, Kräuchi N (2003) Growth response of *Picea abies* and *Larix decidua* to elevation in subalpine areas of Tyrol, Austria. Can J For Res 33:653–662

Li MH, Xiao WF, Wang SG, Cheng G, Cherubini P, Cal XH, Liu XL, Wang XD, Zhu WZ (2008) Mobile carbohydrates in Himalayan treeline trees I. Evidence for carbon gain limitation but not for growth limitation. Tree Physiol 28:1287–1296

Lindsay JH (1971) Annual cycle of leafwater potential in *Picea engelmannii* and *Abies lasiocarpa* at timberline in Wyoming. Arctic Alpine Res 3:131–138

Litaor MI (1987) The influence of aeolian dust on the genesis of alpine soils in the Front Range, Colorado. Soil Sci Soc Am J 51:142–147

Lloyd AH, Graumlich LJ (1997) Holocene dynamics of treeline forests in the Sierra Nevada. Ecology 78:1199–1210

Loris K (1981) Dickenwachstum von Zirbe, Fichte und Lärche an der alpinen Waldgrenze/ Patscherkofel. Mitt der Forstlichen Bundesversuchsanstalt Wien 142:417–441

Loris K, Havranek WM, Wieser G (1999) The ecological significance of thickness changes in stem, branches, and twigs of *Pinus cembra* L. during winter. Phyton Ann Rei Bot 39:117–122

Lugina KM, Groisman PY, Vinnikov KY, Koknaeva VV, Speranskaya NA (2006) Monthly surface air temperature time series area-averaged over 30 – degree latitudinal belts over the globe, 1881–2005. In: Trends: a compendium of data on global change. Carbon Dioxide Information Analysis Centre, Oak Ridge National Laboratory, U.S. Department of Energy, Oak Ridge. http://cdiac.ornl.gov/trends/temp/lugina/lugina.html. Accessed 30 Oct 2010

Lyr H (1996) Effect of the root temperature on growth parameters of various European tree species. Ann For Sci 53:317–323

Lyr H, Hoffmann G (1967) Growth rates and growth periodicity of tree roots. Int Rev For Res 2: 181–236

Malanson GP, Bulter DR, Fagre DB, Walsh SJ, Tomback DF, Daniels LD, Resler LM, Smith WK, Weiss DL, Peterson DL, Bunn AG, Heimstra CA, Lipton D, Bourgeron PS, Shen Z, Miller CI (2007) Alpine treelines of western North America and global climate change: linking organism to landscape dynamics. Phys Geogr 28:378–396

Malanson GP, Brown DG, Butler DR, Cairns DM, Fagre DB, Wlash SJ (2009) Ecotone dynamics: invasibility of alpine tundra by tree species from the subalpine forest. In: Butler DB, Malanson GP, Walsh SJ, Fagre DB (eds) The changing alpine treeline: the example of Glacier National Park, MT, USA, vol 12, Developments in earth surface processes. Elsevier, Amsterdam

Marchand P, Chabot BF (1978) Winter water relations of tree-line plant species on Mount Washington, New Hampshire. Arctic Alpine Res 10:105–116

Marr JW (1977) The development and movement of tree islands near the upper limit of tree growth in the southern Rocky Mountains. Ecology 58:1159–1164

Mattes H (1978) Der Tannenhäher im Engadin. Studien zu seiner Ökologie und Funktion im Arvenwald. Münstersche Geographische Arbeiten, vol 2

Mattes H (1982) Die Lebensgemeinschaft von Arve und Tannenhäher. Eidgenössische Anstalt für das Forstliche Versuchswesen, Ber 241

Matyssek R, Wieser G, Patzner K, Blaschke H, Häberle K-H (2009) Transpiration of forest trees and stands at different altitude: consistencies rather than contrasts. Eur J For Res 128:579–596

McCaughey WW (1994) The regeneration process of whitebark pine. In: Schmidt WC, Holtmeier FK (eds) Proceedings of the international workshop on subalpine *Pinus pinea* and their environment: the status of our knowledge, St. Moritz, 5–11 Sept 1992. USDA Forest Service, Intermountain Research Station, General technical report INT-GTR 309

McDonald GM, Kremenetski KV, Beilman DW (2008) Climate change and the northern Russian treeline zone. Phil Trans R Soc B 363:2285–2299

Mellmann-Brown S (2005) Regeneration of whitebark pine in the timberline ecotone of the Beartooth Plateau, U.S.A.: spatial distribution and responsible agents. In: Broll G, Keplin B (eds) Mountain ecosystems, Studies in treeline ecology. Springer, Berlin

Michealis P (1934a) Ökologische Studien an der Baumgrenze. IV. Zur Kenntnis des winterlichen Wasserhaushaltes. Jahrb der Wiss Bot 80:169–247

Michealis P (1934b) Ökologische Studien an der Baumgrenze. V. Osmotischer Wert und Wassergehalt während des Winters in den verschiedenen Höhenlagen. Jahrb der Wiss Bot 80:37–362

Millard P, Sommerkorn M, Quen-Aelle G (2007) Environmental change and carbon limitation in trees: a biochemical, ecophysiological and ecosystem appraisal. New Phytol 175:11–28

Mooney HA, Wright RD, Strain BR (1964) The gas exchange capacity of plants in relation to vegetation zonation in the White Mountains of California. Am Midl Nat 72:281–297

Müller-Stohl WR (1954) Beiträge zur Ökologie der Waldgrenze am Feldberg im Schwarzwald. In: Janchen E (ed) Angewandte Pflanzensoziologie. Festschrift Erwin Aichinger, vol 2. Springer, Wien

Müterthies A (2002) Struktur und Dynamik der oberen Waldgrenze des Lärchen-Arvenwaldes im Bereich aufgelassener Alpweiden im Oberengadin. Arb Institut für Landschaftsölogie, vol 11. Westfälische Wilhelms-Universität, Münster

Müterthies A (2003) The potential timberline: determination with dendrochronological methods. In: Schleser G, Winiger M, Bräuning A, Gärtner H, Helle G, Jensma E, Neuwirth B, Treydtke K (eds) TRACE tree rings in archeology, climatology and ecology, Proceedings of the Dendrosymposium 11–13 Apr 2002, Bonn

Neuner G (2007) Frost resistance at the upper timberline. In: Wieser G, Tausz M (eds) Trees at their upper limit. Treelife limitation at the alpine timberline, vol 5, Plant ecophysiology series. Springer, Dordrecht

Nicolussi K, Bortenschlager S, Körner C (1995) Increase in tree-ring width in subalpine *Pinus cembra* from the central Alps that may be CO_2 related. Trees 9:181–189

Nierhaus-Wunderwald D (1996) Pilzkrankheiten in Hochlagen, Biologie und Befallsmerkmale. Wald und Holz 10:18–24

Nilsson MC, Höberg P, Zackrisson O, Wang F (1993) Allelopathic effects by Empetrum hermaphroditum on development and nitrogen uptake by roost and mycorrhiza of *Pinus sylvestris*. Can J Bot 71:620–628

Ninot JM, Batllori E, Carillo E, Carreras J, Ferré A, Gutiérrez WE (2008) Timberline structure and limited tree recruitment in the Catalan Pyrenees. Plant Ecol Divers 1:47–57

Noble DL, Alexander RR (1977) Environmental factors affecting natural regeneration of Engelmann spruce in the central Rocky Mountains. For Sci 23:420–429

Oberhuber W (2007) Limitation by growth processes. In: Wieser G, Tausz M (eds) Trees at their upper limit. Treelife limitation at the alpine timberline, vol 5, Plant ecophysiology series. Springer, Dordrecht

Odland A (1996) Differences in the vertical distribution pattern of *Betula pubescens* in Norway and its ecological significance. Paläoklimaforschung 20:42–59

Ohsawa M (1990) An interpretation of latitudinal patterns of forest limits in south and east Asian mountains. J Ecol 78:262–339

Ow LF, Griffin KL, Whitehead D, Walcroft AS, Turnbull MH (2008) Thermal acclimation of leaf respiration but not photosynthesis in *Populus deltoides*×nigra. New Phytol 178:123–134

Patten DT (1963) Light and temperature influence on Engelmann spruce seed germination and subalpine forest advance. Ecology 44:817–818

Paulsen J, Weber UM, Körner C (2000) Tree growth near treeline: abrupt or gradual reduction with altitude? Arct Antarct Alp Res 32:14–20

Paus A (2010) Vegetation and environment of the Rødalen alpine area, Central Norway, with emphasis on the early Holocene. Veg Hist Archeobot 19:29–51

Payette S, Delwaide A (1994) Growth of black spruce at its northern range limit in Arctic Québec, Canada. Arctic Alpine Res 24:40–49

Perkins TD, Adama GT, Klein RM (1991) Desiccation or freezing? Mechanisms of winter injury to red spruce foliage. Am J Bot 78:1207–1217

Pregitzer KS, King JS, Burton AJ, Brown SE (2000) Responses of tree fine roots to temperature. New Phytol 147:105–115

Rada F, Azócar A, Gonzales J, Briceno B (1998) Leaf gas exchange in *Espeletia schultzii* Wedd, a giant caulescent rosette species, along an altitudinal gradient in the Venezuelan Andes. Acta Oecologia 19:73–79

Retzer JL (1974) Alpine soils. In: Ives JD, Barry RG (eds) Arctic and alpine environment. Methuen, London

Richardson AD, Friedland AJ (2009) A review of the theories to explain arctic and alpine treelines around the world. J Sustain For 28:218–242

Richardson AD, Berlyn GP, Gregorie TG (2001) Spectral reflectance of *Picea rubens* (Pinaceae) and *Abies balsamifera* (Pinaceae) needles along an elevational gradient, Mt. Moosilauke, New Hampshire, USA. Am J Bot 88:667–676

Richter M, Diertl K-H, Peters T, Bussmann RW (2008) Vegetation structures and ecological features of the upper timberline ecotone. In: Beck E, Bendix J, Kottke I, Makeshin F, Mosandl R (eds) Gradients in a tropical mountain ecosystem, vol 198, Ecological studies. Springer, Berlin

Rien M, Spengler T, Richter M (1998) Klimaökologische Aspekte in Gebirgen der südwestlichen USA unter besonderer Berücksichtigung der White Mountains. Mitt der Fränkisch Geographishen Ges 45:301–333

Sakai A, Larcher W (1987) Frost survival of plants. Responses and adaptations to freezing stress, vol 62, Ecological studies. Springer, Berlin

Sakai A, Weiser CJ (1973) Freezing resistance of trees in North America with reference to tree regions. Ecology 54:118–126

Schreiber H (1991) Untersuchungen zur Standortdifferenzierung im Waldgrenzbereich am Koahppeloaivi (Utsjoki, Finnish Lapland). Diploma thesis, Westfälische Wilhelms-Universität, Münster

Schulze E-D, Chapin FS, Gebauer G (1994) Nitrogen nutrition and isotope difference among life forms at the northern treeline of Alaska. Oecologia 100:406–412

Schwarz R (1983) Simulationsstudien zur Theorie der oberen Waldgrenze. Erdkunde 37:1–11

Scott PA, Bentley CV, Fayle DCF, Hansell RIC (1987a) Crown forms and shoot elongation of white spruce at the treeline, Churchill, Manitoba, Canada. Arct Alp Res 19:175–186

Scott PA, Hansell RIC, Fayle DCF (1987b) Establishment of white spruce populations and responses to climate change at the treeline, Churchill, Manitoba, Canada. Arct Alp Res 19: 45–51

Shi P, Körner C, Hoch G (2006) End of season carbon supply status of woody species near the treeline in western China. Basic Appl Ecol 4:370–377

Shi P, Körner C, Hoch G (2008) A test of the growth-limitation theory for alpine tree line formation in evergreen and deciduous taxa of the eastern Himalayas. Funct Ecol 22:213–220

Skre O (1993) Growth of mountain birch (*Betula pubescens* ERH.) in response to changing temperature. In: Alden K, Mastrantonio JL, Odum S (eds) Forest development in cold climates, NATO ASI series A 244. Plenum Press, New York

Slatyer RO (1976) Water deficits in timberline trees in the Snowy Mountains of southeastern Australia. Oecologia 24:357–366

Slatyer RO, Noble IR (1992) Dynamics of treelines. In: Hansen A, DiCastri F (eds) Landscape boundaries: consequences for biotic diversity and ecological flows, vol 92, Ecological studies. Springer, Berlin

Smith JMB (1980) Ecology of the high mountains of Guinea. In: Van Royen O (ed) The alpine flora of New Guinea. Cramer, Vaduz

Smith WK, Brewer CA (1994) The adaptive importance of shoot and crown architecture in conifer trees. Am Nat 143:528–532

Smith WK, Carter GA (1988) Shoot structural effects on needle temperature and photosynthesis in conifers. Am J Bot 75:496–500

Smith WK, Donahue R (1991) Simulated effects of altitude on photosynthetic CO_2 uptake potential in plants. Plant Cell Environ 14:133–136

Smith WK, Geller G (1979) Plant transpiration at high elevations: theory, field measurement, and comparison with desert plants. Oecologia 41:109–122

Smith WK, Hughes NM (2009) Progress in coupling plant form and photosynthetic function. Castanea 74:1–26

Smith WK, Johnson DM (2009) Biophysical effects of altitude on plant gas exchange. In: De la Barrera E, Smith WK (eds) Perspectives in biophysical plant ecophysiology: a tribute to Park S. Nobel, Universidad Nacional Autonoma Mexico, CIECO

Smith WK, Knapp A (1990) Ecophysiology of high elevation forests. In: Osmond CB, Pitelka L (eds) Plant biology of the Great Basin and Range, vol 80, Ecological studies. Springer, London

Smith WK, Germino TE, Hancock TE, Johnson DM (2003) Another perspective on the altitudinal occurrence of alpine tree lines. Tree Physiol 23:1101–1113

Smith WK, Nobel PS, Reiners WE, Vogelmann TC, Chritchley C (2004) Summary and future perspectives. In: Smith WK, Vogelmann TC, Critchleyz (eds) Photosynthetic adaptation from chloroplast to landscape, vol 178, Ecological studies. Springer, Berlin

Smith WK, Geronimo MJ, Johnson DM, Reinhardt K (2009) The altitude of alpine treeline: a bellwether of climate change effects. Bot Rev 75:163–190

Sowell JB, Kouitnik DL, Lansing AJ (1982) Cuticular transpiration of whitebark pine (*Pinus albicaulis*) within a Sierra Nevadan timberline ecotone, USA. Arctic Alpine Res 14:97–103

Sparks JP, Ehleringer JR (1997) Leaf carbon isotope discrimination and nitrogen content of riparian trees along an elevational gradient. Oecologia 109:362–367

Steinbjörnsson B, Nordell O, Kauhanen H (1992) Nutrient relations of mountain birch growth at and below the elevational tree-line in Swedish Lapland. Funct Ecol 6:213–220

Stöhr D (2007) Soils – heterogeneous at a microscale. In: Wieser G, Tausz M (eds) Trees at their upper limit. Treelife limitation at the alpine timberline, vol 5, Plant ecophysiology. Springer, Dordrecht

Streule A, Häsler R (2006) Windschutz für junge Bäume in subalpinen Aufforstungen an stark windexponierten Standorten, "Pru dal vent" (Alp Grüm, GR). Eidgenössische Anstalt für Wald, Schnee und Landschaft WSL, Birmensdorf

Strömgren K, Linder S (2002) Effect of nutrition and soil warming on stemwood production in a boreal Norway spruce stands. Glob Chang Biol 8:1195–1204

Stützer A (2005) Bildsequenzen als Zeugen der Vegetationsdynmaik in der subalpinen-alpinen Höhenstufe der Koralpe (Kärnten/Österreich). Wulfenia 9:89–104

Suni T, Berninger F, Markkanen T, Keronen P, Rannik U, Vesala T (2003) Interannual variability and timing of growing season CO$_2$ exchange in a boreal forest. J Geophys Res 108:4265–4273

Susiluoto S, Perämäki M, Nikinmaa E, Berninger F (2007) Effects of sink removal on transpiration at the treeline: implications for the growth limitation hypothesis. Environ Exp Bot 60:334–339

Sutinen M-L, Arora R, Wisniewski M, Ashworth E, Strimbeck R, Palta J (2001) Mechanisms of frost survival and freeze-damage in nature. In: Bigras FJ, Colombo SJ (eds) Conifer cold hardiness, vol 1, Tree physiology. Springer, Dordrecht

Sutter E, Amann F (1953) Wie weit fliegen vorratssammelnde Tannenhäher? Ornithol Beobachter 50:89–90

Sveinbjörsson B, Kauhanen H, Nordell O (1996) Treeline ecology of mountain birch in the Torneträsk area. Ecol Bull 45:65–70

Tasanen T, Norokorpi Y, Sepponen P, Jutunen V (1998) Monitoring timberline dynamics in northern Lapland. In: Tasanen T (ed) Research and management of the northern timberline region, Proceedings of the Gustav Sirén symposium, Wilderness Center Inari, 4–5 Sept 1997, Research papers 677. The Finnish Forest Research Institute, Vantaa

Thorn CE (1978) The geomorphic role of snow. Ann Assoc Am Geogr 68:414–425

Tinner W, Kaltenrieder P (2005) Rapid response of high-mountain vegetation to early Holocene environmental changes in the Swiss Alps. J Ecol 93:936–947

Tomback DF (1977) The behavioral ecology of Clark's Nutcracker (*Nucifraga columbiana*) in the eastern Sierra Nevada. PhD thesis, University of Santa Barbara

Tomback DF, Sund SK, Hoffmann L (1993) Post-fire regeneration of *Pinus albicaulis*: height-age relationships, age structure, and microsite characteristics. Can J For Res 23:113–119

Tomback DF, Holtmeier FK, Mattes H, Carsey KF, Powell ML (1994) Tree clusters and growth form distribution in *Pinus cembra*, a bird-dispersed pine. Arctic Alpine Res 25:74–381

Tranquillini W (1959) Die Stoffproduktion der Zirbe an der Waldgrenze während eines Jahres. 1. Standortsklima und CO$_2$-assimilation. Planta 54:107–129

Tranquillini W (1964) Photosynthesis and dry matter production of tress at high altitudes. In: Zimmermann MH (ed) The formation of wood in forest trees. Academic, New York

Tranquillini W (1973) Der Wasserhaushalt junger Forstpflanzen nach dem Versetzen und seine Beeinflussbarkeit. Centralblatt für das Ges Forstwes 90:46–52

Tranquillini W (1976) Water relations and alpine timberline. In: Lange OL, Kappen L, Schulze E-D (eds) Water and plant life, vol 19, Ecological studies. Springer, Berlin

Tranquillini W (1979) Physiological ecology of the alpine timberline. Tree existence at high altitudes with special reference to the European Alps, vol 31, Ecological studies. Springer, Berlin

Tranquillini W (1980) Winter desiccation as the cause for alpine timberline. In: Benecke U, Davis MR (eds) Mountain environments and subalpine tree growth, New Zealand Forest Service Technical Paper 70

Tranquillini W (1982) Frost drought and its ecological significance. In: Lange OL, Nobel PS, Osmond CB, Ziegler H (eds) Encyclopedia of plant physiology 12B, Physiological plant ecology II. Springer, Berlin

Tranquillini W, Turner H (1961) Untersuchungen über die Pflanzentemperaturen in der subalpinen Stufe mit besonderer Berücksichtigung der Nadeltemperatur der Zirbe. Mitt der Forstlichen Bundesversuchsanstalt Mariabrunn 59:127–151

Tranter M (1991) Controls of the composition of snow melt. In: Davies TD, Tranter M, Jones HG (eds) Seasonal snowpacks: processes for compositional change, Proceedings of the NATO advanced research workshop on Processes of chemical change in snowpacks, Maratea, July 1990, Ecological sciences 18, Berlin

Tranter M, Jones HG (2001) The chemistry of snow. Processes and nutrient cycling. In: Jones HG, Pomeroy JW, Walker DA, Hoham RW (eds) An interdisciplinary examination of snow-covered ecosystems. Cambridge University Press, Cambridge

Troll C (1973) The upper timberlines in different climatic zones. Arctic Alpine Res 5:A3–A18

Tryon PR, Chapin FS III (1983) Temperature control over root growth and root biomass in taiga forest trees. Can J For Res 13:827–833

Tuhkanen S (1980) Climatic parameters and indices in plant geography. Acta Phytogeogr Suec 67: 1–110

Turner H (1958) Maximaltemperaturen oberflächennaher Bodenschichten an der subalpinen Waldgrenze. Wetter und Leben 10:1–12

Turner H (1968) Über "Schneeschliff" in den Alpen. Wetter und Leben 20:192–200

Turner H, Streule A (1983) Wurzelwachstum und Sprossentwicklung junger Koniferen im Klimastress der alpinen Waldgrenze, mit Berücksichtigung von Mikroklima, Photosynthese und Stoffproduktion. Wurzelökologie und ihre Nutzanwendung. Internationales symposium Gumpenstein, 1982. Bundesanstalt Gumpenstein

Van Gradingen P, Grace J, Jeffree CE (1991) Abrasive damage by wind to the needle surface of *Pinus sylvestris* L. and *Picea sitchensis* (Bong.) Carr. Plant Cell Environ 14:185–193

Vittoz P, Rulence B, Largey T, Freléchoux (2008) Effects of climate and land-use change on the establishment and growth of cembran pine (*Pinus cembra* L.) over the altitudinal treeline ecotone in the central Swiss Alps. Arct Antarct Alp Res 40:225–232

Walther G-R, Beißner S, Pott R (2005) Climate change and high mountain vegetation shifts. In: Broll G, Keplin B (eds) Mountain ecosystems, vol Studies in treeline ecology. Springer, Berlin

Wardle P (1968) Engelmann spruce (*Picea engelmannii* Engel.) at its upper limit on the Front Range, Colorado. Ecology 49:483–495

Wardle P (1974) Alpine timberlines. In: Ives JD, Barry R (eds) Arctic and alpine environments. Methuen Publishing, London

Wardle P (1985) New Zealand timberlines. 1. Growth and survival of native and introduced tree species in the Craigieburn Range, Canterbury. N Z J Bot 23:219–234

Wardle P (2007) New Zealand forest to alpine transitions in global context. Arct Antarct Alp Res 40:240–249

Wardle P, Coleman MC (1992) Evidence for rising upper limits of four native New Zealand forest trees. N Z J Bot 30:303–314

Weih M, Karlsson S (1999) The nitrogen economy of mountain birch seedlings: implications for winter survival. J Ecol 827:211–219

Wesche K, Cierjacks A, Assefa Y, Wagner S, Fetene M, Hensen I (2008) Recruitment of trees at tropical treelines: *Erica* in Africa versus *Polylepis* in South America. Plant Ecol Divers 1: 35–46

Wieser G (1997) Carbon dioxide gas exchange of cembran pine (*Pinus cembra*) at the alpine timberline during winter. Tree Physiol 17:473–477

Wieser G (2000) Seasonal variation of leaf conductance in a subalpine *Pinus cembra* during the winter months. Phyton Ann Rei Bot 40:185–190

Wieser G (2002) The role of sapwood temperature variations within *Pinus cembra* on calculated stem respiration: implications for upscaling and predicted global warming. Phyton Ann Rei Bot 42:1–11

Wieser G (2012) Lessons from the timberline ecotone in the Central Tyrolean Alps: a review. Plant Ecol Divers 5:127–139

Wieser G, Bahn M (2004) Seasonal and spatial variation in woody-tissue respiration in a *Pinus cembra* tree at the alpine timberline in the Central Austrian Alps. Trees 18:576–580

Wieser G, Kronfuss G (1997) Der Einfluß von Dampfdruckdefizit und mildem Bodenwasserstreß auf den Gaswechsel junger Fichten (*Picea abies* [L.] Karst.). Centralblatt für das Ges Forstwes 114:173–182

Wieser G, Stöhr D (2005) Net ecosystem carbon dioxide exchange dynamics in a *Pinus cembra* forest at the uppertimberline in the Central Austrian Alps. Phyton Ann Rei Bot 45:233–242

Wieser G, Tausz M (eds) (2007) Trees at their upper limit: treelife limitation at the alpine timberline, vol 5, Plant ecophysiology series. Springer, Dordrecht

Wieser G, Hammerle A, Wohlfahrt G (2008) The water balance of grassland ecosystems in the Austrian Alps. Arct Antarct Alp Res 40:439–445

Wieser G, Matysssek R, Luzian R, Zwerger P, Pindur P, Oberhuber W, Gruber A (2009) Effects of atmospheric and climate change at the timberline of the Central European Alps. Ann For Sci 66:402

Wieser G, Oberhuber W, Walder L, Spieler D, Gruber A (2010) Photosynthetic temperature adaptation of *Pinus cembra* within the timberline ecotone of the central Austrian Alps. Ann For Sci 67:201

Wilson C, Grace J, Allen S, Slack F (1987) Temperature and stature: a study of temperatures in montane vegetation. Funct Ecol 1:405–413

Yanagimachi O, Ohmori H (1991) Ecological status of *Pinus pumila* scrub and the lower boundary of the Japanese alpine zone. Arctic Alpine Res 23:424–435

Chapter 11
The Future of Trees in a Changing Climate: Synopsis

Nancy Grulke and Michael Tausz

Abstract Trees as long-lived stationary organisms are particularly challenged by rapid changes in their environment, most importantly the recent changes in the Earth's climate which proceed at unprecedented rates. This volume addressed the main characteristics of tree life with a view to explain and evaluate how the life functions of trees interact with the environment and environmental changes, and how understanding of physiological functions helps to assess how trees may be able to adapt to future conditions. Chapters address specific aspects of tree physiology, such as transport of water and assimilates, water relations, carbon dynamics and allocation, and environmental interactions such as defence chemicals, mycorrhiza, environmental limitations to growth, and impact of pollution. The relevance of each of these processes and aspects of tree life for environmental adaptation is highlighted.

In this book, the authors have reviewed, described, and evaluated the capacity for long-lived, stationary organisms, trees, to respond to rapid change in their physical environment and a novel, anthropogenically-driven change in atmospheric chemical composition: air pollution. Current projections predict changes in the Earth's climate at unprecedented rates, causing serious concerns for the survival of slow evolving, relatively immobile species such as trees (e.g. Allen et al. 2010; Anderegg et al. 2013). Changes in atmospheric chemistry by air pollutants has already proven

N. Grulke (✉)
Pacific Northwest Research Station, USDA Forest Service, Prineville, OR 97754, USA
e-mail: ngrulke@fs.fed.us

M. Tausz
Department of Forest and Ecosystem Science, The University of Melbourne, Creswick, VIC 3363, Australia
e-mail: michael.tausz@unimelb.edu.au

M. Tausz and N. Grulke (eds.), *Trees in a Changing Environment*,
Plant Ecophysiology 9, DOI 10.1007/978-94-017-9100-7_11,
© Springer Science+Business Media Dordrecht (outside the USA) 2014

dangerous for trees and forests in the past (e.g. Shepperd and Cape 1999), and air pollutants remain an intrinsic part of anthropogenic activities that drive climate change. Because these combined changes are unprecedented, the authors in this book used their expert knowledge and insight to develop qualitative conclusions within the context of many unknowns. Collectively, they have addressed two main systems essential for tree life: (1) water handling capacity from the perspective of water transport, the coupling of xylem and phloem water potential and flow, water and nutrition uptake via likely changes in mycorrhizal relationships, and control of water loss via stomata and its retention via cellular regulation; and (2) within plant carbon dynamics from the perspective of within-canopy distribution of photosynthetic resources, environmental limitations to growth, allocation to defences, and changes in partitioning to respiration. In several of the chapters, and specifically the last two, all of these processes are integrated into tree ecosystem response to environmental change.

11.1 Carbon and Nutrient Acquisition and Optimization Processes

Hikosaka et al. (2014) synthesized research analysing the complex within-canopy distribution of resource investment into different parts of the photosynthetic machinery. They pointed out a number of acclimatory adjustments that optimize carbon gain on a whole tree basis. Acclimation potential is required and, to some extent, implemented in tree species to respond to seasonal or sudden changes in light environments and temperature. The optimal resource allocation to different parts of the photosynthetic machinery is different for different sets of environmental conditions. The authors show convincingly how optimization can confer competitive advantages that outweigh any investments, and how observed acclimation to light gradients or typical seasonal temperature changes were close to predicted optimum patterns. A particularly interesting result showed that some species apparently 'overinvest' in shade leaves, which gives them a greater acclimation margin in the case of a sudden exposure to high light, such as after a formation of gaps in a forest canopy. One way of acclimating fully developed leaves to a sudden increase in light is to increase the number of chloroplasts. But this is only effective if there is enough vacant space along cell walls for chloroplasts to occupy. Some species therefore build thicker leaves (with vacant space) that can accommodate an increase in chloroplast volumes. The authors distinguish "optimistic" species that put resources towards the possible gap formation event, and "pessimistic" (or maybe conservative) species that optimize their investments according to present conditions only.

With respect to the change in climate it is particularly noteworthy that the theoretical optimal resource distribution also changes with increasing atmospheric CO_2 concentrations, from major investments into CO_2 fixation (Rubisco) towards the light reactions (photosystems). Although plants appear to acclimate to growth in high CO_2, the extent to which resource allocation changes in response to elevated CO_2 is less than optimization requirements. This suggests that current individuals are unable to acclimate sufficiently to rising CO_2 and that their resource allocation may be significantly sub-optimal in 50–100 years time – well within the lifespan of individual trees.

Once carbon is assimilated, it must be re-allocated and redistributed to the whole plant, a process dominated by phloem transport. Hölttä et al. (2014) investigated their thesis that phloem transport is driven by three factors: photosynthate production (availability), tree water relations, and phloem structure. The driving factors of sucrose transport in phloem are discussed within the context of pressure, osmotic potentials, and sieve and tracheal diameters of both phloem and xylem, as well as transport distance.

The classical Münch flow hypothesis that requires active processes only at the sources and sinks, is not fully supported by the most recent published experiments. Low temperature and metabolism blocks both reduce phloem transport, and leakage and reloading of solutes have been discovered. On the other hand, osmotic gradients have been shown to generate sufficient pressure gradients within the phloem to support Münch's hypotheses. Of particular interest is that sugar transport is faster than the movement of (labeled) individual molecules from source to sink, suggesting intermediate 'stations' of unloading/reloading which accelerate signaling and communicate source and/or sink strength within the whole plant.

Hölttä et al. (2014) present a mathematical representation of phloem transport rate as a function of the difference in turgor pressure between source and sink, transport distance, sugar concentration, and phloem conductivity, cross-sectional area, and viscosity, as well as outline which values are available or are uncertain or unknowable. They argue that it is not pressure potential differentials that drive phloem transport, but the flow rate, which is determined by plant metabolism, i.e., the rate of assimilate production or remobilisation. Hölttä et al. (2014) also offer intriguing arguments for the interdependency of sugar transport in phloem on one hand, and the transpiration movement of water from roots to leaves on the other. Not only are the two systems linked by turgor potentials, but ionic strength and ion species content in the xylem influence phloem function.

Perhaps one of the most important aspects of this chapter is the role of drought in decreasing photosynthate production and loading, decreasing phloem water potential (Ψ_{phl}), and its influence on xylem water potential (Ψ) and gradients. Because the water potential of phloem and xylem is linked, trees under drought stress have, by necessity, a metabolic cost to phloem transport through the requirement for a concomitant decrease in phloem water potential. The authors estimate that if whole tree water potential is decreased by half, sugar concentration in the phloem must double. They further estimate that this would represent 1 week's worth of photosynthate, which would then not be available for other plant needs, such as maintaining root metabolism. This helps us understand the significance of whole tree plant drought stress in increasing need for photosynthate, at a time when assimilation is restricted by reduced stomatal apertures. The carbon cost of tree drought stress is rarely considered from this viewpoint.

In the third chapter, Allen et al. (2014) address how environmental change affects mycorrhizae, and the symbiotic interchange between mycorrhizae and plants. They discuss three areas of effects: the role of increasing (soil) temperature on respiratory metabolism of both mycorrhizae and roots; the role of increasing soil temperature on soil water availability and likely mycorrhizal community composition and functioning; and the change in mycorrhizal carbon and nutrient metabolism as a function of increased carbon allocation to roots, exudates, and plant nutrient needs in an elevated CO_2 environment. The authors evaluated likely ecosystem responses to global change using two approaches. The first approach was to model ecosystem production components using the model DayCent, and to compare results to measured net ecosystem production (NEP), canopy transpiration, and fine root dynamics. The second approach used a physical model (Fick's), and measured CO_2 concentrations to estimate soil respiration (Rs), assuming that mycorrhizal respiration comprises 50 % of soil respiration, and that every root tip was mycorrhizal (fine root NPP was assumed to be proportional to increasing mycorrhizae numbers).

All GCMs predict future increasing air temperature, but the translation of that increase to soil temperature is imperfect due to differences in soil type, structure, and moisture content. Increased air temperature will increase evapotranspiration from the soil, decreasing water availability to plants, but elevated CO_2 will increase plant water use efficiency (WUE). However, their model did not account for increases in tropospheric oxides (O_3, NO_x) which can negate CO_2-induced increases in WUE by depressing photosynthesis (editor's comment). That being said, the range of ecosystem component response to different GCM scenarios was a 5–10 % increase in almost all ecosystem components (e.g., NEP, NPP, fine and coarse root production, and Rs) under both increased temperature and precipitation, in contrast to a 10–15 % decrease in the same attributes with an increase in temperature and a decline in winter precipitation. Changes in temperature and water availability will likely affect the growing season length and production of mycorrhizal fungi (in this case for arbuscular mycorrhizae, AM) and fine roots differently. The repercussions of these outcomes at the plant : root : mycorrhizae, and ecosystem levels are unknown, but are discussed.

With an increase in atmospheric CO_2 concentration, the need for plant nutrition will increase, which will lead to a concomitant increase in dependency on mycorrhizae. In an elevated CO_2 exposure field experiment conducted in a chaparral shrubland, CO_2-enrichment decreased plant N content, but root and AM fungal mass increased. A change in fungal community composition was also observed. In a nitrogen-fixing shrub, both shrub N fixation and uptake, and transfer of N by ectomycorrhizae (EM) to the plant increased. Of note, production and soil ecosystem functioning was severely compromised by the extreme N limitation that developed between 600 and 750 ppm CO_2 (a mid-2100 scenario; IPCC).

Measuring belowground ecosystem responses to environmental change is difficult, and the ability to adequately predict composition changes in fungal communities even more so. Fungi can tolerate a wide range of environmental conditions, but they are short-lived and competition under the new environmental conditions may lead to rapid, dramatic shifts in composition. Using a model developed from known physiological attributes of different fungal taxa to shifting ratios of C, N, and P, a wide range of outcomes were observed. Although increased diversity of mycorrhizal fungi generally increases plant production, the model output suggested a wide range in response, from supporting to refuting that generality. The model output varied primarily due to the physiological attributes of the initial fungal community taxa. Experimental results from AM and EM fungal associates in both the field and in laboratory experiments were also difficult to generalize. Although we know that new mycorrhizal types developed during periods of rapid changes in atmospheric CO_2 concentrations in geologic history, changes in fungal communities may be dis-synchronous with responses to that same environmental change of the long-lived plant associate. The current and future atmospheric chemical composition differs from any known in the past (due to anthropogenic pollutants, editorial comment). However, long-term CO_2-enrichment experiments with woody plants may be the best view that we have into future forest ecosystem responses.

11.2 Stomata, Osmoregulation and Water Transport

One of the topics that has been underrepresented in physiological modeling efforts is dynamic stomatal responses to rapidly changing environmental conditions, the topic of chapter three (Kaiser and Paoletti 2014). To date, all stomatal models and the ecosystem exchange models that build on them rely on physiological responses under steady state conditions. However, environmental conditions are rarely stable, and there is growing evidence that theoretical relationships between photosynthesis and stomatal conductance to water do not hold under moderately high ozone concentrations and above (>70–90 ppb). This chapter describes the mechanism by which this decoupling – of photosynthesis and stomatal conductance – may occur. These discrepancies from theory affect estimates of ozone flux into leaves, but also estimates of plant water use under moderate ozone concentrations. The focus of this chapter was to review the current understanding of dynamic stomatal responses to environmental and internal conditions, with special consideration for stomatal oscillations, which have been used to tease apart feedback and feed-forward controls of stomatal conductance.

The authors analyze the two primary (negative) feedback regulations on stomata, that of guard cell sensitivity to CO_2 and guard cell sensitivity to transpirational water loss, as a means of understanding the orogeny of stomatal oscillations. The authors discuss in depth the possible sensing mechanisms of transpirational feedbacks, as well as the roles of physiological and physical gain. A third, feed-forward mechanism, the role of air humidity alone as a stimulant of guard cell reaction independent of the transpiration stream, is also discussed. Evidence for this feed-forward mechanism was found for localized water loss (guard cells plus subsidiary cells, 'hydropassive movement'), but not guard cell water loss alone. It is likely that both feedback regulations occur with different time lags for initiation, and at different rates (gain), which contribute to stomatal oscillations. Identifying the feedback and feedforward mechanisms of stomatal control have been elusive to date.

Hydropassive movement has confounded understanding feedback and feedforward mechanisms due to two factors: the threshold of guard cell turgor for movement (with cellular water potential that has zero gain up to a certain point before the feedback mechanism can influence aperture), and 'Wrong Way Responses' (WWR, or the Iwanoff effect). With responsive enough equipment, WWR are often observed in response to *any* rapid environmental perturbation. With a rapid increase in leaf to air water vapor deficits, an increase in stomatal conductance is observed (increasing not decreasing transpiration), lasting for only minutes, which acts as a positive but destabilizing feedback to the transpirational feedback control loop. A mechanistic explanation for this response has been developed but not supported by experimental evidence.

Ozone exposure results in a rapid, sharp, transient decline of stomatal conductance, likely induced by a nearly simultaneous increase in reactive oxygen species (ROS). There was a 'recovery' of stomatal conductance, but then a slow, controlled decline in gs occurred with continued ozone exposure. Interestingly, the stomata did not respond to additional high ozone concentrations without a 'rest.' This may explain sluggish stomatal responses to environmental changes as has been reported in the literature. Sluggish stomatal responses have important consequences for setting and understanding the biological effectiveness of air quality regulations: a single, very high ozone concentration exposure may negate foliar uptake of subsequent O_3 peaks, depending on the time for stomatal desensitization to occur.

The authors conclude that stomatal oscillations in and by themselves are not particularly adaptive to maintaining an optimum water use efficiency, but they are a by-product of feedback mechanisms that have evolved for stomata to respond rapidly to rapid environmental changes. Oscillations may permit small adjustments and lower error about an optimum water use efficiency than would have been achieved by transpirational feedbacks alone.

In Chap. 5, Merchant (2014) presents a concise review of regulatory control of plant osmotic water potential under chronic drought stress that leads to acclimatory responses that permit survival, as well as acute drought stress, that permit short term survival. It seems that the acclamatory value of osmotic regulation does not provide an adaptive capability of responding to, for example, the increasing frequency, duration, and extremity of water deficits expected in some parts of the world with climate change. Although most literature available describes osmotic control of leaf water potential, the coupled water potential of xylem and phloem is an essential part of water, nutrients, and sugar transport within the whole tree (Höltt ä et al. 2014).

Only four classes of solutes have been shown to regulate osmotic water potential in plant cells: inorganic ions, carbohydrates, organic acids, and polyols. Of these classes, inorganic ions are perhaps the most discussed, and among ions, potassium plays the largest role. The direct role of potassium in osmoregulation is not clear, but it is implicated in osmotic sensing in membranes. The limitation of inorganic ions as cellular osmoregulators is the disruption of the both the chemical and electrical balances within the cell. Also, the cell has to isolate toxic ions, such as sodium and chloride, from ions functional as osmoregulators. Carbohydrates, especially non-reducing sugars (sucrose, oligosaccharides), are also commonly cited as cellular osmoregulators. They are present in quantity, and continue to be produced and accumulate to some extent after drought-induced stomatal shut down as well as after cell growth ceases. Because they are metabolically active, carbohydrates are not stable osmoregulators. Amino acids and their derivatives accumulate and lower osmotic potential in plant tissues in response to abiotic stresses. However, accumulation of organic acids are generally correlated with reductions in growth (which occurs with moderate drought stress), and breakdown of cellular processes (implying severe drought stress). It is unclear whether an increase in organic acids is an adaptive, or a survival-preserving response. Organic acids are stable osmoregulators. Polyols (reduced aldose and ketose sugars) are also regularly described as stress metabolites in plant tissues. As non-charged, inert and thus stable solutes, they are good candidates as osmoregulators. Their production uses excess photochemical energy from photorespiration, which occurs with drought stress and stomatal closure. These compounds have strong structural integrity within a taxon, and as such, could be used to select for drought tolerance for expected future environmental changes.

In the chapter on long-distance water transport in trees, Richter and Kikuta (2014) present a sound background on the biophysical fundamentals of water transport (SPAD, the soil-plant-atmosphere continuum; CTT, cohesion-tension theory), clarifying how component and total potentials vary (as in the case for differences among leaves at the same canopy height), their rate of change, and for biological components, the metabolic cost of adjustment. They particularly investigate when water potentials are unknowable, such as frictional water potential in non-steady state conditions, or whether trees are actually in equilibrium with soil water potential in the rooting zone at pre-dawn, or the level of tension that develops in bole xylem. The authors compare trees and herbaceous plants to elucidate similarities in their conservative use of water.

The topics that need to be investigated further include the total length, diameters, turnover rates, and total resistance to water transport of fine roots; extent, frequency, and persistence of embolism and cavitation in stems (and especially in roots); friction in bottlenecks (e.g., due to anatomical anomalies anywhere in the transport system); and resistances to water transport within leaves (for example, where does water pass from capillary veins into mesophyll cells?). Richter and Kikuta (2014) outline potential changes likely to occur with climate change for the primary environmental factors that influence long distance water transport in trees. Despite expected changes in rainfall, air temperature and humidity, wind-speed and direction, and radiation loads modified by changing cloud cover, the authors conclude that trees have long endured periods of prolonged drought stress, and elaborate mechanisms are already in place to maintain functionality of water transport under expected future environmental conditions. Although trees have persisted through many extreme droughts in the past, the authors caution that still not all species will persist due to competition with better adapted species.

11.3 Air Pollution

Matyssek et al. (2014) provide a thorough review of both anthropogenic and biogenic air pollution as a component of climate change. They identify the key components as CO_2, O_3, N_2O, and halogenated hydrocarbons. They point out that biogenic volatile organic compounds (BVOCs) also contribute to global warming. Forests and shrublands account for ~90 % of BVOC emissions. Anthropogenic VOCs emitted from refineries and combustion of carbon-rich fuels play a much smaller role. Of the greenhouse gases, O_3 is the most phytotoxic and the most likely to negate the elevated CO_2-induced increase in C sequestration, thus potentially accelerating global warming. Understanding its effects on forests is critical to anticipate and prevent forests from being converted to a source, rather than a sink for carbon. Although excess nitrogen deposition tends to be regional because its aerial transport distance is limited, its effects are also of note because it initially promotes forest growth, but as it accumulates in the soil, it becomes toxic, has long residence times (>100 years), and has significant, deleterious effects on many ecosystem components. For these reasons, the effects of O_3 to a greater extent, and excess N deposition to a lesser extent on forests are the focus of this chapter. A well-executed schematic is presented for NO_2 and O_3 effects at the whole tree level.

Combustion of fossil fuels results in the release of NO_x and N_xO, and over-fertilized agricultural lands also release NH_3. Where N deposition exceeds ecosystem use or even luxury consumption, N builds in the soil, leads to nitrogen saturation, and results in acidified, eutrophic soil systems. The human consequences are increased run-off of nitrates which deteriorate stream water quality, soil respiration, as well as increase denitrification, a process that releases nitrous oxide (a greenhouse gas). An imbalance in soil nutrients has multiple effects on trees including reduced water uptake and radial growth due to Ca and Mg limitation and its effects on photosynthetic pigments, and increased herbivory. Tree mortality may increase from these belowground effects.

Ozone uptake effects at the cellular, leaf, and whole plant are described as are differences between juvenile vs. mature trees. The interactive effects of O_3 action on trees is discussed with regard to both physical conditions (elevated temperatures, drought, radiation loads, nutrition), gaseous interactions (VOCs, CO_2), other biotic factors (genetics, competition, symbionts), and global dimensions (long-distance transport, biomass burning). Lastly, the authors discuss future issues around O_3 pollution and climate change in the near term (next several decades).

We have a good understanding of cellular defenses against gaseous oxidants. They begin to be decomposed as soon as they come in contact with the apoplastic water and plasmolemma. This decomposition requires energy, both to construct and recycle the antioxidants. O_3 uptake has a cascade of effects at the whole tree level: leaf loss is accelerated and leaves not necessarily replaced, WUE decreases, and C allocation is altered, with root mass especially reduced. In their discussion of mature tree response to O_3, the authors utilize a unique whole stand, experimental O_3 exposure of European beech and Norway spruce. Of note, tree bole volume was reduced in beech trees (44 %), but was undetectable at the base. The shape of the bole was modified significantly, such that the younger part of the bole had less diameter growth; the spruce bole volume was unaffected. Also unexpected was that fine root growth was enhanced, not depressed in beech, likely due to O_3-mediated cytokinen destruction in the leaves, accompanied by lack of suppression of cytokinens in fine roots.

For the interaction of O_3 and temperature, examples of the depressive effects of low temperature on plant defense systems are enumerated. If O_3 is transported into higher elevations or latitudes, this could render these treeline species more sensitive. In hotter, drier years, O_3 concentrations are generally higher. However, from the example of the whole stand O_3 exposure, severe drought stress reduced foliar O_3 uptake relative to a more humid growing season due to reduced stomatal conductance. Overall, drought overrode the effects of O_3 exposure. Direct effects of O_3 exposure on sluggishness of stomatal responses could exacerbate physiological drought stress. The key point is that both drought and O_3 exposure will likely limit forest C sequestration. Although modeling cloud cover under current conditions is still elusive, the authors argue that with higher temperatures and greater drought developing, it is likely that plants will experience higher irradiance loads. The increased production of O_3 with irradiance is well known. There have been few direct tests of the interaction between O_3 exposure and plant nutrition. Of these, N amendment/deposition generally, but does not always ameliorate O_3 effects. Only at low nutrient availability does O_3 exposure lower stomatal conductance (inadvertently limiting O_3 uptake) in Norway spruce. This is one example, but the interactive effects of CO_2 and O_3 combined is likely species-specific. In conclusion, the authors warn that forest ecosystems will be far different than what we are familiar with under future climate change and air pollution.

11.4   Secondary Metabolites

Secondary plant metabolism is affected by environmental stressors such as CO_2, air temperature, O_3, UV radiation, and drought stress, as reviewed in Goodger and Woodrow (2014). Different environmental stimuli elicit different metabolite responses. For example, CO_2, O_3, UVB, and drought stress increase phenolics. Elevated temperature increases foliar content and emissions of VOCs (terpenes). However, little is known about the effects of flooding. Plant type, age, and genetic makeup can also modify the tree's response to environmental conditions. Secondary plant compounds are sinks for plant carbon, as well as for defense against herbivory and pathogens.

Elevated CO_2 variably affects plant carbon and nitrogen supply and use within the plant, and along with it, variable effects on plant secondary metabolites. Different species had different strategies for balancing growth and chemical defense. To date, CO_2-enrichment effects have been investigated on only 11 tree genera. Of these, four were among the most common globally, but species from tropical forests were not well represented. With increasing carbohydrate supply, generally both phenolics (phenylpropanoids) and tannins (hydrolyzable) also increase. Terpenoids were less responsive to elevated CO_2. There were few data on elevated CO_2 effects on nitrogen-based defense chemicals (e.g., alkaloids, cyanogenic glycosides). The primary effect of increased global temperatures is likely to be increased volatile emissions (terpenoids and hydrocarbons). Ozone exposure elicits defense responses, and changes in plant secondary metabolites such as flavonoids, phenolics with antioxidant properties, and lignin properties have been observed to follow, but not tannins. Mature trees were more responsive than seedlings or saplings. As with other interactive effects, elevated O_3 and CO_2 together negated (simply) or synergistically exacerbated effects on secondary plant metabolites.

With a linear increase in air temperature, there was an exponential increase in emissions. Because isoprene emissions increase as temperatures rise, and the gain of photosynthesis is likely lost with increased temperatures, the authors suggest that accelerated isoprene emission may negate CO_2-stimulated increases in photosynthesis. A similar argument arises from mono- and sesquiterpene emissions from foliage. There was inconsistency in the direction and magnitude of changes in phenolics (e.g., condensed tannins, flavonoids) with increased temperature, likely due to differences in their biogenic pathways. Responses to combined elevated CO_2 and temperature were independent, offset by temperature, or apparent only in higher temperatures (as summarized from Zvereva and Kozlov 2006). Overall, the greatest effect of elevated temperature on secondary plant metabolism may well be the phenological mismatch between host and insects, which is suggested as true for both increased temperature alone, and for both elevated temperature and CO_2 combined.

Although legislation has reduced CFC release, UV-B radiation is still expected to increase through mid-century due to stratospheric O_3 degradation. Exposure affects foliar UV-B-absorbing compounds, especially phenolics, with linked effects on terpene metabolism. With exposure, aliphatic and aromatic amino acids, and hydrolyzable tannins were increased as well, although complexities of responses abound: most secondary plant compounds increased, but in some species with exposure to UV-B, others declined. Similar to elevated CO_2 effects, UV-B affects plant carbon and nitrogen balance, and so changes in allocation to secondary plant metabolites are also expected. There have been few interactive studies conducted with UV-B. However, in a review of terrestrial ecosystems, UV-B sensitivity was reduced with concurrent drought. Drought suppresses activity of some enzymes, reduces photosynthesis limiting C more than nutrients, but also increases oxidative stress through reduced stomatal apertures. In Douglas fir, there was a parabolic relationship in condensed tannins from mild, to moderate, to severe drought stress.

Drought stress was found to depress volatile emissions of monoterpenes but emissions then increased during recovery with watering, or to increase isoprene emissions with short term drought stress.

Clearly more research is needed on these topics, especially across low, medium to high stress, and short term vs. chronic stress. Stresses that affect carbon and nutrient supplies and allocation within the plant generally affect secondary plant metabolism and metabolites. The authors summarize the following: elevated CO_2 consistently increases phenolics; elevated temperature increases isoprene and may also increase mono- and sesquiterpene emissions and foliar terpene concentrations; drought stress can increase phenolics, and mild stress can increase terpene emissions; water-logging increases phenolic concentrations; elevated O_2 increases phenolics and diterpene resin acid, as does UV-B. Throughout, nitrogen containing metabolites have rarely been studied. These effects imply multi-trophic level repercussions with atmospheric and climate change.

11.5 Limits to Tree Growth and Treelines

Rossi et al. (2014) provide an informed analysis of wood growth, partitioning cell division, enlargement, cell wall formation, and lignification into distinct periods with associated, driving environmental factors. Different tree species show differing periods of the onset and conclusion of each stage, which may give us insight as to which species will be able to respond favorably and/or competitively to climate change. For example, an earlier onset of cambial cell division, a prolonged plateau of maximum growth activity, and delayed lignification may indicate greater capability for plasticity, and could be more competitive under climate change with increased temperature, longer growing season, and potentially more responsive to fertilization (elevated CO_2 and N deposition: editorial comment).

Despite wide study of its occurrence, the exact mechanisms that shift plants from active growth to dormancy, and from dormancy to quiescence (in the latter state, inducible with first day length, then secondarily with temperature or other appropriate environmental conditions) are unknown. There are two hypotheses for temperature control of the onset and egress from dormancy: a gradual phenomenon caused by insufficient assimilated carbon, vs. a threshold effect caused by a reduction in carbon investment in structural growth. Once inducible, bud beak may be accelerated by warmer spring temperatures.

There is recent support for carbon allocation limitation by low temperature, supporting the 'sink' hypothesis proposed by Körner (1998). Interestingly, xylogenesis is active in the spring at very low temperature (air and stem temperatures of 2–4 °C and 4 °C, respectively). Cell expansion should not occur before root temperatures exceed 6 °C, the temperature at which there is a significant increase in cell membrane permeability to water. In their chapter, Rossi et al. (2014) provide evidence for a minimum air temperature of 6 °C for active xylem growth in Fig. 9.3 for treeline species (*Larix decidua*, *Picea abies*, and *Pinus cembra*), but 8–9 °C for

Abies balsamea, *Pinus leucodermis*, and *Pinus uncinata*, all mountainous or northern temperate zone species. The authors suggest that temperatures below 6 °C affect water transport and temperature-dependent mechanisms of cell growth in all plant tissues, however a minimum temperature for cell division vs. cell elongation or enlargement may differ (editorial comment). Rossi et al. propose that all northern tree species studied achieve maximum xylem growth rates at the same time: at the apex of day length (June 21), not at maximum temperature. With that consideration, the earlier the onset of xylem cell differentiation, the longer the greater the potential for growth through the year. This is quite important for understanding future responses to climate change: despite increasing temperatures, bole growth of trees as they occur in situ will be ultimately limited by day length and thus limit the first half, not that last half of a climate change increase in growing season length.

Over the long term, growth of northern forests is currently constrained by total photosynthetic gain and nutrients from the soil. Snow cover limits not only water availability for stem and foliar growth, but also root activity for uptake of nutrients. Warmer global temperatures will reduce spring snow cover, increase soil warming, and presumably speed up all subsequent biological growth processes. However, despite known warming at northern latitudes and higher elevations, tree rings are thinner, not thicker. The effect of increased temperature may have resulted in greater respiratory costs, which have reduced whole tree carbon gain and thus allocation to wood growth. Despite shifts in the timing of snow melt, the authors suggest that tree growth and nutrient availability will still be out of sync in a future, warmer climate.

In the acidic soils of northern latitudes, availability of ammonium and organic N are directly related to the duration of root growth. Although boreal trees store 1–2 years of nitrogen (N) for future growth, the uptake of N is asynchronous with soil N mineralization and root uptake, which occurs after peak growth. With this asynchrony in mind, over the long term, plant N reserves may become depleted, plant growth slowed, or ceased. Root uptake may replenish nutrients for the next year's growth, but climate change is expected to exacerbate N limitation of northern forests. Increased nitrogen has a wide range of effects on tree growth, from increasing photosynthetic gain and foliar growth, altering tissue C:N ratio, increasing lignin content of earlywood, as well as decreasing wood density affecting wood mechanical properties. Anthropogenic increases in N deposition appear to have stimulated growth of northern forests, and are likely to have cascading effects on the availability of N and its timing for especially spring growth, as well as select for species capable of responding to these inputs.

Under future climate scenarios, xylem growth is predicted to increase by about 30 days than under current conditions. The limitations of photoperiod on growth, asynchronies in plant growth demands and nutrient availability, drought stress, increased temperatures on total plant carbon balance, and increased herbivory are all challenges requiring more investigations into the future.

Scientists have always been intrigued with controls of abrupt changes in vegetation from one type to the other, and that of treeline has had exceptional interest.

There have been many advances in knowledge and mechanistic work on resource acquisition and growth at treeline since the explosion of research on the topic in the late 1960s. What is unique in the presentation of Wieser et al. (2014), though, is a thorough handling of limitations on both young and mature trees, as well as addressing changes in both CO_2 *and* water vapor with elevation. Changes in water vapor with elevation has significant repercussions on tree transpiration rate. The authors address limiting factors for: seedlings to mature tree growth at treeline; soil water and nutrient uptake; how limitations on trees differ from that of other low stature vegetation, and finally, mechanisms of treeline advancement, a logical assumption of response to climate change.

The newly proposed causes of treelines are either: limitation to growth due to limitations on carbon metabolism (e.g., glycolysis) not photosynthesis; or, the abiotic stresses on tall tree growth are excessively severe, especially in winter. With regard to the first cause, treeline species would not be able to take advantage of an increase of atmospheric CO_2, because they are already end-product limited. However, the same may not be true for seedlings as most gas exchange research has been conducted on mature trees at treeline. Also, specific leaf mass and leaf nitrogen content is greater in treeline trees, such that they have a morphology and N storage capacity suited to take advantage of elevated CO_2. Although net photosynthesis and stored non-structural carbohydrate content are similar between lower and upper elevation trees, the authors note significant increases in net assimilation and non-structural carbohydrate concentrations of what would be considered saplings (larch) to experimentally enriched CO_2 exposure. No such responses were observed in an evergreen species (pine). With regard to the second cause, there is much support for overall temperature limitation at treeline, such as accelerated tree growth at treeline near warm springs (accompanied by increased length of growing season), and in years of higher spring temperature. Barring impediments to seedling establishment, treelines should expect to advance with greenhouse gas-induced warming. With a few exceptions, there has been little field evidence for this over the last 50 years (editorial comment). Wieser et al. posit that drought, insect and disease outbreaks, fire, and anthropogenic disturbance as well as nutrient limitation and/or dys-synchrony in growth vs. nutrient availability may impose additional strong limitations to treeline expansion.

In addition to systematic decreases in temperature (adiabatic lapse rate) with elevation, atmospheric pressure is lower, decreasing the concentrations of CO_2, but decreasing the water holding capacity of the air. Despite this interplay of atmospheric pressure and temperature, the changing pressure is a stronger driver than that of temperature, and the evaporative capacity of the air at higher elevations increases, not decreases, necessarily increasing transpiration, and thus decreasing leaf water use efficiency. Although not enough is known about how higher air temperatures will be translated to higher soil temperatures, there is no question that low soil temperatures currently limit spring water availability for plant growth at high elevation and latitudes. Global warming will be translated to higher soil temperatures, but whether soil nutrient availability will still be out of sync with plant requirements remains to be discovered. At high latitudes, soil water logging will still be limiting, but at high elevations, lower soil water availability will likely develop except in depressions. Low soil water availability at high elevation will impose additional limitations on shallow-rooted germinating tree seedlings.

Leaf temperatures of trees at treeline average 4 °C higher than air temperature, but also experience more extremes: under low wind-speeds or on the lee side, sunlit leaf temperatures can exceed 10 °C greater than air temperature. However, night-time leaf temperatures can be as much as 10 °C lower than that of the air with radiative cooling. Reduced height growth in treeline trees has been attributed to a shorter growing season as well as increased wind damage, as the boundary layer is much closer to the ground surface at treeline than in the tall tree forest. Particle abrasion (mineral or ice) significantly erodes cuticles, increasing plant water loss at treeline. However, once the cuticle is compromised, it is unlikely that the needle or leaf would be retained long due to desiccation and death (editorial comment).

The authors stress that environmental limitations on seedlings must be understood before predictions of treeline response to climate change can be predicted. In forest and treeline alike, seedlings have the highest mortality of all tree age classes. Similar to the forest, the greatest survival of tree seedlings is in locations with not too much nor too little sun exposure. High radiation loads were associated with mortality due to photoinhibition of photosynthesis. Excessively high temperatures, desiccation, and low radiation microsites were associated with poor root growth, poor mycorrhizal infection, and greater infection with parasitic snow fungi. In seedlings as in trees at treeline, there was a greater carbon sink limitation with elevation. Regional and local climatic fluctuations affect carbon gain and growth in mature trees at treeline.

Microclimate and soil conditions affect carbon gain and survival of treeline seedlings. Seedlings establishing at a continental treeline were more likely within a krummholz or low shrub mat than in the open and exposed to both excessive sunlight and longwave radiation loss at night. Seedling recruitment is poor in the treeline ecotone. For example, germinability of seed produced at treeline was 1 % vs. 24 % in the subalpine belt: the upper limit for reproductive viability is lower than the physiological limit to tree growth. Reproduction at and above treeline was likened to a series of hurdles: carbon gain, reproductive bud formation (1 year; carbon gain plus appropriate hormonal ratio at bud), viable seed production (1 year), germination (1+ year), establishment (3 years [based on high mortality in the first 3 years]), and growth to the next size class. Conditions that promote seed germination may be different than those that promote establishment. These processes occur over a minimum of 6 years at treeline, and thus rely on favorable environmental conditions at each stage. Six consecutive favorable years is unusual at treeline, and may still be unusual with climate change. However, wind and bird dispersal of viable seeds cannot be ignored for establishment beyond current treeline. The most favorable microsite for seedlings to become emergent is in the lee of mats with lower wind-speeds and ice abrasion. Although higher temperatures and longer growing seasons may be favorable for regeneration, establishment, growth, and winter hardening, drought stress expected with warmer and long growth seasons with climate change and low nutrient availability may limit treeline advance.

11.6   Conclusion

The work by the authors of this book volume will help us understand tree responses to the complex, concurrent effects of environmental stresses imposed by climate change, and its source, air pollution. Whilst statements about the likely capacity of trees for adaptation to environmental changes on a short time scale relative to their lifetime must necessarily remain vague, the often new and intriguing approaches presented, and the challenges to dogma pursued in the chapters, permitted qualitative conclusions about tree responses to an altered, future physical and chemical environment.

References

Allen CD, Macalady AK, Chenchouni H, Bachelet D, McDowell N, Vennetier M, Kitzberger T, Rigling A, Breshears DD, Hogg EH, Gonzalez P, Fensham R, Zhang Z, Castro J, Demidova N, Lim J-H, Allard G, Running SW, Semerci A, Cobb N (2010) A global overview of drought and heat-induced tree mortality reveals emerging climate change risks for forests. For Ecol Manag 259:660–684

Allen MF, Kitajima K, Hernandez RR (2014) Mycorrhizae and global change. In: Tausz M, Grulke NE (eds) Trees in a changing environment. Springer, Dordrecht

Anderegg WRL, Kane JM, Anderegg LDL (2013) Consequences of widespread tree mortality triggered by drought and temperature stress. Nat Clim Chang 3:30–36

Goodger JQD, Woodrow I (2014) Influence of atmospheric and climate change on tree defence chemicals. In: Tausz M, Grulke NE (eds) Trees in a changing environment. Springer, Dordrecht

Hikosaka K, Yasumura Y, Muller O, Oguchi R (2014) Ecophysiology of resource allocation and trade-offs in carbon gain under changing environment. In: Tausz M, Grulke NE (eds) Trees in a changing environment. Springer, Dordrecht

Hölttä T, Mencuccini M, Nikinmaa E (2014) Ecophysiological aspects of phloem transport in trees. In: Tausz M, Grulke NE (eds) Trees in a changing environment. Springer, Dordrecht

Kaiser H, Paoletti E (2014) Dynamic stomatal changes. In: Tausz M, Grulke NE (eds) Trees in a changing environment. Springer, Dordrecht

Körner C (1998) A re-assessment of high elevation treeline positions and their explanation. Oecologia 115:445–459

Matyssek R, Kozovits AR, Schnitzler J-P, Pretzsch H, Dieler J, Wieser G (2014) Forest trees under air pollution as a factor of climate change. In: Tausz M, Grulke NE (eds) Trees in a changing environment. Springer, Dordrecht

Merchant A (2014) The regulation of osmotic potential in trees. In: Tausz M, Grulke NE (eds) Trees in a changing environment. Springer, Dordrecht

Richter H, Kikuta S (2014) Ecophysiology of long-distance water transport in trees. In: Tausz M, Grulke NE (eds) Trees in a changing environment. Springer, Dordrecht

Rossi S, Deslauriers A, Lupi C, Morin H (2014) Control over growth in cold climates. In: Tausz M, Grulke NE (eds) Trees in a changing environment. Springer, Dordrecht

Shepperd LJ, Cape JN (1999) Forest growth responses to the pollution climate of the 21st century. Springer, Berlin

Wieser G, Holtmeier F-K, Smith WK (2014) Treelines in a changing global environment. In: Tausz M, Grulke NE (eds) Trees in a changing environment. Springer, Dordrecht

Zvereva EL, Kozlov MV (2006) Consequences of simultaneous elevation of carbon dioxide and temperature for plant-herbivore interactions: a meta analysis. Glob Chang Biol 12:27–41

Index

M. Tausz and N. Grulke (eds.), *Trees in a Changing Environment*,
Plant Ecophysiology 9, DOI 10.1007/978-94-017-9100-7,
© Springer Science+Business Media Dordrecht 2014